MEASURING MANHOOD

Measuring Manhood

Race and the Science of Masculinity, 1830–1934

Melissa N. Stein

University of Minnesota Press
Minneapolis
London

Published by the University of Minnesota Press
111 Third Avenue South, Suite 290
Minneapolis, MN 55401-2520
http://www.upress.umn.edu

Library of Congress Cataloging-in-Publication Data

Stein, Melissa N.
Measuring manhood : race and the science of masculinity, 1830–1934 / Melissa N. Stein.
Includes bibliographical references and index.
ISBN 978-0-8166-7302-5 (hc)—ISBN 978-0-8166-7303-2 (pb)
1. Racism—United States—History. 2. Masculinity—United States—History. 3. Sexism—United States—History. 4. Individual differences—Social aspects—United States—History. 5. Individual differences—Political aspects—United States—History. 6. Science—Social aspects—United States—History. 7. Sociobiology—United States—History. 8. United States—Race relations—History. 9. United States—Social conditions—1865–1918. 10. United States—Social conditions—1918–1932.
I. Title.

E184.A1S793 2015
305.800973—dc23 2014043033

Printed in the United States of America on acid-free paper

21 20 19 18 17 16 15 10 9 8 7 6 5 4 3 2 1

For my family—past, present, and future

Contents

Making Race, Marking Difference

IN A 1994 SPEECH BEFORE the Rutgers Council of the American Association of University Professors discussing college admissions and diversity programs, Rutgers University president Francis L. Lawrence remarked that minority students lacked the "genetic hereditary background" to excel on standardized tests—a comment for which he later apologized, insisting that he misspoke.[1] Similarly, Harvard University president Larry Summers suggested at a 2005 conference on diversity in science careers that "socialization and continuing discrimination" were perhaps less responsible for the small representation of women in the field than women's own biologically determined capacities (or incapacities). That is, women simply had less of an innate aptitude for certain scientific professions than men.[2] Summers's comments were eerily reminiscent of another respected Harvard man more than a century prior: Dr. Edward Clarke, a professor in the university's medical school and an opponent of higher education for women. In his tremendously influential 1873 publication, *Sex in Education, or, A Fair Chance for the Girls,* Clarke insisted that women's bodies were not designed for intellectual pursuits, and as a result, higher education would impinge on their true biological function—reproduction—by diverting blood from their wombs to their brains, in turn jeopardizing not only their own health but also the future of the (white) race.[3] Going back farther still in the hallowed halls of Harvard, we find Louis Agassiz, a revered Swiss-born professor of zoology and prominent racial theorist. Agassiz was famous in the mid-nineteenth century for his contributions to natural science and innovations in teaching. He was also known among his contemporaries for his marriage to Elizabeth Cabot Cary, famous in her own right as a writer and advocate of women's education, who, following her husband's death, led the charge to open Harvard's doors to women. She ultimately succeeding in 1879—the same decade her colleague Edward Clarke published *Sex in Education*—and later became the first president of Radcliffe College, Harvard's "sister" school. But Agassiz remains most well known today for his vocal support for the theory of polygenesis, which claimed that the "races" originated separately and that each race's unique biological makeup had not and would not change over time—a theory that in the antebellum period in particular often served as a scientific justification for America's racial hierarchy.

This is intended not to disparage Harvard University (or Rutgers, my own graduate alma mater) but rather to show how *mainstream* biological determinism is in regards to both race and gender that one might hear the ethos of "biology is destiny" espoused at even the most esteemed institutions in the United States and also how *persistent* such ideas have been since the nineteenth century. In the case of both university presidents in the 1990s, the response to their comments—particularly from women's and minority rights groups—was swift and very public. However, neither of the men lacked defenders, some of whom were quick to assert that the comments were well supported by science, seemingly an entity unto itself that cared little for liberal agendas and only concerned hard facts. Indeed, though more likely to be challenged today (particularly when articulated as baldly as Lawrence and Summers), the belief that race and gender, as well as sexuality, are biological categories never completely disappeared. In fact, such beliefs have seen a resurgence in the United States in the twenty-first century, albeit often in more implicit ways and with different cultural or political aims than the long nineteenth century—the ways in which studies of the "female brain," for example, continue to be employed toward questions of educational policy and classroom practice.[4]

In regards to sexuality in particular, television and print news media regularly report on scientific studies seeking a biological basis of homosexuality. From the search for the "gay gene" to a widely reported story that linked homosexuality to the ratio of index to ring finger lengths, scientists have continued to scrutinize every part of the body for clues to sexual orientation.[5] In "Born Gay? How Biology May Drive Orientation," a lengthy article in the *Seattle Times* that described the finger length studies, among other scientific investigations on homosexuality and heredity, science reporter Sandi Doughton clearly articulated the sociopolitical context—and applicability—of such work: "The social and political implications of the research are impossible to ignore, leading to unease on both sides of the gay-rights debate. If science proves homosexuality is innate, is there any basis to deny gays equal treatment—including the right to marry? But if scientists unravel the roots of sexual orientation, will it some day be possible to 'fix' people who don't fit the norms or abort fetuses likely to be born gay?"[6] While much has changed in American culture since the turn of the twentieth century, the scientific debate Doughton depicted in 2005 shared more than a passing resemblance to medical scientists' considerations of the "congenital" or "acquired" nature of homosexuality nearly a hundred years earlier; for example, in an article in the *Medico-Legal Journal,* respected New York gynecologist and sexologist Bernard Talmey sought to determine whether one of his patients demonstrated "a case of inherited or acquired homosexuality" and found that indeed the man's "entire clan was sexually tainted."[7] Likewise, the concern Doughton raises about biologically deterministic sexual science today underscores that the "nature or nurture" question in regards to homosexuality also continues to have eugenic implications.[8]

The dawn of the twenty-first century has also seen a resurgence of scientific interest in biology and race. Biomedical researchers have devoted considerable energy to comparing disease rates across racial groups, and in 2005, the Food and Drug Administration approved the first race-specified drug, BiDiL, marketed by the pharmaceutical company Nitromed exclusively to African Americans for heart failure. In addition to ethical concerns and uncertainty about its effectiveness, scientific and public debates over the controversial drug have frequently raised the question "Is race biological?"[9] Advances in DNA research have also precipitated scientific investigation into race and racial histories. In conjunction with the African American Roots Project, which provides similar services for members of the general public, Harvard professor Henry Louis Gates Jr. combined genealogical research and DNA mapping to trace the ancestry and family histories of famous African Americans in a series of television documentaries. In the nineteenth century, racial theorists used science to deny black people a history, any claim to civilization, and even a place in the human family. There is certainly a poignant irony, then, in science being used to help African Americans reclaim their past. To be sure, the cultural politics driving this work on DNA or on disease are quite different from the nineteenth- and early twentieth-century racial science described herein. Still, any slippage between race as a social category and biological entity in scientific endeavors, no matter how noble the intentions, raises important questions and concerns—not the least of which is that, as this book demonstrates, any conclusion implying racial difference of some kind could have a life well beyond the scientist's intent.[10]

This book describes how race became the purview of science and the processes by which race was constructed as a biological phenomenon with far-reaching social, cultural, and political resonance. In so doing, it tells the story of an overlapping, interdisciplinary group of scientists who asserted their relevance and authority by offering expert advice on America's most pressing issues. These scientists often used gender and sex difference to buttress their claims about racial difference, but the mechanisms by which they did so constantly shifted according to what was at stake in that specific historical moment. But throughout the nineteenth and early twentieth centuries, their conclusions about human difference helped naturalize and institutionalize sociopolitical difference—and hierarchy—in America. For these scientists, the physical body both reflected and determined the character of the social body.

But why should we care about a hodgepodge group of long-dead scientists, whose work in some cases seems so absurd and blatantly biased by contemporary standards that it is all too easy to dismiss? For one, they are still making race—and gender—trouble today. Scientists who find evidence of biological differences between the races, or between the sexes, continue to make front-page headlines and spark debate among talking heads on television shows targeted to mainstream audiences, while the countless studies that challenge those findings or the implications drawn from them receive far less public attention. From the "gay gene" and the "female brain" to minority students' insufficient "genetic hereditary background," we are still dealing with the legacy of scientific ideas that coalesced in the nineteenth century. This project, then, seeks to unpack their roots, as well as their impact in U.S. history.

More specifically, I am concerned with not only ideas and discourses but also how the public received such ideas and how they shaped the way real people were treated. Accordingly, this book examines a variety of sources, both textual and visual, to interrogate racial scientists' engagement with social and political issues as well as the incursion of scientific thought into political culture. I first survey the ways in which gender and sex were implicated in the rise of race as a subject of *scientific* inquiry and then focus on a number of key moments from the heyday of biological determinism—roughly concurrent with the long nineteenth century—that are particularly illustrative of the complex intersections of race, gender, and sexuality in medico-scientific thought; the sociopolitical import of human bodies; and the ways in which scientists sought to establish and defend their authority over Americans' most pressing concerns.

The centrality of race in public and scientific discourse reached an apex at the turn of the twentieth century, a time during which the body was imbued with enormous social meaning and significance. Within the multiple and intersecting discourses about bodies—evolution, hysteria, pathology, sexual difference, masculinity—race remained, to borrow Evelyn Brooks Higgenbotham's useful terminology, the metalanguage that wove those discourses together.[11] Race determined how bodies would be interpreted and also how they would be treated as a result. Moreover, as historian Laura Briggs notes, "Race is not an attribute that inheres in bodies, but rather attaches itself to bodies through the ideological and material work of things like law, medicine, science, economy, education, literature, social science, public policy, and popular culture. Race in turn fundamentally shapes both the subject and content of those diverse fields."[12] As we shall see, race provided the raison d'être for whole fields of scientific inquiry and a space for scientists to lay claim to social and political relevance.

However, race itself was still being *made* as a category of medico-scientific analysis during much of the nineteenth century, and early scientists of race relied on more familiar categories of human difference to conceptualize and describe it. *Gender*—a metalanguage in its own right and one already in use as a biological means through which the populace could be parsed and understood—functioned as a rhetorical tool for scientists to employ in constructing race. That is, while it is common for scholars to describe intersecting differences as mutually constitutive, when it comes to the role scientists and physicians played in making and naturalizing race, it would be more precise to say gender difference (which predated the shift during the early modern period in the meaning of the word "race," from lineage and bloodlines to the more recent definition regarding skin color and physical features) offered scientists a convenient reference point and foundation through which to frame their ideas about *race* as a nascent category of difference similarly ordained by nature itself.[13] At a time when most Americans found it difficult to imagine human beings without regard for gender or sex—categories to which many social and political rights were attached—racial scientists often maintained that blacks deviated from white standards of both gender performance and bodily sex to call into question their very humanity.

Race in Early America

During the nineteenth century, racist ideologues increasingly drew upon science for legitimacy and authority, whereby popular ideas about the roles and capacities of men and women were naturalized as biologically determined when science came of age in earlier centuries.[14] Racism was not new, but as George M. Fredrickson points out in his seminal text *The Black Image in the White Mind,* the type of racism that gained strength throughout the nineteenth and into the early twentieth centuries was "a rationalized pseudoscientific theory positing the innate and permanent inferiority of nonwhites." Fredrickson distinguishes this new scientific racism from the "protoracist" prejudice and discrimination of prior centuries—that is, what might be better interpreted as xenophobia.[15] His use of the term "pseudoscience" is somewhat misleading, however. Largely forgotten today, ethnology, or the "science of race" as defined by its early champions, was familiar to many in the nineteenth century. In his influential 1854 tome *Types of Mankind,* Alabama physician Josiah Nott declared ethnology to be "eminently a science for American culture." Indeed, proponents of ethnology generally read into the body ideas about race that already permeated the national imaginary. Racial science shaped and was shaped by American race relations and political rhetoric. Scientists frequently used gender and sex difference to bolster their claims about racial hierarchy and police the category of citizenship, but precisely how sex and race intersected in scientific thought changed significantly over time, corresponding to specific sociopolitical concerns among the scientists.

Though Nott was careful to define the term "ethnology" for his mid-nineteenth-century readers, the principles of scientific racism were by then at least a half-century old, even if ethnology as a defined field was still developing. In the early colonial period in North America, European colonists initially distinguished themselves from Africans and Native Americans not on the basis of color distinctions but rather through asserting religious and cultural differences. However, since cultures could adapt and people could convert, it soon became clear that society needed to be organized around differences that were both more visible and permanent if Europeans were to build and maintain their own power, authority, and dominance. Color filled the need for a means of conferring status that was immutable and readily apparent, and color thus gradually became the foundation through which difference was constructed and maintained. Over time, Americans no longer considered themselves Anglo-Christians, distinguished from the heathens in their mix, but as "white." An inchoate and amorphous category, whiteness was defined not so much by what it was as by what it was not—Native American or African. More important, though, in the developing nation leading up to and after independence, "white" was a category with increasingly tangible rewards, including the rights and privileges of citizenship.[16]

The birth of the United States represented a pivotal and complex moment in race making. The language of equality, protection from tyranny, and the "rights of man" used in American Revolutionary rhetoric seemed to conflict with the reality of a slave society—an incongruity that eighteenth-century African Americans certainly noted. However, many historians have argued that the Founding Fathers actually intended their vision of a democratic society to apply only to white, propertied men. Most notably, Thomas Jefferson, who wrote so eloquently about the ideals of democracy in the Declaration of Independence, was himself a slave owner and yet publicly expressed deep ambivalence over slavery. Despite conflicted sentiments over the institution, Jefferson proposed in his widely read text *Notes on the State of Virginia,* first published in the early 1780s, that blacks were most likely naturally and irreconcilably inferior. In his *Notes,* Jefferson, an early theorist of scientific racism, applied the principles of animal husbandry to a lengthy and strange discussion of racial mixture and expressed clear distaste for black physiognomy, which he deemed an "eternal monotony" and an "immovable veil of black which covers all the emotions of the other race."[17]

A virtual microcosm of the racial tropes and themes that later characterized ethnology, Jefferson made a number of claims in *Notes* about black people specifically and about racial difference more generally that would continue to echo in scientific writing on race well into the twentieth century. Notably, gender played a key role in Jefferson's assessment of racial difference and hierarchy. "A black, after hard labour through the day, will be induced by the slightest amusements to sit up till midnight, or later, though knowing he must be out with the first dawn of the morning. They are at least as brave [as white men], and more adventuresome. But this may perhaps proceed from a want of forethought," he wrote. His argument here foreshadowed paternalistic defenses of slavery by antebellum scientists; black men were suited to hard labor but impetuous and incapable of acting in their own best interests. Even areas in which they seemed to compare favorably with white men—"they are at least as brave, and more adventuresome"—instead underscored racial difference and gender deviance. Bravery in white men was simply that, while apparent bravery in black men stemmed from their recklessness, impugning their manhood rather than recommending it.

Most important among his many claims about race, though, was that racial differences were not cultural but rather natural and rooted in the body itself. This belief in particular drove racial science and remained a central tenet in the field even as other concerns, debates, and frameworks in the field changed over time. Though Jefferson expressed some uncertainty as to exactly from where in the body race sprung, he was an early proponent of the idea that it was both biological and pervasive. Jefferson also assumed racial difference held tremendous importance. "This unfortunate difference of colour, and perhaps of faculty, is a powerful obstacle to the emancipation of these people," he maintained. Moreover, he found the black and white races so different that he suggested that they might be entirely different species. Near the end of his discussion of slavery and race in *Notes,* Jefferson offers a call to arms for further study into racial difference, writing, "It is not against experience to suppose, that different species of the same genus, or varieties of the same species, may possess different qualifications. Will not a lover of natural history then, one who views the gradations in all the races of animals with the eye of philosophy, excuse an effort to keep those in the department of man as distinct as nature has formed them?"[18] Scientists began to take up Jefferson's call shortly thereafter and more widely still in the nineteenth century.

Scientific Bodies

But it is also important to note that Jefferson himself was writing at a critical moment in the development of the biological and natural sciences. The eighteenth century was the Age of Enlightenment, the Age of Reason. Science gained a new place of prominence and respect in Europe and North America, with particular emphasis on understanding the natural world and, in the wake of Swedish naturalist Carl Linnaeus's taxonomy system, classifying it as well. As Londa Schiebinger demonstrates in *Nature's Body: Gender in the Making of Modern Science,* gender profoundly shaped how scientists viewed and understood that natural world. Plants took on gendered human characteristics in botany texts that read like romantic novels, for example. Human beings, of course, were a part of the natural world, and scientists were also increasingly interested in biological differences between men and women and the implications of anatomy for their place in the social order. Thomas Laqueur's *Making Sex: Body and Gender from the Greeks to Freud* examines an eighteenth-century shift from a one-sex model of sexual difference in medical and cultural discourse, whereby women were viewed as imperfect, underdeveloped men rather than a separate sex, to a two-sex model with women and men positioned as distinct and opposing sexes. This shift illustrates Laqueur's assertion that the body, and seemingly immutable anatomy itself, is viewed through a cultural lens. In both models, anatomists and medical scientists interpreted the body and sex difference in such a way that they reinforced the existing cultural and political ideology about gender roles in society. In other words, the sexed body both reflected and determined the gendered sociopolitical order, regardless of the era in which it was conceived. For Laqueur, sex, like gender, has been—and continues to be—made.

Other scholars have similarly demonstrated the ways in which the scientific body is inextricably tied to the historical and cultural context in which scientists interpret it. In examining the changing medical treatment of intersexuality in England and France in the nineteenth and early twentieth centuries, historian Alice Dreger reveals how scientists and physicians have at times quite literally constructed sex. With advancements in medicine and surgical technology, physicians attempted to manipulate sexually ambiguous bodies to fit into increasingly rigid gender roles—though the people inhabiting those bodies were often more resistant than they expected.[19] Echoing Dreger, as well as Schiebinger and Laqueur, Lisa Moore and Adele Clarke's "Clitoral Conventions and Transgressions: Graphic Representations in Anatomy Texts, c. 1900–1991" argues that anatomy, or specific parts of the body, can take on enormous cultural significance and reflect the gendered assumptions of both the scientists and the society in which they are writing.[20] Anne Fausto-Sterling's *Sexing the Body* unpacks various theories of sex difference from scientific, medical, philosophical, and historical texts and critiques what she sees as the false dichotomy between nature and nurture, asserting that the boundaries between the body and the world outside are more permeable than we think. Most notably, Sterling proposes that sex, like gender, be read as a spectrum or fluid continuum. The body is a site of knowledge production, she maintains, and science is the tool with which it is produced—though never in isolation from cultural influences. Moreover, she notes that the more feminists have attacked the norms and limitations of gender, the more scientists have been determined to find and prove biological differences of sex in an attempt to naturalize social distinctions or inequalities. As I demonstrate, scientific studies of race in the long nineteenth century followed a similar trajectory. While scientists continuously drove American discourse on race in the nineteenth century, they were also in many regards reactionary. As nonwhite races—and women—resisted the sociopolitical limitations they faced and challenged white male dominance, scientists rallied to ground their power in biology.

Jefferson's *Notes on the State of Virginia* was very much a product of the eighteenth-century scientific genealogy that Laqueur and Schiebinger describe, but it also anticipated the direction science would take in the nineteenth century. With a well-established model of using the body to naturalize gender roles and justify social divisions based on sex, scientists began to utilize the logic of "biology is destiny" in regards to race. Echoing the notion of gender differences rooted in biology that could never be legislated away, scientists in Jefferson's wake moved increasingly away from environmental concepts of race and spoke instead of the immutable and "natural" corporeal differences of African Americans in particular, differences that irrevocably determined their social inferiority. As the male–female binary was increasingly bifurcated and "legitimized" by scientists, so too was the binary of black–white. Grappling with the kinds of questions Jefferson had raised, the field of ethnology—which its adherents defined as the "science of race"—developed gradually over the first decades of the nineteenth century and coalesced in the United States in the 1830s with the work of Samuel Morton and the elaborate racial taxonomy he created based on his enormous collection of human skulls, discussed in chapter 1.

Europe had long seen a booming scientific literature on race dating to the early modern period, prompted by the diverse new peoples Europeans continuously encountered during the age of exploration and colonization.[21] Numerous scholars have pointed out that these explorers seemed particularly fascinated by Africans' sex organs and sexual practices.[22] Europeans' scientific and popular fascination with sexual aspects of race later culminated in—and was exemplified perhaps most famously by—the tragic story of Saartjie Baartman, a young South African woman who was brought to London in 1810 and publically exhibited across Western Europe as "the Hottentot Venus," a living specimen of sexualized exoticism and racial difference.[23] In the late eighteenth and early nineteenth centuries, European scientists also increasingly looked to the newly formed United States of America as a perfect laboratory for their theories about race, while at the same time, American science was coming into its own and a growing number of U.S. scientists added their voices to the transatlantic conversation, with Morton among the most publicized. Indeed, contemporaries and modern scholars alike have characterized Morton as the founding father of the American school of ethnology, which was largely synonymous with the theory of polygenesis (the separate origins of the races) but also characterized by its proponents applying biology and history to issues specific to the U.S. racial context, such as chattel slavery and political debates over what to do with indigenous peoples within the new nation's borders. For this reason, then, I mark the 1830s as the primary starting point for this book, with the turn toward a distinctly American scientific literature on race.

Ethnology, Gender, and American Science

From the start, ethnology was not a field on the fringes of the scientific establishment. On the contrary, its proponents held degrees or professorships from universities like Princeton, Harvard, the University of Pennsylvania, and Yale; maintained memberships in established and well-regarded organizations like the American Medical Association, the American Philosophical Society, and the Academy of Natural Sciences of Philadelphia; and gave lectures in such venues as the Smithsonian and the Royal Society of London. Nineteenth-century practitioners of ethnology should be termed *scientists,* not to lend credence to their long-discredited claims, but rather because their contemporaries regarded such work as legitimate and their qualifications and authority were rarely questioned, even by those who might have taken issue with their conclusions.

Though some racial theorists self-identified as ethnologists or spoke directly to their desire to see ethnology recognized as a distinct discipline, particularly in the early to mid-nineteenth century, racial science was a truly interdisciplinary endeavor with wide-ranging social and political concerns. A myriad of scientific disciplines were brought to bear on questions of race in the nineteenth century, including biology, comparative anatomy, medicine, linguistics, Egyptology, anthropometrics, and, later in the century and into the twentieth, social hygiene, criminology, cultural anthropology, eugenics, and sexology as well. Moreover, many individual scientists bridged several of these fields in their professional interests or approaches. An abiding interest in human difference united these diverse scientific endeavors.

Ethnological writers were also engaged in a national conversation on race. To be sure, debates in the larger culture over slavery's efficacy informed many of the questions early ethnologists raised, but scientific critiques or endorsements of the institution did not break down neatly along regional lines, nor did such differences in political opinion preclude individual scientists from admiring—and citing—each other's conclusions on race more generally. That is, while ethnology developed as a field of scientific inquiry amid mounting political and cultural debate over slavery, individual ethnologists had complicated relationships with the institution. Numerous practitioners of ethnology themselves owned slaves and supported the institution personally and professionally; at the same time, some of the most adamant proslavery voices in ethnology were Northerners who did not themselves own slaves but defended racialized slavery in medico-scientific terms. Other proponents of ethnology in the North commented little on or even explicitly opposed the institution but seemed to welcome the national boost to their professional status when proslavery scientists or politicians cited their conclusions on racial difference. Still others criticized slavery as an abuse of a biologically inferior race, but they neither imagined nor promoted political equality between the races if slavery were to be abolished. Most white ethnological scientists across the political spectrum and regional divides insisted that biological distinctions between the races not only explained existing social and political hierarchies in America—and thus determined which races were worthy of citizenship—but also *necessitated* the social order.

While we do not typically think about scientists as mediating citizenship, that is precisely what the "scientists of race" attempted to do. As sociologists Michael Omi and Howard Winant remind us in their examination of racial formation in the second half of the twentieth century, "The state from its very inception has been concerned with the politics of race. For most of U.S. history, the state's main objective in its racial policy was repression and exclusion. Congress' first attempt to define American citizenship, the Naturalization Law of 1790, declared that only free 'white' immigrants could qualify. . . . Historically, a variety of previously racially undefined groups have required categorization within the prevailing racial order."[24] The U.S. "scientists of race" provided such categorization and spoke directly to debates over who should be included—or, more often, excluded—from the category of citizen, bolstering their claims with biological assertions and scientific "evidence." And in turn, at various moments in history when new groups sought to lay claim to citizenship rights, political leaders drew on scientific literature on race to justify their policies. Many scientists, for example, argued that African Americans were naturally servile, their bodies built for labor and their minds incapable of the reasoned thought needed for political involvement or leadership. Meanwhile, they touted the accomplishments of white men throughout history as the builders of great civilizations; the conquerors; the artists; and the writers of political doctrine, poetry, and prose. In the antebellum and Reconstruction eras, scientists often compared the "lower races" to women, particularly in intellectual capacity; their similarities to each other, in conjunction with the vast chasm that separated both from white men, illustrated that neither white women nor nonwhites were capable of political participation. Scientists in the late nineteenth and early twentieth centuries revised such characterizations in a far more ominous way, casting black men as a dangerous, menacing presence and white women as vulnerable outside the confines of domesticity, but that too functioned to underscore that neither had a place in American public life.

White men alone had a special biological mandate to full citizenship. Some scientists insisted they were merely categorizing the races and avoided commenting directly on politically charged issues like slavery or the franchise. However, their characterizations of the races as fundamentally different nonetheless had broad implications recognizable to their audience as well. In nineteenth- and early twentieth-century America, biology *was* destiny: different bodies inherently denoted different social roles—and by then, race and gender had replaced property ownership as the defining characteristics of U.S. citizenship. Historians Allison Sneider and Evelyn Nakano Glenn both detail the checkered history of citizenship in the nineteenth and early twentieth centuries, demonstrating that its definition and parameters were constantly being challenged—and ultimately reshaped—by women's suffragists as well as the larger sociopolitical developments of U.S. expansion, imperialism, immigration, and debates over free labor.[25] Faced with such challenges, racial scientists continuously reinforced the de jure and de facto definition of citizenship as white and male. Consequently, the contested nature of citizenship—political as well as medical—is a theme that threads throughout this book and continues to resonate in U.S. politics in the twenty-first century. Michelle Alexander's groundbreaking book *The New Jim Crow: Mass Incarceration in the Age of Colorblindness* reminds us that the criminal justice system has replaced slavery and Jim Crow segregation as a means of controlling black men and maintaining their second-class citizenship status.[26] Meanwhile, the election, and subsequent two-term presidency, of Barrack Obama saw many white voices rushing to declare America "postracial," others questioning his ability to lead utilizing rhetoric strikingly similar to ethnological scientists writing during Reconstruction, and still others (the so-called Birthers) challenging his U.S. citizenship entirely despite all evidence to the contrary, based in part on his racial heritage.

Likewise, racial science was essentially teleological. Rather than the strictly objective disciplines scientists often purported them to be, nineteenth- and early twentieth-century scientific endeavors on race took their cue from and reflected existing cultural ideologies and hierarchies. As feminist biologist Anne Fausto-Sterling reminds us, "'Science' is created by scientists." This seemingly simple statement underscores that *who* does science shapes the scientific knowledge that they produce. "Scientists are not neutral machines. They are social beings who have grown up in a particular society and have, through the process, acquired a particular framework or way of viewing the world," Fausto-Sterling notes. Further, "if science is made by non-neutral scientists, then it follows that science itself is not neutral. . . . Science is a construct created by a particular society which offers that society a particular view of the world. It changes and grows in response to societal changes, even ones which it may itself stimulate."[27] Similarly, as historians such as Laqueur have demonstrated in regards to sex and gender, scientists' expectations about race shaped what they saw in the body.

As a result, racial science was rife with claims that today seem contradictory and paradoxical, revealing the nature and extent of scientists' situational logic. For example, racial theorists often claimed that black men had longer arms proportional to white men, which they interpreted as a sign of the former's physical proximity to apes. But at the same time, they insisted that white men's greater amount of body hair was emblematic *not* of their own simian similarities but of their manliness. Beards could be particularly evocative for racial scientists. New York physician and proslavery propagandist John Van Evrie, for instance, argued that white men had fuller beards than all other races. It is important to note, however, that the beard was a marker not just of racial difference for Van Evrie but also of manhood. Gender alone, though, as an analytic category, does not fully encompass or explain the implications of Van Evrie's claims, or many similar examples in ethnology tracts. For Van Evrie and many of his contemporaries, black men were irrevocably excluded from the category of citizen or political actor not simply because of their allegedly deviant gender performance, presumably a potentially adaptable behavior, but because their very biology excluded them from the male *sex*.

Putting Science to Work

From its beginnings in the early nineteenth century, racial science was an applied science. These scientists considered a range of political and social issues in their work, including slavery, the Reconstruction amendments, education, sexual variance, and violent crime. They examined Civil War soldiers for both health defects and racial characteristics, penned letters to the Freedmen's Bureau and editorials to newspapers, gave public speeches on America's "race problem," endorsed political parties or even specific candidates, testified in court in "sexual perversion" trials, served on vice commissions, treated (and mistreated) patients, and made recommendations on lynching. Racial science very much had a public life outside the scientific and medical establishment. As physician and prolific racial scientist G. Frank Lydston articulated in 1912, "The medical doctor is the best social doctor."[28]

Accordingly, these scientists also directed their work toward ever-widening audiences. More and more over time, scientists articulated their conclusions about race and its sociopolitical implications in mass-produced pamphlets, university and medical school classrooms, mainstream periodicals, public lecture halls, and newspapers. And racial science indeed gradually *reached* larger segments of the public, in part because science and medicine had come of age, but also because the circulation of professional knowledge(s) in American culture had as well. The church and the increasingly professionalized sciences vied for cultural authority, while the nineteenth century also saw campaigns to bring both religion and education to the masses. With the growth of the publishing industry that produced (and mass produced) books and periodicals at ever-cheaper prices, the proliferation of newspapers, tent revivals, lecture circuits, and traveling ministers, a far greater number of American citizens than ever before had access to information from a diversity of sources. The Chautauqua movement of the late nineteenth and early twentieth centuries, which brought an array of performers, religious speakers, scientists, and specialists of varying stripes to the public, embodied this trend toward mass education. Scientific theories about race reached a wide audience through such lecture circuits and print media; ethnology was a science *for* the people as much as it was *about* people.

These scientists often saw themselves as reformers, upholding the natural order and curing society of its ills. While their claims seem anything but progressive today—for example, some scientists' impassioned defenses of slavery or other scientists' advocacy of castration as a humane alternative to lynching at the turn of the century and into the 1920s—it is vitally important to examine racial scientists in their historical context, not to excuse them, but to better understand the culture that produced them and the role they played in American society. Politicians and cultural commentators cited racial science to buttress and legitimize their claims and policies, while scientific theories about race manifested themselves in popular visual culture, from political cartoons to campaign posters. In the nineteenth and early twentieth centuries, racial science was mainstream science.

It was never uncontested domain, however. Quite the contrary, even as scientific racism took shape over the course of the nineteenth century, so too did a vibrant counterdiscourse challenging it. Some white scientists challenged key tenets of racial science—or even, on occasion, the whole enterprise—as being built on tenuous logic or as inherently biased. More poignantly, throughout the long nineteenth century, African American intellectuals disputed scientific claims about black inferiority, argued for their place in the human family, and laid claim a proud history of great civilizations that their white counterparts tried to deny them—in other words, they offered their own theories about race and what it meant. Inheriting their intellectual mantle, Walter White, leader of the National Association for the Advancement of Colored People (NAACP), insisted that racial science was among the causes of racial violence and strove to dismantle scientific racism as an integral—but often overlooked—part of his antilynching work in the 1920s and 1930s. I end my story there, with White's antilynching work, the last spectacle lynching in the United States in 1934, and the gradual decline of biological models of racial difference during this period as the influence of Franz Boas and cultural anthropology grew. The book, then, seeks to place racial science in a broader conversation in nineteenth- and early twentieth-century America, to unpack not just its genealogy but also its reception and social function. At turns embraced and contested, science was at the center of debates over the meanings of race, gender, and sexuality and the nature of power and privilege in the United States.

Race and Sex: Bodies under Threat

In the nineteenth and early twentieth centuries, the corporeal body—specifically, the sexed and sexualized body—was the primary site on which scientists examined the moral character and intellectual capacity of those Americans already outside the boundaries of the white, heteronormative nexus of social power and "respectability."[29] The body held tremendous meaning for racial and sex differences, meanings that constantly shifted according to the specific context and historical moment that led a scientist to "read" it. This project, then, pays close attention to change over time, but it also reveals the persistence of certain ideas about race in American science in the nineteenth and early twentieth centuries—namely, that racial differences were biological and permanent and social hierarchies were a reflection of the natural order. In the antebellum period, scientists looked at bodies across racial lines to answer the "origins question" of whether the races constituted separate species, and some looked to naturalize slavery by interpreting black bodies specifically as biologically designed for servitude. In the late nineteenth century, evolutionary theory provided an ideological framework in which racial scientists compared and ranked bodies along a hierarchical ladder of human perfection in which white men always occupied the apex. And in the context of locating racial character and destiny on the physical body, the same scientists also began to examine the bodies of homosexuals for proof of their biological deviance.

Indeed, as the nineteenth century drew to a close, ethnological scientists were increasingly interested in sexual activity as it related to the nation's "race problem." Concerned with a vast category they termed "sexual perversion," America's most prominent racial scientists also became the country's first sexologists. Under this framework of "sexual perversion," scientists were troubled by two manifestations in particular:

homosexuality among whites, which they saw as threatening the white race from within, and black male sexual aggression, which threatened the white race from outside. During the late nineteenth and early twentieth centuries, racial scientists offered solutions to the sexual threats they saw facing American society, threats that were also highly racialized.

Scientists' growing concern with homosexuality at the turn of the century was largely a product of their concern with the bodily integrity and sociopolitical status of the white race in the face of enormous social change in America. Similarly, more than any specific scientific discovery with regard to sexuality, scientists' reconceptualization of homosexuality as a medical pathology rather than a behavior, with supposedly correlated mental and corporeal imperfection, reflected their understandings of race and gender as biological entities. Scientists mapped these interrelated categories of difference—race, gender, sexuality—onto the body, as physical manifestations of existing social power dynamics and cultural hegemonies.

This book is informed by and in dialogue with a number of historical and scholarly literatures. Part of an emerging scholarship that examines the intersection of race, gender, and sexuality, it brings together histories that are often presented separately—including U.S. history, the history of race and ethnicity, African American history, cultural and intellectual history, the history of sexuality, women's and gender history, and the history of science and medicine. Though grounded in historical research and methodologies, my work also draws on and speaks to a number of interdisciplinary fields, including critical race studies, feminist science studies, gender studies, and queer studies, as well as the rich array of scholarship on the body within and across a variety of disciplines. It follows in the tradition of analyzing gender, race, and sexuality as cultural constructs—though one suspects that historians have only uncovered the tip of the proverbial iceberg when it comes to understanding the complex relationships among these constructed categories of difference.

The first two chapters provide a gendered analysis of scientific racism from the antebellum era through the turn of the century—a framework all too often missing from scholarship on American "sciences of race." Seminal texts by George Fredrickson, William Stanton, Steven Jay Gould, John Haller Jr., and Winthrop Jordan have made undeniably valuable contributions to understanding racial thought in nineteenth-century America and continue to be widely cited today.[30] However, most of these foundational works that address scientific racism were written prior to the rise of women's history as a subfield or, to borrow a phrase from historian Joan Scott, before gender became accepted as "a useful category of historical analysis."[31] The role of gender in ethnological understandings of racial difference, particularly during the mid-nineteenth century, is largely absent from these works. At the same time, this book is also indebted to more recent scholarship that has begun to move the historiography on scientific racism in exciting new directions. Nancy Stepan's work on gender, race, and science in England and Latin America, for example, has opened the door for a gendered analysis of scientific racism in the United States, which this book seeks to do.[32] And Mia Bay's study of black thought about white racialist science, and white people more generally, reminds us of the challenges of confronting hegemonic ideologies and just how high the stakes of ethnology were.[33]

More than simply a corrective to earlier scholarship on scientific racism then, close attention to gender in medico-scientific thought on race challenges traditional understandings of the cultural politics of ethnology and demonstrates that the tone, focus, methodology, and intended audience of American racial science changed considerably over time, reflecting larger transformations in U.S. politics and society. As it did, scientists asked different questions about the sociopolitical implications of race and looked to different parts of the body for evidence of racial difference. Put simply, ethnology in the 1830s and 1840s tended to focus on more gender-neutral features like skulls, hair, and skin (itself all the more striking considering European scientists' fascination with Africans' sexualized bodies and behaviors earlier in the century and prior), while in the years surrounding the Civil War, when scientists took up the profoundly gendered issues of slavery and citizenship, they also featured secondary sex characteristics such as beards and breasts in their bodily analyses. And after Reconstruction, American racial scientists even more overtly engaged not just with gender but also with sexuality and sexualized bodies. Amid frequent discussions of miscegenation and "race suicide," as well as the rise of eugenics—a field that quite literally embodied the nexus of the "sciences of race" and the "sciences of sex"—they added reproductive organs to the list of physical characteristics they investigated for evidence of racial difference. That is, sexual features were *socially* relevant to ethnological scientists' readings of race at the turn of the century in ways they were not earlier in the nineteenth century. While which body parts most received racial scientists' attention changed over time, to some extent so too did what each part signified. One thing changed very little, however: no matter where in the body white racial scientists looked and when, they interpreted what they saw through the lens of pervasive difference rather than similarity.

These developments in ethnology were profoundly interconnected in ways not yet explored by historians of nineteenth-century America. While George Fredrickson has highlighted the increasingly proslavery stance of many ethnological theorists beginning in the 1850s, historians have not examined the impact of this development on gender in scientific considerations of race, for example.[34] Indeed, historical accounts of racial science before the late nineteenth century have paid little attention to gender. Scholarship on racial thought at the turn of the century, on the other hand, has examined the construction of black men as sexual predators during that period and analyzed the centrality of sex and gender in evolutionary theory; however, this scholarship largely attributes scientists' attention to gender to the rise of social Darwinism rather than an extension of earlier developments. The first two chapters, in contrast, reveal that gender *was always* integral to scientific understanding of race from the start, but the function it served changed over time with shifts in the nation's cultural and political landscape. Consequently, they detail a series of "call-and-response" moments in which scientists reacted to and offered their expertise on these shifts while underscoring ethnology as an interracial site of contestation and race making.

Chapter 1 traces the rise of biological paradigms of racial difference and the development of ethnology as a field of scientific inquiry. It argues that early ethnology was entirely male dominated; it focused on men, presumed a male audience, and was produced by male scientists. For example, the question of racial origins—the dominant concern in ethnology prior to the Civil War—was framed in terms of male lines of descent. Scientists debated whether the black and white races alike descended from Adam (rather than Adam and Eve) or if the black race could be traced to Ham, the cursed son of Noah—a racial theory that remains alive and well among modern white supremacist organizations like the Aryan Nation. At the same time, black intellectuals directly challenged the tenets of scientific racism, particularly the theory of polygenesis that placed African Americans outside of the human family, and created their own models of race—though these too were focused on men. When racial scientists did discuss women, it was usually in the context of their considerations of slavery, which often pointed to slave women's reproduction as indicative of the system's overall health and viability. Ethnology's adherents never presented a united front on America's "peculiar institution," however, and prior to the 1850s and 1860s, when the issue heated up in U.S. political culture in response to the increasingly powerful abolitionist movement and debates about the expansion of slavery into the Western territories, it was common for racial scientists to ignore the issue entirely. As this chapter reveals, gender was central to ethnological arguments both for and against the institution. As with other issues in antebellum ethnology, scientists focused more often on men, but "man" itself was a normative category that was subject to little scrutiny. With few exceptions, it was not until the Civil War that scientists began to explicitly analyze manhood—its physical parameters and political import—vis-à-vis questions of race.

Chapter 2 describes this transitional moment in American racial science, when scientists became deeply invested in defining what manhood meant and to whom it applied.[35] During the Civil War, military scientists conducted large-scale anthropometric studies of union troops, quite literally measuring manhood along racial lines, and attempted to ascertain what, and who, constituted the ideal soldier in both body and character. Their interest extended beyond the military context, however, and they also sought to qualitatively and quantitatively define the "average man" and, by extension, the best man. During Reconstruction, ethnological scientists turned their attention from the question of black men's fitness for military service to their fitness for full citizenship, a political category attached to men alone. Many argued against black male suffrage by insisting they were not really "men" in body or behavior, and scientists often compared black men and women as having similarly limited intellectual capacities to illustrate that neither deserved the vote. With Emancipation and black and women's suffrage posing a challenge to the traditional American definition of citizenship, scientists rallied to reassert the citizen body as inherently both white and male. The economic and political competition that the freed black population had posed during Reconstruction was met with violent backlash in the period that followed, as the South was "redeemed" for white supremacy in the late nineteenth century. More generally, white scientists were concerned by reproductive competition and threat as well, that the nation's balance of power could come down to a contest of numbers and that miscegenation represented a menace to American civilization—a sharp contrast to antebellum racial theories that minimized the realities of sexual exploitation under slavery by writing off mixed race people themselves as inconsequential, sterile, and destined to die off. In this context, with the integration of evolutionary theory and social Darwinism into American racial science, ethnological critiques of black men's capacity for citizenship became more sexualized in the late nineteenth and early twentieth centuries. The American obsession with race and reproduction, in popular and political discourse as well as scientific thought, during this period intersected with the nation's imperial project, debates over immigration, anxieties about the free black population, and the eugenics movement.[36] This chapter situates that obsession within a larger chronology of scientific thought in which scientists continually adapted their use of gender and sex in conceptualizing race to new political and cultural circumstances and looked to different parts of the body for racial truths with each shift.

At a time when many Americans believed that reproduction and scientifically strategic breeding would determine the fate of civilization and white supremacy, turn-of-the-century scientists were quite preoccupied with sex. They subsumed under a vast umbrella category a polyglot of sexual acts, crimes, and social trends—real and imagined—they perceived as threatening the natural order. And indeed, American society *was* changing in the wake of Emancipation and the advancements of Reconstruction; the waves of immigrants flooding U.S. cities and the workforce; and women's rights advocates pushing for educational opportunities, the franchise, and the ability to control their own reproduction. For their part, scientists, representing a range of subfields, medical specialties, and professional disciplines, attempted to combat this gender and racial disorder—which challenged their own position at the top of a white patriarchal hierarchy—by controlling sex. Chapters 3 and 4, then, examine the ways in which scientists sought to maintain their authority and cultural relevance by speaking to the nation's intersecting anxieties around race and sex and promoting "practical," medically based solutions for the nation's social ills.

Scholars are just beginning to explore the intersections of race and sexuality—and queer sexualities in particular—in U.S. history. Lisa

Duggan, Siobhan Somerville, Kevin Mumford, Julian Carter, Sharon Patricia Holland, and Nancy Ordover have raised important questions about these intersections in American culture, revealing exciting new insights by applying queer theory to studies of race.[37] Duggan, Somerville, Carter, and Ordover also underscore the racial politics driving scientific studies of sex and the pathological models of same-sex sexuality that emerged at the turn of the century, a topic I take up at length in chapter 3, which extends and deepens such analyses by demonstrating that racial science and sexology were not separate fields in America. They not only shared similar methodologies and concerns, as previous scholars have pointed out, but they literally shared the same key scientists. Moreover, although turn-of-the-century scientists disagreed over the distinctions or overlap between homosexual acts, gender deviance, and physical hermaphroditism, scientific assessments of the causes of "sexual perversion" frequently broke down along racial lines. "Sexual perversion" among nonwhites and the lower classes was characterized as vice, indicative of the physical and moral degeneracy of the group. In contrast, scientists usually read cases of "sexual perversion" or inversion in middle- and upper-class whites as symptomatic of individual pathology or disease, which many attributed to "overcivilization." While African Americans, immigrants, and the lower class threatened moral corruption through the spread of their sexual "vice," white homosexuals from respectable families threatened "race suicide."

Meanwhile, chapter 4 argues that emasculation was central to both scientific and popular discourse on lynching as well as its practice. The waves of lynchings and the demonization of black men as rapists during the late nineteenth and early twentieth centuries are well known. But less well known are scientific responses to lynching. For more than three decades, numerous American scientists recommended surgical castration as an alternative to lynch violence. Far from marginal, these prominent scientists had a profound impact on American society outside the medical establishment. They lent scientific support for the "black beast rapist" trope but positioned themselves as progressive reformers offering a medical solution to the problem, spurned on in part by increasingly negative publicity on lynching itself. In so doing, they sought to place the authority over America's "race problem" in the hands of scientists rather than mobs. Meanwhile, lynch mobs often included castration as a crucial part of—rather than a substitute for—the lynching spectacle. Rather than an anomaly, the "castration remedy" represented a culmination of scientists' intersecting concerns with race and sex, as well as their role as "social doctors" fixing America's ills.

Not everyone thought science offered the cure to America's issues around race, however. In the 1920s and 1930s, NAACP antilynching activist Walter White identified racial scientists as part of the problem, not the solution. Sharing an intellectual genealogy with nineteenth-century black thinkers who challenged the core tenets of ethnology, he pointed to the impact of what was by then a century of scientific racism on the way living, breathing human beings were treated in American society. As demonstrated in chapter 5, White made debunking biological theories of black inferiority a key component of his antilynching work with the NAACP and sought strategic alliances with sympathetic scientists who could help him turn science against itself. That he was able to build such alliances with prominent *white* scientists in particular was a testament to his tenacity, to be sure, but also signaled a crucial moment in American history in which the logic of racial science was increasingly under attack. Throughout his work, White critiqued the black rapist trope that so often served to justify lynching, but he shrewdly robbed this sexualized rhetoric of the legitimacy it had long received from science by attributing it only to ignorant mobs rather than physicians' pens. While he was not the first black thinker to suggest that scientific racism precipitated racial violence, White had something his predecessors did not: powerful white friends, political influence through his leadership in the NAACP, and a wide audience that crossed the color line.

Above all, this book is a story about how categories of human difference are created, maintained, and contested and the role of science in that process. Throughout much of the nineteenth century and into the twentieth, American scientists were engaged in a complex examination of race, sex, gender, and sexuality. These categories were perpetually in flux, and the body was the site on which they were built. Ultimately, they were not just categories of difference, however; they were, and often continue to be, categories of sociopolitical exclusion. Accordingly, several interrelated scholarly and historical concerns have driven this work, including the function of science in American society; the development of racial science as a field and its authority over sociopolitical issues and debates; the racialization of homosexuality and the concurrent rise of sexology; challenges to scientific claims about racial inferiority; the social construction and intersection of race, gender, and sexuality; and the body as a cultural entity and site of contestation. But while scholars generally talk about these categories as social constructs, they continue to have undeniable cultural and material salience in America.

· CHAPTER 1 ·

"Races of Men"

Ethnology in Antebellum America

HAVE THE WOOLLY-HEADED RACES OF men ever produced one, even only one man[,] famous as either a lawgiver, statesman, poet, priest, painter, historian, orator, architect, musician, soldier, sailor, engineer, navigator, astronomer, linguist, mathematician, anatomist, chemist, physician, naturalist, or philosopher?" asked John Campbell in his 1851 tome, *Negro-mania: Being an Examination of the Falsely Assumed Equality of the Various Races of Men.* An Irish-born, highly successful book publisher, Campbell was the first secretary of Philadelphia's Social Reform Society, the founder of several similar organizations, including the Social Improvement Society, and head of the city's typesetters union. A civic leader, he had also become enmeshed in the volatile racial politics of his adopted nation and posed the previous loaded question in what he called his "text book for white men on the subject of negromania."[1]

Campbell's interest in publishing a book on race first took shape after he attended an eight-week discussion series at the Social Improvement Society considering the question "Can the Colored races of men be made mentally, politically and socially equal with the white?"[2] Participants included influential Philadelphians from across the racial spectrum, including prominent abolitionist Robert Purvis, the Charleston-born son of a cotton broker and a free black woman. Sympathetic to working-class whites, Campbell had long been an outspoken proponent of the labor movement—the same political beliefs that had prompted his immigration to the United States from Great Britain as a young man. He came away from the discussion series unsettled by the arguments for racial equality made by Purvis and other speakers. Eager to address the question and keenly aware that not all white men had equal access to the enormous body of scholarly literature on race, he wrote *Negro-mania,* a mass-market publication contesting the "falsely assumed equality" between the "various races of men." The longtime resident of the Philadelphia's Germantown neighborhood, a hotbed of abolitionist activity by the mid-nineteenth century, Campbell took little interest in slavery prior to his publication of *Negro-mania.* But as slavery became increasingly contested, the staunch supporter of the rights of white working men was alarmed.

Campbell feared abolition would have disastrous implications for white labor. Cotton drove Northern industry and all the various trades that supported manufacturing, but the "indolent" black race—the only race capable of laboring in the Southern heat in his estimation—would not work picking cotton unless forced. Campbell predicted that the industry would collapse nationwide, leaving hardworking white "mechanics, bricklayers, carpenters, machinists, [and] miners" without gainful employment and suddenly having to support a lazy, thieving, and entirely dependent African American population.[3]

Though they chided him for his writing style and lack of original contribution to the field of ethnology, Campbell's evolution on the issue of slavery and his critique of the misguided and impractical abolitionist fervor surrounding him in the "City of Brotherly Love" earned him praise from the editors of the *Southern Quarterly Review* in a lengthy book review and essay.[4] By the time of Campbell's writing, ethnology, or the scientific study of race, had become an established field of scholarly inquiry in the United States and an influential source of expertise on social and political issues for a variety of both pro- and antislavery readers and leaders across the country.

Largely a synthetic text, *Negro-mania* summarized the arguments of prominent ethnologists and recounted the field's central debates. Today, the book can also be read as summary of antebellum ethnology's methodological approach and theoretical underpinnings. First, as illustrated in the book's title, Campbell's approach to questions of race was narrow in scope. For all that he subtitled his book—*Being an Examination of the Falsely Assumed Equality of the Various Races of Men*—the title *Negro-mania* accurately reveals that his primary interest was in the black race specifically. More striking still is the book's gendered focus. Although *Negro-mania* ostensibly studied human groups composed of men and women, its assessment of the history and capacity of the black race speaks only of men. Campbell's reference to "races of men" and his use of male pronouns may at first appear simply to reflect popular linguistic convention, in which the word "men" is used to mean humans more generally.

But as Campbell's list of esteemed male occupations in which blacks had presumably failed to succeed at any point in history suggests, *men* were indeed his primary subject. Like the scientists whose work he describes, Campbell presents ethnology as being for and about men.

On the whole, the antebellum-era scientific work on race contains far fewer references to women than men. It should not, however, be interpreted as a gender-neutral discourse but rather one in which men were rendered normative and race was conceptualized primarily in terms of male lineages and patriarchal destinies. More overtly visible in some contexts than others, gender was always a part of the story of scientific race making, but its points of emphasis and foci shifted across time and space. Tracing the uses of gender in scientific arguments about race complicates and deepens our understandings of both the cultural politics and the agendas of the individual scientists who wrote about ethnology and of the sociopolitical climate of antebellum America more generally.

Especially in its discussion of gender, nineteenth-century scientific thought about race changed over time, revising itself to meet the ideological needs of a nation divided first over slavery and later over the status that blacks would occupy as a free people. Manhood, citizenship, and gender were interlinked in both these debates but in different ways; close examination of those differences reveals how nineteenth-century experts on race adapted their scientific work on the subject to meet the changing times in which they lived. Never divorced from the social questions of its day, ethnology was a product of men who sought to bolster the field's influence in an era when science more generally was enjoying increasing authority in American society. And many aimed to build national reputations for themselves individually through their contributions to a cutting-edge new scientific field that spoke to societal concerns.

With these developments in mind, we must ask why the antebellum-era scientists who wrote about ethnology focused so heavily on the bodies on men. Scientific writers were far more engaged with questions of gender deviance and sex difference in the postbellum period and the turn of the century, when they were also more likely to study female bodies as well as male. But in the decades prior to the Civil War, ethnologists largely ignored such subjects, instead exhibiting a steadfast preoccupation with less sex-specific characteristics like skulls, brains, bones, skin, and hair as they constructed a hierarchy among men and debated such hotly contested issues as slavery and the origins of the races.

Ethnology was a field of scientific inquiry premised on the belief that the physical body revealed the intellectual, moral, and political capacities of its owner and, by implication, those of his or her race writ large. The slavery question formed the backdrop for antebellum ethnology and likewise shaped its major themes and gendered points of emphasis. Men emerged as its central subject in the context of a scientific discourse largely dedicated to establishing hierarchal taxonomies of race that drew on biblical accounts of the descent of man through a series of ancient patriarchs. Even ethnologists who questioned the shared origins of the races were likewise preoccupied with the male lineage of the races—they questioned whether blacks were "sons of Adam." At stake in such debates was the question of whether slavery was natural and whether blacks would ever qualify for freedom. In documenting that blacks were the product of a long history of divinely ordained subjugation and that it was a status to which they were naturally suited, some ethnologists hoped to resolve this question for once and for all. In contrast, others maintained that all men descended from common ancestors and that the status of the races had varied over time and in different locations—and thus all were capable of enacting their "manhood rights" within the right circumstances.

Race in this literature was largely a matter of male descent. Ethnologists traced the history of each race through the deeds, accomplishments, and values of its men. Furthermore, though there was much disagreement in the field whether the races shared a singular origin, ethnologists on both sides of the most ubiquitous debate in antebellum ethnology framed their arguments in terms of male lineages in which women seemingly played no role. Antebellum ethnologists occasionally discussed women, but when they did, it was generally in relationship to patriarchal institutions of power, male property ownership, or masculine concerns. For example, numerous ethnologists invoked women's bodies—particularly the productive and reproductive bodies of black women—when defending slavery as a "healthy," naturally expanding institution under benevolent plantation paternalists. The focus on women's bodies in discussions of slavery specifically—and, to a lesser extent, of racial mixture—was a minor theme in antebellum ethnology, however.

Men were both the producers and the primary objects of this scientific discourse, and they wrote at a time when women had little claim to institutional power in America: it was the past, present, and future of the male sex that held the most consequence. In developing a new field of scientific inquiry, these racial theorists constructed a world inhabited by men, in which gender difference was subsumed and supplanted by the rubric of race.

Nineteenth-century experts on race, the men who wrote about ethnology, held a range of political beliefs and came from a variety of personal and professional backgrounds, but most held medical degrees and a number taught biology, anatomy, physiology, archeology, linguistics, natural history, or related topics at universities or medical schools. Though some posited that the races differed in only superficial ways and shared the same potential in comparable environments, most were proponents of scientific racism rather than human equality, and ethnology's advocates were well represented on both sides of the Mason-Dixon Line. In fact, many of them lived in the North, which contained far more medical and scientific institutions than the South. Although now more associated with the Declaration of Independence and the birth of a nation that espoused freedom, liberty, and equality, Philadelphia was perhaps the premier site for ethnological work. The city was home to a wealth of publishing companies and medical schools as well as an active abolitionist movement, all of which fueled the production of ethnological work espousing racial difference—though individual scientists varied considerably in how they characterized the social and political implications of such differences. Such apparent contradiction was hardly unique to Philadelphia though. In the nineteenth century, republican ideology and racist doctrine often shared the same proponents, with some racial theorists even arguing that democracy for white men was dependent on the enslavement of blacks.[5] Antebellum writers of ethnology included working-class men as well as scientific scholars and blacks as well as whites. Indeed, black thinkers who responded to ethnological arguments about racial hierarchy—especially and most poignantly those claims that black people did not belong to the human family—faced the unenviable task of defending their race within a field that rendered them incapable of rational thought. But ethnology saw virtually no female proponents, even though women of this era were beginning to make headway in other fields of science such as medicine and astronomy.[6] Before the Civil War, ethnology was predominantly a science by and about men.

Who Was Born of Adam . . . and Whatever Happened to Eve? The "Great Question" of Antebellum Ethnology

More than any other topic in ethnology prior to the Civil War, the issue of the original unity or diversity of the races was both contentious and ubiquitous. A quarter of all scientific publications on race between 1830 and 1859 were explicitly focused on the question of racial origins, while many texts focused on other themes were shaped by or employed toward the debate.[7] For example, studies of crania regularly provided "empirical evidence" on the plurality of racial origins, and discussions of racial mixture and hybridity frequently intersected with the origins issue as well. Numerous racial theorists claimed that the offspring of black and white parents could produce no "permanent stock," thus proving the races to be separate species entirely. Like most discussions in antebellum ethnology, racial scientists framed their considerations of racial origins in terms of men. They often pondered whether black men were the sons of Adam or if all the races had descended from one father, not whether the races descended from Adam and Eve. Similarly, they charged that African Americans were the "sons of Ham" as evidence of their long-standing inferiority and position of servitude. The "Curse of Ham" myth derived from an odd biblical story in which Noah punished his son Ham for viewing Noah's naked body by declaring Ham's son Canaan to be "the slave of slaves." All of Canaan's descendants were likewise cursed to a life of servitude.[8] Just as surnames and property were passed down patrilineally in nineteenth-century U.S. society, so too was racial lineage an issue of male descent.

A brief review of the racial origins debate illuminates ethnology's development as a field, and one in which masculinity was simultaneously central and implicit.[9] In his 1863 text *The Races of the Old World: A Manual of Ethnology,* New York social worker and minister Charles L. Brace described the debate over "the Unity or Diversity of Origin of Mankind" as "the great question at the basis of the Science."[10] Brace was precisely right in his identification of the "great question" at the heart of ethnology. Indeed, up to the early 1860s, questions about the original unity or diversity of the races constituted both the driving force and central theme of most ethnological writing (see Figure 1.1). Likewise, discussions of this debate have dominated historical scholarship on scientific racism.

Throughout the first half of the nineteenth century, ethnological arguments about race centered on the debate between polygenesis and

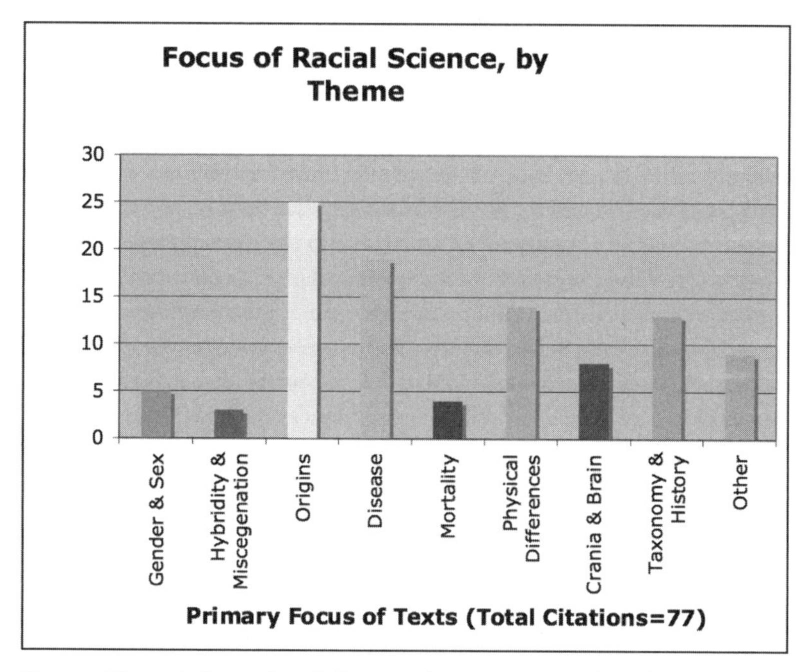

Figure 1.1. Thematic focus of antebellum racial science texts. *As shown here, the debate between polygenesis and monogenesis dominated antebellum ethnology, with 25 percent of texts on race published during this period explicitly focusing on the origins question. Moreover, even publications that were purportedly focused on other topics usually addressed the question somewhere in the body of the text, at least in passing; studies that compared cranial shape or capacity in particular were often employed toward an argument about the origin of the races. Meanwhile, few ethnology texts during this period focused explicitly on gender or racial mixture, which came to be of major scientific concern later in the century (graph by author; data based on the* Index-Catalogue *of the National Library of Medicine; see appendix for discussion of methodology).*

monogenesis. Presbyterian minister and president of Princeton University (then known as the College of New Jersey) Samuel Stanhope Smith made a case for monogenesis, or the shared origin of the races, in 1787 with a speech presented before the American Philosophical Society in Philadelphia. With its publication shortly thereafter and an expanded version reprinted in 1810 titled *Essay on the Causes of the Variety of Complexion and Figure in the Human Species,* Smith became an early and widely respected authority on race. Smith's professional training was in theology rather than medicine or science, though he employed a naturalistic explanation for skin color variation among human beings. He argued that all races were members of the same species and shared a common ancestry. Current physical differences resulted from environmental factors, particularly climate and the divergent lifestyles of "savagery" and "civilization." Unlike most proponents of ethnology, Smith was almost entirely gender neutral in discussion of the effects of climate and lifestyle on human complexion. He used terms like "mankind" in adherence to the masculine generic language convention of the day but inferred through the context of his words neither an implicit nor explicit focus on men alone. However, like most contemporaries who subscribed to the theory of monogenesis, he argued that other races had degenerated from the white, superior race—the human norm by which all others were defined as deviant. Smith believed that blacks could become equal to whites, subject to the same environmental and lifestyle conditions, but only by literally turning white through subsequent generations—through both adaptation to the environment and mixture with whites.[11]

Although Smith's tacit acceptance of miscegenation raised some eyebrows, his theories regarding the origin of color differences between the races went largely unchallenged until the publication of *Thoughts on the Original Unity of the Human Race* in 1830 by North Carolina–born Charles Caldwell. After obtaining his medical degree from the University of Pennsylvania in 1796, Caldwell taught natural history at the university and maintained a practice in Philadelphia before moving to Kentucky in 1819 to help develop two medical schools, first at Transylvania University in Lexington and then at the University of Louisville. Caldwell had begun to attack Smith's argument in a series of essays dating back to 1811, and some of Caldwell's proponents even credited his fiery essays with hastening Smith's death, a charge Caldwell himself denied.[12] Caldwell argued for polygenesis, or the separate creation of the races as distinct species. He drew on biblical chronology and asserted that the "superior" white intellect could not be due simply to differences in environment but rather must be an innate "gift of nature."[13] Answering the vocal critics who rejected claims of the separate origins of the races as heretical because they diverged from the biblical account of Genesis, he positioned himself as a man of science and reason who respected theology but also sought truth, which "can never prove unfriendly to sound religion." At the same time, scholars have described Caldwell as a man who seemed to court controversy.[14]

Moreover, the gender implications of Caldwell's emphasis on reason and intellect were twofold. For ethnological theorists like Caldwell, reason and intellect indicated the superiority of white men as a group over both women and the "lower races." And by positioning himself specifically as eminently reasonable in contrast to his critics, he simultaneously questioned both their scientific soundness and their manhood. In the nineteenth-century gender ideology of separate spheres, emotion and religiosity were increasingly associated with women and the feminine realm, contrasted with masculine rationality and the male domain of science. Though women could lay little claim to religious institutional power in the United States, many Americans believed them to be more spiritual beings, which afforded them a position of some authority and moral influence over their husbands, families, and communities and was supposed to serve as a kind of balance to men's rational and scientific natures.[15] Put simply, in antebellum America, science was a masculine pursuit for the modern man.

As historian George Frederickson reveals, while several European texts in favor of polygenesis were well received among Southern intelligentsia, the theory of separate origins of the races did not become widely accepted in the United States until the 1840s and 1850s with the emergence of the American school of ethnology.[16] This school of thought was precipitated in large part by two important texts, published in 1839 and 1844, respectively: *Crania Americana* by Dr. Samuel Morton (1799–1851) of Philadelphia and *Crania Aegyptiaca,* which Morton coauthored with British-born Egyptologist George Gliddon (1809–57). A physician by training and naturalist by trade, Morton shared Caldwell's emphasis on intellect as a physiologically determined marker of racial difference and cited his lesser-known contemporary as influencing his belief in polygenesis. Morton's grave marker in a lush garden cemetery full of wealthy and famous Philadelphians likewise echoed Caldwell's valorization of the "truth" of science even amid religious controversy; it reads, "Wherever Truth Is Loved or Science Honored, His Name Will Be Revered."[17]

Morton was born in 1799 to an Irish Protestant father, who died when Morton was only a year old, and a Quaker mother, who sent her three children to Friends schools and later remarried a Philadelphia merchant. His Quaker upbringing may have influenced his later attitudes toward slavery; though his scientific conclusions about race bolstered the proslavery logic of many of his colleagues, he always stopped short of overtly defending the institution himself. From a young age, he showed considerable interest in history, natural science, and medicine, and he too obtained his degree from the University of Pennsylvania in 1820 before heading off to the University of Edinburgh for further medical education. He also learned early on the importance of fostering professional friendships and alliances, and he quickly earned a reputation as both gentlemanly and serious in studies.[18] He attended dissection demonstrations in his free time and developed an interest in comparative anatomy from one lecturer's use of skulls as a teaching tool as well as studying the university's enormous zoological collection.[19] While studying in Edinburgh, Morton's new friend George Combe exposed him to another scientific field that would prove equally influential on his future career as a scientist of race: phrenology, for which Combe was internationally famous. As Ann Fabian explains in her fascinating analysis of the cultural politics of studying skulls in nineteenth-century America, *The Skull Collectors: Race, Science, and America's Unburied Dead,* even though phrenology had its share of detractors by the mid-nineteenth century, in earlier decades, it played a key role in establishing the "materialist premise that the brain was the organ of the mind," a concept central to ethnology. And she notes, "It also offered a beginning for understanding cranial location and helped establish human skulls as a favored site for investigation. During these same years, naturalists in Europe and the United States began to work on the idea that differences among the races of men could be set down as differences in the shapes and sizes of skulls." While Combe resisted applying phrenology and the comparative study of skulls to questions of race, Morton would come to build his career on precisely such applications, which in turn influenced the field of ethnology for decades.[20] Morton returned to Philadelphia in 1824, and in so doing, he "took his Edinburgh lessons (on collecting, empiricism, anatomy, and phrenology) home to a country deeply divided along racial lines, and those racial divisions shaped his intellectual world," Fabian adds.[21] In a city with a glut of physicians, Morton supplemented his income by lecturing on anatomy at local medical schools, and he quickly began to turn most of his attention to the Academy of Natural Sciences, which he was effectively running by 1825. The academy housed an ever-expanding and varied collection of specimens, including animals, plants, fossils, and rocks, but Morton soon focused on studying differences among human beings, his real passion and his most famous legacy.[22]

In that capacity, Samuel Morton collected a dizzying number and variety of human skulls and subjected each to a series of measurements, from which he drew inferences about the intellectual abilities of the skull's deceased owner and the tribe or race he represented. Whereas phrenology typically read the bumps on the head for clues about personality traits and abilities, craniometry or craniology measured and/or weighed the capacity of the skull: the bigger the capacity, the bigger the brain it must have held and thus the bigger the intellect and reason.[23]

Indeed, the human head held tremendous significance for scientists studying race throughout the nineteenth and early twentieth centuries, although the "truths" they thought it revealed changed over time in accordance with new social and political contexts. Constituting 15 percent of scientific texts on Native Americans and 8 percent of all other texts on racial science in the period between 1830 and 1859, studies focused primarily on brains or skulls loomed larger than their numbers in terms of prominence and influence. The head received far more scientific emphasis than any other part of the body during that period. An 1860 article in the *American Journal of Science and Arts* offered an explanation why the head was so important to ethnology's adherents. "The seat of those faculties which lie at the base of all the peculiarities of human races," the head bore "essentially and intimately upon their manners and customs, all their institutions, their religious impulses, their capacity for civilization, and the development to which it has attained."[24] Moreover, Morton's crania studies in particular continued to figure in the work of many racial scientists who followed him as evidence of black inferiority—well after the theory of polygenesis, of which his measurements had been offered as evidence, had fallen out of favor.

Morton amassed his enormous collection of human skulls largely by appealing to friends and associates—including diplomats, military officers, Western settlers, missionaries, explorers, and other naturalists and physicians—to procure skulls for him through their work or travels. Many donated skulls to Morton as a professional courtesy or friendly gesture, while he paid a fee for others. Though poor health, family, and work did not allow Morton himself to travel much outside of Philadelphia on head-hunting missions, so to speak, he still managed to collect more than a thousand skulls through his carefully developed network of benefactors. These proxy collectors (grave robbers, in many cases) in turn had gathered the skulls from battlefields, burial grounds, prisons, asylums, and poor houses, among others. By the time of his death, his collection represented skulls from all seven continents and every racial category—as defined in Morton's nineteenth-century terms—including a dizzying array of North and South American aboriginal groups.[25] Morton measured the internal capacity of each skull—as a reflection of the size of the brain it had previously housed—with a precision widely admired by his contemporaries, filling the cavity with mustard seed (later switching to lead shot) and pouring the fine material back out into a graduated cylinder.[26]

Despite his stated commitment to precision and scientific objectivity, there were numerous flaws with Morton's methodology. Particularly notable among them was that he failed to adjust his data for differences of sex, as evolutionary biologist Stephen Jay Gould reveals. Morton's skull collection contained a mix of male and female crania, and he did not attach any additional significance to the skulls according to their sex; indeed, it was race rather than gender for which he read the skulls. He made little effort to distinguish the sex of his skulls, which Gould argues skewed Morton's results, not to mention the conclusions he drew from them. His sample of nonwhite races included more female skulls, usually smaller in size than male skulls, and he—and the many theorists who cited his work—read the smaller average skull size he calculated for those races as evidence of their inferiority. In other words, he read sex difference within a race as proof of physiological and intellectual difference between the races.[27] Ironically—and largely unintentionally—then, it was bodies of women that were equated with racial inferiority in this particular instance.

Nonetheless, Morton's *Crania Americana* received favorable reviews when it was published and continued to be enormously influential throughout the nineteenth century.[28] Though the extravagantly illustrated volume was costly, which limited sales to private readers, it was reprinted numerous times and widely cited by scientists as evidence of a biologically based intellectual hierarchy among the races for many decades, well beyond its original context as part of the antebellum debate over polygenesis. Morton often resorted to giving away copies, though wealthy Southern planters were notable among the private readers who *could* afford to buy the tome—and they did, eagerly latching onto his conclusions about racial difference in their defense of slavery, even if Morton had not defended the institution outright himself.[29] The data had a life of its own from the start, and other racial theorists were quick to reference it as fact rather than repeat such an exhaustive empirical undertaking themselves. The scientific methodology of many ethnological writers during and after Morton's lifetime consisted primarily of casual observation of the living racial subjects around them (or in some cases, patients in their medical practices) and citing Morton's data on skull size.

Not only did other physicians and scientists frequently reference his crania studies; they also often represented the public face so to speak of the racial sciences outside of the scientific world. For example, one 1848 political cartoon uses phrenology—one of the most popular of the cranial sciences outside the scientific establishment itself—in its satirical attack on presidential candidate Zachary Taylor (see Figure 1.2).[30] In the background of the cartoon are several heads or skulls lined up on shelves like

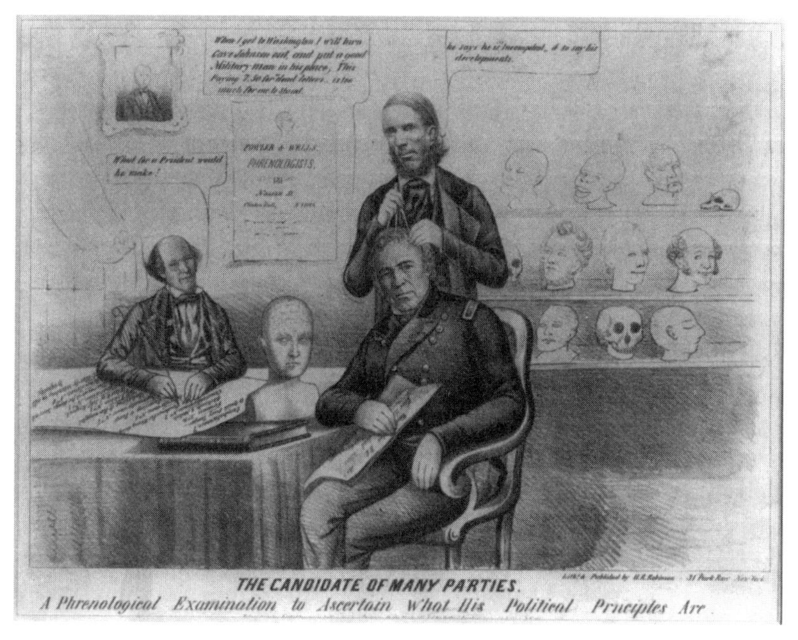

Figure 1.2. The public face of racial science. *As demonstrated by this 1848 cartoon lampooning presidential candidate Zachary Taylor, the basics of racial science—examining the head for clues as to the mind within and making comparisons between various human "types"—would have been familiar to a broad audience. Notably, both the scientist and his human subjects are again male ("The Candidate of Many Parties: A Phrenological Examination to Ascertain What His Political Principles Are," Library of Congress, Prints and Photographs Division,* American Political Prints, 1766–1876 *[LC-USZ62–19668]).*

specimens. All the heads appear to be male, unlike Morton's own mixed-gender crania collection. Furthermore, each of the white men was a figure recognizable to the cartoon's audience, including Martin Van Buren, James Watson Webb, and Henry Clay. The black man, however, is unidentified and anonymous, although a recognizable black man would not have been difficult to find, as Frederick Douglass was speaking across the northern United States during the same decade. The anonymous and racialized heads on the shelf are necessary components to make the scene recognizable to the audience as a typical snapshot of scientific inquiry into human mental characteristics; the unexpected faces of the famous men are what constitute the joke. Though a satire to be sure, the cartoon nonetheless reflects an important and perhaps unintentional truth about the preponderance of such science in midcentury America: both the scientist and his subjects were male.

Morton and his contemporaries used the human head as an ethnological tool in racial taxonomies and in determining the intellectual capacities of the races, but the most prominent use of craniology in the antebellum period revolved around the question of the original unity or diversity of the races. Though Morton himself did not engage the issue as extensively as the racial scientists he influenced, he did note that differences in skull size and shape between the races seemed to date to antiquity and speculated that correlated differences in mental and moral character were ancient as well.[31] For Morton, the physical features of each race were the same today as they had been in antiquity, which implied that the races had originated separately. In his estimation, these features varied across five races, which he ranked in order of skull size and corresponding intellectual ability: Caucasian, Mongolian, Malay, American, and Ethiopian. Interestingly, when listing the physical characteristics of each race, Morton discussed the propensity for beard, or lack thereof, as among those characteristics but did not generally discuss female anatomical features. (Also interesting is the fact that Morton himself is shown clean-faced in all his existing portraits.) For Morton and many of his contemporaries in ethnology, racial taxonomy meant classifying and differentiating between the men of the world.

Despite claims of scientific objectivity, Morton did not shy away from commenting on issues of tremendous political import, reading the character and fates of Native American and African American men through his collection of skulls. Of the former, he posited that biology offered an explanation for their well-established "inaptitude for civilization": "His moral and physical nature are alike adapted to his position among the races of men, and it is as reasonable to expect the one to be changed as the other. The structure of his mind appears to be different from that of the white man, nor can the two harmonise [*sic*] in their social relations except on the most limited scale."[32] As with much of antebellum ethnology, though it may first appear that Morton is simply using "man," "men," and "his" in the conventional generic sense, the context reveals that he means men as a sex specifically. Morton refers to social relations, intellect, and capacity for civilization, all of which were associated with the masculine realm at this point, indicating that Native American and white women fell outside the scope of his racial comparison.

Earlier in his career, Morton tended to deal with the question of the origin of the races indirectly for the most part while also providing support for the polygenesis argument. Indeed, he often stated that his task was merely to categorize human beings, not to speculate on how such diversity came to be.[33] But in his study of human skulls, he had concluded that each race had changed little, if at all, in regards to physical characteristics and, by implication, mental abilities. That conclusion, along with his extensive measurements, provided countless racial theorists who followed him with the "empirical" data they needed to argue for the original and permanent diversity of the races, and by the end of his life, he himself had become increasingly explicit about his belief in polygenesis. Consequently, his contemporaries and historians alike have long regarded Morton as the founding father of the American school of ethnology, nearly synonymous with polygenesis by the mid-nineteenth century.

Moreover, Morton's work also provided scientific support for proslavery apologists such as Josiah Nott, a physician with a private practice in Mobile, Alabama, who became the most vocal and venomous advocate of the new American ethnology. Unlike Morton, however, proving the separate origins of the races was his stated goal.[34] First published in Philadelphia in 1854, *Types of Mankind,* an anthology of ethnological work, was coedited by Nott and George Gliddon, the Egyptologist and former U.S. consul in Cairo, Egypt, who had previously collaborated with Samuel Morton. *Types of Mankind* also contained contributions from Swiss-born naturalist and outspoken polygenesist Louis Agassiz and the late Morton himself, to whom the volume was dedicated. Nott heralded Morton as "the Founder of the American School of Ethnology," which the anthology sought to represent.[35] The dedication to Morton was no idle sentiment; the imprint of Morton's work could be found all over *Types of Mankind,* a tome so popular it was in its eighth edition by 1857, just three years after its original publication. Making frequent reference to Morton's crania studies in support of polygenesis, it also drew heavily on Egyptian history and art, which both Nott and Gliddon interpreted as showing the ancient Egyptians to be white but their slaves black. Thus they argued that blacks were innately inferior and that the relationship between the white and black races had long—perhaps always—been that of master and slave.

Ardently committed to using science to naturalize slavery and American inequality, Josiah Nott was perhaps the most well-known advocate of ethnology during his lifetime and certainly remains the proponent of scientific racism most often cited by historians today. Raised in Columbia, South Carolina, and the son of a wealthy lawyer and planter (himself originally from Saybrook, Connecticut), Nott received his medical training from esteemed schools in the urban North as well as Europe.[36] Openly hostile to blacks throughout his career, Nott revealingly referred to ethnology as "niggerology" in an 1845 letter to proslavery South Carolina planter James Henry Hammond.[37] Nott differed from many antebellum-era adherents to ethnology in both his unabashed racism and his open commitment to the proslavery cause. Pure science was the subject that many such writers claimed to study—including Nott at times. Indeed, *Types of Mankind* began with a decidedly more formal definition of ethnology than the coarse characterization of the field contained in his private letter to Hammond. In this influential anthology representing the American school of ethnology, Nott looked across the Atlantic to assert ethnology's legitimacy: "Mr. Luke Burke, the bold and able Editor of the *London Ethnological Journal,* defines Ethnology to be 'a science which investigates the mental and physical differences of Mankind, and the organic laws upon which they depend; and which seeks to deduce from these investigations, principles of human guidance, in all the important relations of social existence.'"[38] Nott adopted this working definition of ethnology for the expansive volume to follow and emphasized that the driving questions and social import of ethnology were far reaching.[39] Nott did not envision ethnology merely as science for science's sake, despite his claims elsewhere to political disinterest. Rather, it was an endeavor with considerable appeal to and relevance for important *men,* including "the philanthropist, the naturalist, and the statesman." The line between science and politics in nineteenth-century America was thin indeed.

Like Nott, Charles Brace, a personal friend of Charles Darwin, introduced *The Races of the Old World* by defining ethnology for his audience, though he did so in a way that underscores the fact that not all ethnological writers shared Nott's desire to see the discipline be driven by contemporary political interests. He stated, "Ethnology, according to its literal derivation, means the Science of Nations; but in the more comprehensive classification, nations have been divided according to descent or race, and the word has come to mean the Science of Races."[40] I borrow Brace's definition of ethnology as the "science of race," or more precisely, the science*s* of race, to encompass the broad array of scientific disciplines focused on race in the nineteenth century, even if the scientist himself did not use the term.[41]

Together, Nott, Morton, and Brace also illustrate the diversity of the *scientists* of race. Southern born and raised and later educated in New York, Paris, and Philadelphia, Nott would maintain close relationships with other physicians and scientists in the North and Europe throughout his life, often traveling outside the South for both business and personal visits with medical colleagues. As Nott biographer Reginald Horsman noted, "In no way did he feel himself intellectually isolated or out of the main lines of European and transatlantic thought"—nor did his contemporaries view his work as outside the scientific mainstream even when they disagreed with his claims.[42] He settled in Mobile, a booming coastal city by the time of Nott's arrival and the center of Alabama's slave trade. Though the city's cotton industry thrived on slave labor, it was also home to a sizable free black population as well as a small but visible population of mostly impoverished Choctaw Indians. Nott would come to base many of his scientific conclusions about race in general and on race mixing in particular on his observations of the groups inhabiting his city—and the various mixtures between them—as he encountered them in the streets and in his medical practice.[43] He saw ethnology as a science that proved racialized slavery was natural and the servile position of the black race a transhistorical phenomenon rooted in their specific biological makeup. Nott himself owned nine slaves and in the late 1840s opened a thriving slave infirmary that catered primarily to young black men and later served as a military hospital during the Civil War. Nott also prided himself on an extensive and varied private practice that included some of the most wealthy and powerful elites in Mobile and saw him perform a range of medical services from gynecology and obstetrics to general surgery to autopsies in suspicious death cases. Widely respected and sought after as a physician in Mobile, he also developed a national reputation for his expertise in yellow fever and disease among slaves.[44] Consequently, Horsman observed, "Slavery and white-black relations were entwined with Nott's private life, his practice, and the life of his city and his region."[45]

In contrast, Charles Brace inhabited a New York City in which immigration and urban poverty shaped many of his observations about race. Like John Campbell, he was deeply involved in the social reform movements sweeping urban centers in the Northeast and helped found the Children's Aid Society in New York in 1853, with which he remained engaged until his death in 1890. Celebrated by most of his contemporaries as a hero (with the exception of Catholic clergy and relief workers who criticized the anti-immigrant and anti-Catholic sentiments that undergirded his policies) and described by scholars today as both sinner and saint, Brace is most famous today for the "orphan trains" he facilitated through his work with the Aid Society. The trains transported impoverished children, usually orphans or runaways from immigrant—and predominantly Catholic—families, to live with rural Anglo-Protestant families in the West, ostensibly to save them from a life of crime and destitution, though evidence suggests at least some were received in their new homes as little more than young servants.[46] Brace was born in 1826 in Litchfield, Connecticut, to an old New England family that included many religious and political leaders. His father was a teacher and later the principal of the Female Seminary of Hartford, and though his mother died when he was fourteen, he was very close to his sister and was surrounded by unusually educated women for the time period on account of his father's work. An eager student of history and philosophy from a young age, Brace pursued a degree in theology from Union Seminary and soon after began his work with New York City's poorest. Though he did not have a medical or scientific degree, his work in the city's slums sparked his interest in race and society. Nor was Brace the only racial theorist who had a theological rather than medical background (as we have also seen with one of ethnology's earliest adherents, Samuel Stanhope Smith), which further underscores the fact that nineteenth-century ethnology often blended medico-scientific and religious expertise rather than pitting them against each other. In addition to his day-to-day work as a minister and social worker, he also soon became a prolific writer on such issues as immigration, poverty, crime, and race, most notably with the influential and widely circulated books *The Races of the Old World* (1863) and *The Dangerous Classes of New York* (1872). He saw the former, which he subtitled *A Manual of Ethnology*, as a dispassionate primer of the field's central discoveries, separated from any defense of slavery, which he vehemently opposed and which had become almost completely imbricated into most American ethnological work by the time of his writing.

In contrast, Nott naturalized the racial status quo as something transhistorical and unchangeable by human action or law, describing the world as the eminent domain of the white race, which was destined "to conquer and hold every foot of the globe where climate does not interpose an impenetrable barrier." He further asserted, "No Philanthropy, no legislation, no missionary labors can change this law; it is written in man's nature by the hand of his creator."[47] This often cited passage is notable not just for its explicitly political interpretation of the past in defense of the present but also as an example of gendered language in antebellum ethnology. While "man's nature" could appear to mean "human nature," the overall language and tone of the passage reflects traits and domains associated with men in antebellum America, revealing that Nott's subject was indeed specifically male.

Again, when antebellum racial scientists spoke of a race or races, they usually meant men specifically, although that was not always immediately apparent. Much of the literature in the field was written in what psychologists and linguists refer to as the "masculine generic"—that is, "man" and "mankind" to mean all of humankind. The fascinating title of one early ethnology essay in particular—"An Account of a Female of the White Race of Mankind, Part of Whose Skin Resembles That of a Negro; with Some Observations on the Causes of the Differences in Color and Form between the White and Negro Races of Men" (1818)—clearly demonstrates the linguistic specificity of women and the generality of "men."[48] Other titles, such as *Types of Mankind* (1854), *Natural History of the Prognathous Species of Mankind* (1857), *The Classification of Mankind* (1850), and *Races of Men* (1850), seem to be simply conforming to this linguistic convention but were indeed primarily about men.[49] As feminist scholars have argued, the masculine generic functions to reinforce men and masculinity as normative while simultaneously rendering women invisible.[50]

As both a linguistic convention and a scientific framework in ethnological writing, the masculine generic reinforced man as normative at the very same time that such writing interrogated the category of race with regard to the differences "between men." In other words, early proponents of ethnology used "man" or "men" in this generic sense *and* limited their discussion to men as a sex.

For example, in *Types of Mankind,* Nott provided a detailed taxonomy of the races of the world and often characterized a particular race by the thickness of its beards. Notably hirsute himself (see Figure 1.3), he deemed this mark of manliness as a marker of white superiority—even though he was ostensibly describing each race as a whole. More typically, though, racial scientists' discussions of the body usually centered on skin or nonreproductive organs such as lungs or kidneys, skulls, and hair common to

Figure 1.3. The scientist as specimen. *In the influential 1854 collection of American ethnology,* Types of Mankind, *Alabama physician and proslavery racial theorist Josiah Nott highlighted beard thickness among the physical characteristics that distinguished the races of the world from each other—races he believed originated separately and with different purposes ordained by nature. In so doing, not only did he characterize each race as a whole by a feature specific to men, but he also used a feature that he himself possessed as a marker of both ideal manliness and racial supremacy ("Josiah C. Nott," courtesy of the National Library of Medicine,* Images from the History of Medicine *Collection, Portrait No. 5063).*

both sexes.[51] Nonetheless, the sociopolitical import they read into these body parts was male centered. White men like Nott studied the races to assess their capacity to govern or lead armies, as John Campbell's introductory question underscored.

In addition to focusing overwhelmingly on men, antebellum ethnology was also quite often literally black and white in its approach to race. Nott, for example, was prone to characterizing ethnology as "the nigger business."[52] Nott's choice of words is indicative of antebellum ethnology's overwhelming focus on the black race. As Figure 1.4 demonstrates, 54 percent of all racial science texts published from 1830 to 1859 were explicitly focused on African Americans, as compared to 21 percent for Native

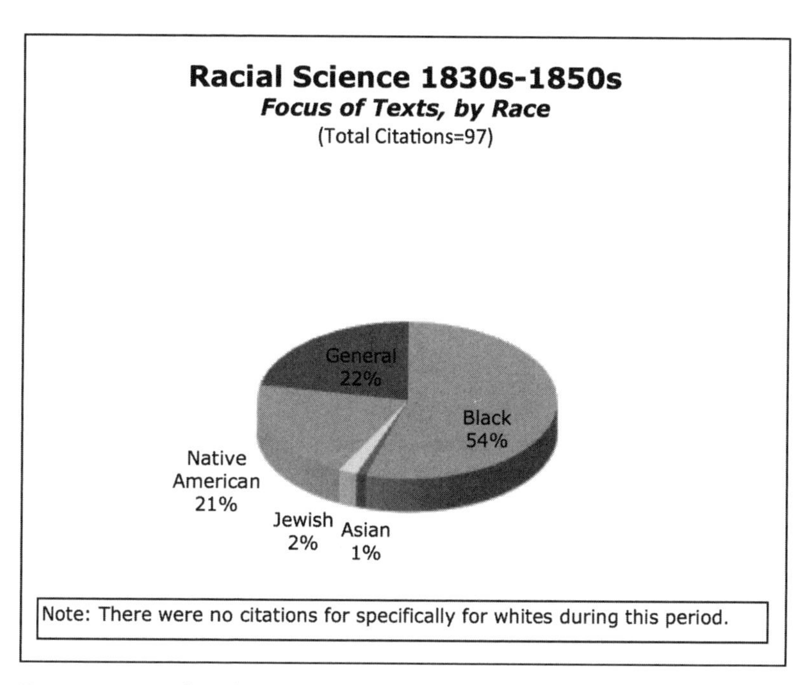

Figure 1.4. Focus of racial science texts by race, 1830–59. *The majority of ethnology texts in the antebellum period focused on the history and characteristics—physical and otherwise—of the black race. Though ostensibly about the history of the races, American ethnology was largely driven by concerns about race relations in the present, much of which centered around the social role and destiny of black people specifically. Notably, there were no citations specifically on whites alone during this period (graph by author; data based on the* Index-Catalogue *of the National Library of Medicine).*

Americans, the race emphasized in the next largest number of American ethnological publications.[53] Nott neatly summarized the preoccupations of the field in his *Two Lectures, on the Natural History of the Caucasian and Negro Races* (1844) when he declared, "The Anatomical and Physiological differences, between the Caucasian, the Malay, Mongol, Indian and Negro races, have elicited a great deal of scientific research, and I might very well write an octavo on these points alone. Time, however, compels me to restrict my lecture to a parallel between the Caucasian and Negro races."[54] Many other American ethnological texts in the antebellum period and beyond began with similar disclaimers. Furthermore, even those texts that were ostensibly on race in general, as indicated by their titles (22 percent of texts published between 1830 and 1859), usually focused most of their attention on the black race specifically or on comparing the black and white races after first delineating and characterizing the races of the world more broadly.[55]

In the black-and-white world of antebellum ethnology, the white race was normative whereas the black race was scrutinized for evidence of difference and deviance. John Van Evrie, a New York physician, publisher, and proslavery pamphleteer born in 1814, even glibly proposed that the African existed for the sole purpose of illuminating the white race's unique virtues and that "his wonderful capacity of imitation—that striking quality which those ignorant of his nature have often mistaken for real capacity—is a positive proof that God designed him to exist in juxtaposition with the superior race."[56]

Part of the reason for racial scientists' overwhelming focus on both men and the black race seemed to stem from three interrelated beliefs: (1) that racial difference was pervasive throughout the human body; (2) that such differences were most prominent between the black and white races, commonly believed to represent opposite poles on a racial scale of perfection; and (3) that racial characteristics were more pronounced in men than women. Indeed, one racial theorist, prominent Louisiana physician Samuel Cartwright, went so far as to claim in 1857 that the African race's darker color infiltrated "the whole inward man," inside and out, including organs and viscera.[57]

Though invoked by scholars and cultural critics today more than any other proponent of ethnology as exemplifying the outlandishness of nineteenth-century scientific claims about race, Samuel Cartwright was taken far more seriously during his own lifetime, with a professional pedigree and experience that rivaled other top physicians. His assertion in 1857 that race pervaded every part of the body right down to the mucus membranes, for example, remained a consistent theme in ethnology throughout its reign, often stated in nearly identical language over time and among different authors, including Morton in 1830, Van Evrie in 1866, and Daniel Garrison Brinton (a Yale-educated physician, editor of the *Medical and Surgical Reporter,* and professor of ethnology and archaeology at the Academy of Natural Sciences in Philadelphia) in 1890.[58]

Cartwright was born in Fairfax County, Virginia, in 1793 and, like Charles Caldwell, studied medicine at the University of Pennsylvania—both, ironically, under the tutelage and mentorship of the famous Dr. Benjamin Rush, an ardent abolitionist. He soon returned to the South to practice medicine, first in Huntsville, Alabama, then Natchez, Mississippi, before settling in New Orleans in 1848 with his family. Twice he served as a military surgeon and physician under General Andrew Jackson and later for the Confederate Army during the Civil War, recruited for his work on diseases endemic to the Deep South. While practicing medicine in New Orleans, he also worked as a professor at the University of Louisiana and began to build his national reputation as an expert on the "peculiar diseases" of African Americans, particularly slaves.[59]

As can be seen in Cartwright's work, a mixture of biology, anthropology, comparative anatomy, and religion characterized nineteenth-century writing on race. When he spoke of blacks as "prognathous," for example, he referred to facial angle, specifically a jutting jaw. Racial scientists often attached this characteristic to people of African descent to illustrate their alleged similarity to apes, as seen in an illustration (Figure 1.5) from *Races of Men* (1850) by British anatomist Robert Knox. In contrast, his frequent reference to "Canaanites" and "Cushites" drew on the biblical history of the races.

Meanwhile, Cartwright's explanation of how race permeated the body invoked gender, but only to underscore the difference between the races of men, declaring, "The black color is not so deep in the female as in the male, nor in the feeble, sickly negro as in the robust."[60] Cartwright's binary language implied a correlation between female and feeble and between male and robust, oppositions that reflect common nineteenth-century ideas about gender. But more important, for Cartwright, race literally seemed to be located more wholly in men. His readers agreed, even when they called out for more evidence in support of his other claims. In an

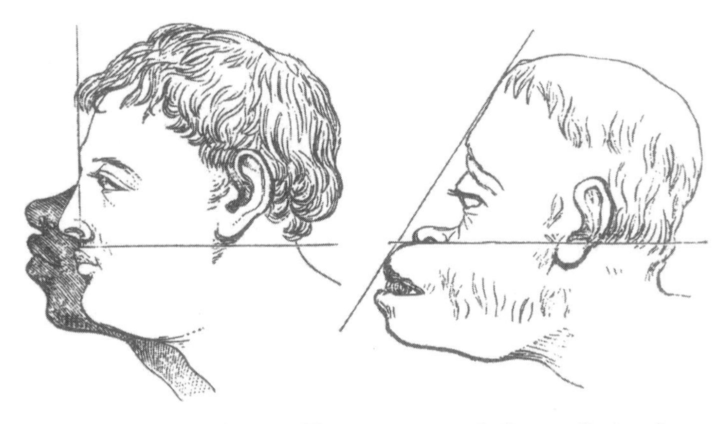

[*Profile of Negro, European, and Oran Outan.*]

Figure 1.5. Ethnology's images. *The image is quite typical of ethnological illustrations in the first half of the nineteenth century in several regards. First, its emphasis on the shape of the head as a marker of racial difference; second, the comparison between black people and animals, especially simian species; and finally, the use of men to represent the race as a whole. This particular drawing was supposed to illustrate the "prognathous" facial angle of orangutans and the black race, in contrast to whites (Knox,* The Races of Men*).*

article that analyzed Cartwright's theories on racial mixture, the editors of the *Georgia Blister and Critic* noted, "The genital organs are the last parts of the body that lose all trace of the Negro. The scrotum of the brightest mulatto is found almost black."[61] Strangely, while the article was otherwise about ovary cells, the editors did not follow up their example about the "almost black" scrotum with a corollary about female genitalia. Even when ostensibly discussing female anatomy, scientists were most interested in the impact of race on men. Moreover, here race was not simply more prominent in the bodies of men; it permeated the very body parts that distinguished them *as* men.

Though they came to the study of race with a range of political agendas, employed diverse methodologies, and focused on different bodily characteristics to bolster their claims, virtually every racial theorist in the antebellum era addressed the debate between monogenesis and polygenesis, in a variety of venues, and always with reference to men. In an article titled "Diseases and Physical Peculiarities of the Negro Race," first published in the May 1852 issue of the *New Orleans Medical and Surgical Journal,* Samuel Cartwright exemplified the interplay between religion, politics, and science in antebellum ethnology. After delineating some of the allegedly distinct anatomical features of the black race that made blacks natural slaves, he turned his attention to the question of unity. Pondering the male lineage of the white race, he queried, "Does [the African] belong to the same race as the white man? Is he a *son* of *Adam?* Does his peculiar physical conformation stand in opposition to the Bible, or does it prove its truth? These are important questions, both in a medical, historical and theological point of view."[62] Likewise, he argued that all three perspectives must be employed to ascertain the answer.[63]

Cartwright found a philosophical opponent in Charles Brace, though they shared a similar language of gender. Brace challenged Cartwright's claim that the unity question could only be answered by examining science, theology, and history together; for Brace, "the subject is purely scientific." While a monogenesist, Brace shared Cartwright's male framework of racial descent and devoted a lengthy chapter of *The Races of the Old World* to "the great question" of ethnological science. In sharp contrast to Nott and Cartwright, Brace was careful to divorce the science of racial difference from possible social and political application, but he did not reject the basic premise of racial hierarchy. And ironically, he employed the same paternalistic framework of race relations and male focus as the proslavery colleagues he critiqued. In Brace's discussion of the "races of men," humanity constituted a "brotherhood" with God as its "Father." While "the moral Brotherhood of man does not depend on community of descent," this community did not appear to include women, for it is apparently possible for mankind to have "one parent" rather than one pair of parents.[64]

Not all northern adherents to ethnology shared Brace's views on slavery or origins, but across regional and political lines, they framed race and racial descent in terms of men. A New Yorker like Brace, John Van Evrie argued that the actions of monogenesists, despite their words, proved the "fact" that the races were distinct species with separate origins. That is, if monogenesists like Brace truly believed that blacks "descended from the same parentage" and differed from whites in color alone, then there would be no reason for their aversion to intermarriage and race mixture. "But on the contrary, the reverend and distinguished gentleman who has ventured to declare that the belief that the Negro is a being like ourselves, is essential to Christianity, would infinitely prefer the death of his daughter to that of marriage with the most accomplished and most pious Negro in existence!" he averred.[65] Here, "a negro" meant a black man specifically. Invoking the controversial image of a white woman with a black man (which would become even more volatile as the century drew to a close), he used the specter of miscegenation as a challenge to those who would defend the African's place in the human family.

Racial mixture also featured prominently in the origins debate in antebellum ethnology. Numerous scientists of the era argued that the offspring between a black parent and white parent were largely infertile and thus incapable of producing a "permanent stock" beyond that first generation. For these scientists, this proved that the two races constituted separate species. In fact, the word "mulatto" derived from the word "mule" or "a sterile hybrid."

Not surprisingly, the subject of race mixing was one of the few contexts in which women—both black and white—were objects of discussion, in contrast to more general considerations of the original unity or diversity of the races. Even though antebellum scientists thought "mulatto" men and women alike were weak and largely infertile, their discussions of sterile hybridity were more likely to target biracial women specifically as bad breeders. For example, H. A. Ramsey and W. T. Grant, physicians and coeditors of the *Georgia Blister and Critic,* asked their readers, "In the cross of the white and negress, do the Ovary Cells diminish with each cross, until the fourth, and then nearly disappear entirely?" The *Blister,* a quasi-medical journal largely dedicated to scientific justifications of chattel slavery, was part of an antebellum-era surge in Southern periodicals, most of which were short lived, privately funded, and geared toward a regional audience.[66] Like Van Evrie characterizing miscegenation between a black man and white woman as an absurdity, the *Blister*'s focus on the "cross of the white and negress" hinted at the reality of interracial sex in antebellum America. More often than not, such mixture occurred between white men and black women in a culture in which white masters legally owned the bodies of their female slaves, and even free black women's rights to their own bodies were frequently ignored in law and practice.

Hoping to collect opinions and anecdotal evidence from the *Blister*'s readership to assist his research, Samuel Cartwright had submitted this revealing query about ovary cells to the journal, but he was less interested in women per se than in uncovering further proof of (permanent) racial difference. For their part, the journal's editors were happy to oblige him, noting that "the question is important" and thus it required "a candid and careful investigation." They admitted they had "presumed an answer, without the necessary data to confirm it." Their presumptive answer presented no information about black women, however. Instead, it made an observation about animals, just as Jefferson had done in the previous century in his own discussion of racial mixture. Ramsey and Grant wrote, "We think it quite probable that the Ovary Cells in the cross of the negress and white, may diminish, until sterility would be the result. Our dissections are not ample enough to determine the point precisely, but we see a cross in the horse and mule, produce sterility and why not in the white and black biped race? We see no reason to question." They concluded by offering their own anecdotal example about a black man they knew, who had previously fathered children before marrying a "fourth cross" wife, who did not "breed," despite being otherwise healthy. Perhaps not surprisingly in a society in which black women, particularly slaves, so often faced sexual exploitation that made the paternity of their children either difficult to ascertain or all too tempting for whites to ignore, white men wanted to believe that "mulatto" women in particular were sterile—at least by the fourth cross.[67]

Ultimately, though, investigations of hybridity usually focused less on women—white or black—than on the question of whether the races constituted different species or variations of the same species. In a two-part lecture before the Academy of Natural Sciences in Philadelphia in 1846, later published as an article in the *American Journal of Science and Arts,* Morton argued that natural laws governing hybridity in animals could be brought to bear on the question of racial origins in human beings. Whereas Morton did not contend that human "mulattoes" were sterile, he maintained that their ability to reproduce did not prove the races to be one species of singular origin either. As Morton's counterargument indicates, many scientists had made "hybridity the test of specific character," arguing that animals of different species were unable to reproduce fertile offspring. For some, "sterile hybrids" were thus proof that the races were distinct species.[68] For Morton, however, arguments for the separate origins of the races were best supported by the distinct and unchanging character of the various races over thousands of years rather than the reproductive capacities of racial "hybrids."[69]

Other experts were not so quick to divorce the issue of racial hybridity from the origins question or to concede that mulattoes could themselves reproduce, but their arguments were similarly geared toward proving long-standing and permanent racial difference. In introducing his 1844 *Two Lectures, on the Natural History of the Caucasian and Negro Races,* Nott discussed the "effect of crossing races."[70] He also thought that animals could shed light on questions of race, but he believed that the natural sciences had not adequately addressed the issue. Elsewhere, Nott conceded that though fertile offspring could be produced from black and white parents, such offspring did not have the fecundity of its parent races and that, over time, it was "the higher type that in the end predominates."[71] However, no amount of infusion of white blood could turn the black race white or enable a mulatto to escape detection, for the skilled eyes of Nott and other racial experts could always "instantaneously trace the Negro type in complexion and feature." After all, ethnology's proponents had positioned themselves as trained scientists who could always "see" the truths of race. And why did the "higher type" predominate but never subsume the lower race? Nott concluded, "The only physiological reason that may be assigned is this: the mulattoes, or mixed-breeds, die off before the dark stain can be washed out by amalgamation. No other rational explanation can be offered."[72] In a text that also offered an explicit defense of slavery—under which the sexual exploitation of slave women by their white masters was actually profitable—it was politically expedient to render racial mixture nonthreatening.[73] Thus Nott argued that the issue of hybridity was of considerable interest to scientists, but he dismissed the human "hybrids" themselves as inconsequential, weak, and ultimately destined to die out.

Nott was correct that his views on hybridity in particular would stand out to other scientists. Echoing his assessment of the importance of examining racial mixture, an overwhelmingly favorable review of *Types of Mankind* first published in the *Philadelphia Medical Journal* and reprinted in the *Georgia Blister and Critic* focused entirely on the "detailed conglomeration of fact upon amalgamation" that could be gleaned from the text. The reviewer included an excerpt from *Types* that summarized Nott's critical conclusions on the issue for a wide audience—conclusions that continued to ground scientific discussions of race mixtures for decades to come:

1. That mulattoes are the shortest lived of any of the human race.
2. That mulattoes are intermediate in intelligence between the black and the whites.

3. That they are less capable of undergoing fatigue and hardships than either the black or whites.
4. That the mulatto women are peculiarly delicate, and subject to a variety of chronic diseases. That they are bad breeders, bad nurses, liable to abortions [miscarriages], and that their children generally die young.
5. That when mulattoes intermarry, they are less prolific than when crossed on the parent stocks.[74]

Nott explained that he first drew these conclusions in 1842 in a short essay on hybridity in animals and humans and that, while he remained convinced of their validity, he must now add an important caveat: they are only fully accurate when applied to "intermixture of the strictly white (i.e. the Anglo-Saxon, or Teuton) with the true Negro."[75] Nott claimed that his experience living in Mobile and New Orleans, where the population included "a preponderance of the blood of the French, Italian, Spanish, Portuguese, and other dark-skinned races," taught him that mixture between these "darker" Europeans and people of African descent had far less, if any, effect on the fertility of the offspring than when Anglo-Saxons and "pure Negroes" mixed.[76] Rather than disproving Nott's claims regarding the black race's separate origin, the more fertile offspring resulting from unions between blacks and non-Anglo Europeans simply drew the whiteness of the latter into question while reinforcing the common ethnological view that "true" whites and "negroes" represented opposite poles on the racial spectrum.

According to Nott, thorough investigation of racial hybridity in the United States was a complicated venture, and mulatto women both added to the difficulties and illuminated vital facts on the subject. Though his "circumstances, personal and professional, [afforded him] ample opportunities for observation," for Nott, "the difficulty arises solely from the want of chastity among mulatto women, which is so notorious as to be proverbial."[77] He further claimed that even when they were married to similarly "hybrid" men, their children were just as likely to be fathered by other men, black, white, or otherwise, as they were by their mixed race husbands. Like many other slaveholders and scientific experts, Nott obscured the frequent sexual exploitation of black and mixed-race women with convenient stereotypes about their promiscuity. While such alleged promiscuity may have prevented Nott from fully tracing the racial lineage of the individuals he encountered, it did not stop the intrepid doctor from going about his business of collecting anecdotal evidence regarding mulatto women as a group—and mixed race people in general—by inquiring of those women he encountered in his medical practice, "either free or slave," about the color, number, health, longevity, and parentage of their offspring. Their answers in turn convinced him of the validity of claims about racial mixture.[78] Sadly, the bleak image that Nott constructed of mixed race people was a characterization shared by many of his contemporaries within and outside the science world.

In sum, antebellum racial theorists presented mulattoes as without purpose in nature—lesser versions of both parent races and thus physically fit neither to rule nor to labor. Gender played some role in this alleged lack of purpose, for "hybrids" also did not fit the sex-defined roles nature had designed; mulatto men were weak and mulatto women were "bad breeders." A short time later, scientists during the Civil War and Reconstruction applied similar arguments not just to "mulattoes" but to black men and women as well. After emancipation, blacks in general would join "mulattoes" as deviating from the gendered schema of nature and being undeserving of the rights and privileges afforded their respective sexes, such as the vote.

Ethnological discussions of mulattoes as weak, sexually compromised hybrids persisted even as the debate over polygenesis, to which considerations of hybridity were initially connected, waned in the 1860s.[79] So too did ethnological scientists continue to focus overwhelmingly on the black and white races in discussions of racial mixture, for as Van Evrie described, "As with all other genera, there is a certain capacity of interunion in the several human species, less, however, in the instance of Caucasians and Negroes than in other races, for these two occupy the extremes of the generic column, the former being at the head, and the latter at the base of this column."[80] This common belief that blacks and whites represented opposite ends of a racial hierarchy was in part why so much of antebellum ethnology in general focused on comparing the two races rather than other races that scientists thought were more closely linked.

Also noteworthy in Van Evrie's words is his emphasis on terms like "genera" and "species," a clear indication that racial taxonomy was never just about categorizing. Applying Swedish naturalist Carl Linnaeus's system of classification to race provided a way for some ethnologists to attempt to reconcile polygenesis with more traditional Creation stories.[81] They claimed that the theory of the separate origins of the races did not fly in the face of the Bible because some races did not fall under the category of "man" whose genesis the Bible describes.

Like the question of origins with which it was often associated, racial taxonomy was framed in terms of men; it was male bodies and male pursuits that defined the parameters and character of the races. One of the most detailed antebellum texts on racial taxonomy, *Types of Mankind,* classified the races by both physical characteristics and degree of civilization and barbarism. Among anatomical features like skin color and facial features, Nott also used male-specific traits such as beards and brute strength to distinguish between the races. Likewise, he viewed world history through the male enterprises of war and domination, maintaining "that human progress has arisen mainly from the war of races" and from "conquests and colonizations."[82]

Linnaean taxonomy allowed racial theorists to try to cloak the ignominy of their claim that nonwhite races constituted separate species from whites as a matter of scientific semantics.[83] Nott was one such proponent, countering theological critiques of polygenesis by noting that there were many separate creations of plant and animal species, so there was no reason for people to be so reluctant to accept that the same could be true of man.[84] Racial taxonomy—the number of races in the world, their geographic origins or migrations, physical and cultural features, and so on—then was irrevocably tied to the question of racial origins in antebellum ethnology. Taxonomy also afforded such experts the appearance of being objective scientists who were simply providing names and order to the natural world, which included human beings. By placing race within a larger context of other differences in nature—not unlike distinguishing between varieties of plants, for example—even proslavery apologist Nott could describe himself as a politically disinterested naturalist.

Although well received in some intellectual circles, many Americans, including a range of Southern religious leaders, had resisted Nott's position on polygenesis, objecting to his blasphemous rejection of the biblical story of Creation despite their overwhelming acceptance of his claim regarding black inferiority.[85] Nott, for his part, seemed to enjoy making anticlerical statements while simultaneously bragging that the very same pastors in Mobile who critiqued his polygenesist beliefs still respected him so much as a medical professional that he remained the private physician to them all.[86] As Horsman points out in his nuanced biography, "Nott could have said much of what he wanted to say without attacking the use of the Bible for scientific purposes," like "most of his scientific friends, both in the North and the South," who had "reached much the same conclusions about black incapacity, but . . . found ways to pay lip service to religious orthodoxy while writing works that in effect challenged the scientific validity of Genesis."[87] Nott remained a provocateur, however. Not surprisingly, then, his critics—most of whom had long reconciled their Christianity with their belief in racial hierarchy and even the moral soundness of chattel slavery—seemed far more bothered by Nott's refusal to show deference to religious authority and scripture than his grim assessment of the black race, which was such a common view it barely received mention.[88]

Despite his critique of traditional interpretations of the Bible, Nott was hardly a devoted scientist defending "fact" and rationality in the face of religious condemnation and oppression. Indeed, he was not even particularly committed to the notion of polygenesis but rather to the "practical fact" of permanent black inferiority, however it had originated. "Whether an original diversity of races be admitted or not, the permanence of existing physical types will not be questioned by any archaeologist or Naturalist of the present day." The "consequent permanence of moral and intellectual peculiarities" was undeniable as well: "The intellectual man is inseparable from the physical man."[89] Nott's argument was representative of nineteenth-century ethnology in its insistence on the "practical fact" of racial hierarchy and the firm belief that one's moral character and intellectual ability is inscribed upon and irrevocably determined by the physical body—a body that was also implicitly male.

As Nott demonstrates, racial taxonomy was a high-stakes venture in antebellum America. Furthermore, just as similar racial language could be found on both sides of the sectional crisis, the theory of polygenesis found an audience and support across the political spectrum, with articles arguing on behalf of the separate origins of the races appearing in both *The United States Democratic Review* and the *American Whig Review*. Notably, though, in each case, such arguments were made by men and implicitly or explicitly targeted a male audience.

"We are MEN": The American School of Ethnology Attacked

Polygenesis—or slavery, with which it was increasingly linked by the 1850s—was by no means universally accepted within or outside the scientific establishment, but its critics were just as likely to frame their arguments in terms of men. Chief among these critics were black intellectuals, who faced the daunting and unenviable task of developing a logical argument to refute the often-illogical racist rhetoric of the day. In arguing for the shared origin of the races, they literally had to assert their place in the human family as part of "mankind." Employing some wordplay of their own, black intellectuals insisted that African Americans too were "men" in the generic sense of the word, but they also defended their status as men in the gendered sense of the word. For example, in his passionate abolitionist text, *Appeal to the Colored Citizens of the World,* North Carolina–born David Walker, the son of an enslaved father and a free black mother, posed a hypothetical question to "every [white] man who has a heart:" "If you will allow that we are MEN, who feel for each other, does not the blood of our fathers and of us their children, cry aloud to the Lord of Sabaoth against you, for the cruelties and murders which you have and do continue to afflict us."[90] Though legally free himself, Walker's hatred of chattel slavery drove him to leave the South, travel, and eventually begin a new life in Boston; the success of the clothing store he opened there allowed him to use his money and influence to help other African Americans, particularly runaway slaves.[91] Addressing the primary audience for his missive—"colored citizens"—Walker appealed to their sense of manhood, querying, "Are we men who have any spirits at all? I know that there are many swell-bellied fellows among us, whose greatest object is to fill their stomachs. Such I do not mean—I am after those who know and feel, that we are MEN, as well as other people."[92] Whites, he added, have "always been an unjust, jealous, unmerciful, avaricious and blood-thirsty set of beings, always seeking after power and authority." But while criticizing white men for masculinity run amok, he also demanded that black men assert their own place in the ranks of manhood. "Are we MEN!!—I ask you, O my brethren, are we MEN?"[93]

Here, Walker evoked male images of the white race similar to those conjured by African Americans' most vocal—and racist—critic, Josiah Nott. But where Nott heralded a conquering white race that was asserting its rightful place in the natural order, Walker, and numerous other black thinkers who also challenged ethnology, denounced whites as vicious warmongers.[94] Both groups were ostensibly speaking of the white race as a whole, but as neither conquering nor warmongering were activities linked to women in the nineteenth century, it would seem that their respective assessments of the white race revolved around its men. Indeed, beyond being insulted by the theory of polygenesis, black writers also took issue with racial scientists' valorization of white manhood.

The black counterdiscourse that developed in response to widespread scientific charges of racial inferiority in the nineteenth century also utilized masculinist language and male-centered arguments about race. "No man is any thing more than a man, and no man is less than a man," avowed former slave James Pennington in 1841 in response to white claims of intellectual disparities between the races. Blacks, he argued, were "inferior in attainment" not as a product of their anatomy or a separate creation but because of racial injustice and the degradation of slavery in America.[95] Arguing for the shared origin of all humanity, as described in the biblical story of Creation, black minister Hosea Easton asserted, "God hath made of one blood all nations of men for to dwell on all the face of the earth. Or, in other words, I conclude it is a settled point with the wisest of the age, that no constitutional difference exists in the children of men, which can be said to be established by hereditary laws."[96] Like many antebellum racial theorists on both sides of the color line, Easton answered the "great question" at the heart of ethnology with reference to "nations of men" inhabited by "children of men." Similarly, in the preface to the third edition, David Walker proclaimed his hope that "all coloured men, women and children" would read his text. Yet on the next page, he nonetheless addressed his *Appeal to the Colored Citizens of the World* to his "dearly beloved brethren and citizens" and consistently referenced his "fellow men" and "brethren" throughout, indicating that Walker anticipated and intended a male audience despite any initial claims of wider appeal.[97]

Famous black abolitionist Frederick Douglass also addressed scientific attacks on black capacities in a commencement address at Western Reserve College in 1854, the same year *Types of Mankind* was published. Douglass was born a slave in Maryland and, unlike most slaves, learned to read as a child, first from his master's wife, then, after she'd been severely reprimanded by her husband for defying the ban on slave literacy, from white children on the plantation. He credited his unusual literacy as awakening his spirit and thirst for freedom, and his eloquent writing and oratory later made him a sought-after speaker and respected leader in the abolitionist movement as well as a number of other humanitarian causes. After he was sold as an adolescent to a particularly cruel master known for "slave breaking," he eventually escaped and settled in a community of free blacks in Massachusetts. There he began his formal involvement with antislavery activism and, with the encouragement of influential white abolitionist William Lloyd Garrison, published an autobiography of his experience under slavery in 1845. By the time of his address at Western Reserve, he had already toured the country giving lectures to packed houses, and the public intellectual had become a celebrity across the United States, not just for his work with the abolitionist movement, but also for his devotion to the cause of women's rights.

His speech at Western Reserve, titled "The Claims of the Negro, Ethnologically Considered," at times conformed to the gender conventions of other black ethnologists as well as white racial scientists, while in other instances completely inverted them. "To know whether negro is a man, it must first be known what constitutes a man," Douglass posited. He bemoaned the fact that black men's "faculties and powers, uneducated and unimproved," were constantly being contrasted with the most educated and accomplished of white men and their likeness instead drawn with animals, an absurd comparison in Douglass's estimation and one easily refuted by their ability to speak and acquire knowledge. He then turned both the comparison and the common use of intellect as a measure of manhood back on African Americans' critics. Noting that horses, dogs, and livestock all recognize the black man's mastery over them and that even they know instinctually that "the negro is a man," he jested that surely white men would not want to exhibit less intellectual acuity than these barnyard beasts in attempting to argue otherwise.[98] He then upended the common scientific convention of attributing intellectual capacity to white men alone even further. He first noted that proponents of ethnology always attributed any demonstration of intelligence in an African American, himself included, to white blood in their ancestry. "There is, however, a very important physiological fact, contradicting this last assumption; and that is, that intellect is uniformly derived from the *maternal* side," he claimed, a rather striking exception within a scientific discourse that nearly universally divorced intelligence from women in every possible sense. Then alluding to the realities of sexual exploitation within America's racial hierarchy, he rightly added, "Mulattoes, in this country, may almost wholly boast of Anglo-Saxon male ancestry."[99] In other words, he and other accomplished mulattoes inherited their intelligence from their mothers, *black women,* countering both the racial and gender assumptions in nineteenth-century science.

Like the white ethnological writers whose theories they were countering, other black intellectuals framed the racial antagonisms in America as a contest of manhood and their own struggle for equality in terms of manhood rights.[100] David Walker invoked a list of male professions in which blacks were not represented, utilizing rhetoric similar to white supremacist John Campbell's work two decades later. "Now I appeal to heaven and to earth, and particularly to the American people themselves," he began, "to show me a coloured President, a Governor, a Legislator, a Senator, a Mayor or an Attorney at the Bar . . . show me a man of colour, who holds the low office of a Constable, or one who sits in a Juror Box, even on a case of one of his wretched brethren, throughout this great Republic!" Unlike Campbell, however, Walker invoked this list not as evidence of black men's inability to hold such positions but as an indictment of the racist system that barred them from the ranks.[101] For Walker, the destiny of a race lay in the *opportunities* and status afforded its men.

The stakes—and the risks—differed greatly for white and black theorists as well. White theorists like Nott used science to defend and naturalize an existing sociopolitical culture that empowered them as white men, whereas black writers were forced to challenge that same culture and its conventional beliefs about race. But challenging authority and the status quo too was a privilege afforded white men alone in nineteenth-century America. When Josiah Nott deliberately antagonized religious leaders and flouted canonical understandings of the Bible, he came through the storm of controversy with both his livelihood and his life. In contrast, David Walker's fiery and widely circulated pamphlet encouraging slaves to rise up against their masters—to *be men*—caused such anger and fear among Southern slaveholders that many believe his mysterious death at only forty-five years of age was the result of foul play.[102]

While white adherents to ethnology endeavored to ground race in biology, black respondents emphasized culture and environment. Like their white counterparts, black racial theorists employed a mix of science, history, and scripture but tended to read less into the corporeal body in their discussions of race. Particularly before the Civil War, black intellectuals often proposed climate-based explanations of racial difference or conceptualized race as a cultural rather than biological phenomenon. Largely eschewing the biological determinism that dominated white ethnological work, many antebellum black thinkers pointed to classical African civilizations like Ethiopia and Egypt to counter white claims that the African had no history or capacity for civilization and highlighted Europe's own barbaric past.

Within such arguments, black theorists like Walker and Easton often critiqued the white race as overly masculine, aggressive, and violent and positioned blacks as a "redeemer race."[103] Though he echoed Walker's assessment of whites as a savage race of warmongers, Easton did not call upon black men to prove their manhood. Instead, he celebrated the feminine qualities he saw as characteristic of the black race, even personifying Africa as a "she," all the more notable since he otherwise relied on the masculine generic language conventions of the time: "Africa never will raise herself, neither will she be raised by others, by warlike implements, or ardent spirits; not yet by a hypocritical religious crusade, saying one thing and meaning another. But when she rises, other nations will have learned to deal justly with her from principle. When that time shall arrive, the lapse of a few generations will show the world that her sons will again take the lead in the field of virtuous enterprise, filling the front ranks of the church, when she marches into the millennial era."[104] Like white ethnological writers, Easton presented the white and black races in sharp contrast but in culture and ideals rather than biology. Furthermore, Easton described the black race as having the potential to ameliorate the worst abuses of white men and to lead not by force but by example. In so doing, Easton invoked an argument that mirrored nineteenth-century discourse on gender whereby women served as a moral influence and counterbalance to the more aggressive pursuits of men. However, as historian Mia Bay argues, black men like Easton redefined manhood to include more "feminine" characteristics.

Many black intellectuals drew heavily on the Bible to defend their shared humanity dating to Creation as well as to address the "Curse of Ham" argument that white scientists used to "prove" that black inferiority was inevitable and divinely determined. In rejecting their alleged place in a cursed bloodline, several black theorists constructed a similarly male-focused alternative lineage for the black race, tracing it to Ham's other son

Cush rather than Canaan. Indeed, regardless of how they addressed the "Curse of Ham" argument, black respondents described racial lineage as concerning the "children of men."

Moreover, the argument by numerous white scientists that the black race perhaps belonged to the same genus as whites but constituted a distinct species was never a matter of mere taxonomical semantics for black writers, or for the black populace as a whole, who were very much aware of claims that they were a separate species. Educated and uneducated blacks alike passionately and poignantly asserted their shared humanity and common origin with whites. Still, as Bay points out, in so doing, antebellum black thinkers were largely unable to escape racial essentialism and struggled to reconcile their own competing claims that the races were both different and equal.[105]

While polygenesis remained controversial within the scientific world, even among white advocates of ethnology firmly committed to racial hierarchy, the theory was met with greater skepticism and opposition still in mainstream periodicals aimed at wide audiences. This was the case for a myriad of reasons, though rarely indicative of a belief in the innate or potential equality between the races. In 1851, the *North American Review,* a popular literary magazine published out of Boston with some of the most well-known writers in nineteenth-century America among its contributors, ran a series of essays from linguists and scientists, including prominent defender of monogenesis James Prichard, under the umbrella title "The Unity of Language and of Mankind." The issue was preceded by a lengthy introduction by the journal's editors. Originally published in the 1847 "Reports of the British Association," the editors of the *North American Review* chose to reprint the articles for an American audience because of "their bearing upon the vexed question of the unity of the human race," which "is daily assuming more and more prominence in the researches and discussions of scientific men and scientific bodies." The important and far-reaching issue of racial origins "may be regarded as physiological, psychological, or genealogical," the *Review*'s editors explained.[106] They also defended polygenesists against knee-jerk accusations of sacrilege and stated that their conclusions were based on legitimate science and sound logic. But ultimately, the editors came down on the side of a single origin for all humanity regardless of race.[107] For them, the proof lay not in physical science or in any evidence that the races were exactly the same in moral or intellectual capacity. In fact, they argued that the races did indeed differ markedly by those measures, but such difference was a matter of degree and not nearly to such an extent to justify their classification as separate species, for there was much more difference between man and lower animals than there was among men across racial lines. Instead, linguistics, which so many tended to write off as "dry and trite," held the answer to "the unity of the human race."[108] Like much of the black response to ethnology, the *Review* deemphasized the body in the "vexed question" of racial origins.

Popular publications also shared black theorists' emphasis on biblical lineages as well as their gender conventions. "Are All Men Descended from Adam?" asked the title of an unsigned critique of polygenesis published in *Putnam's Monthly Magazine of American Literature, Science and Art* in 1855. Men presumably meant all races of humanity and descent was traced not to Adam and Eve but to Adam alone. Like "The Unity of Language and of Mankind," the article in *Putnam's Monthly Magazine* acknowledged differences in the moral and intellectual character of the various races but argued that these differences did not prove separate origins of the races. However, "Are All Men Descended from Adam" did not share the previous article's deference for the polygenesists as worthy if ultimately incorrect men of science and proceeded to counter each component of the typical argument for the plurality of racial origins.[109]

Other publications went further still, attacking not just polygenesis specifically but rather the teleological nature of ethnology as a whole. The popular periodical *Scientific American,* for example, opined, "It appears to us that modern Ethnology is something like spiritualism, neither of them are new subjects, but as treated by their students they develop many new absurdities." It discussed in particular an Irish-born doctor named McElheran currently residing in the United States who defended the often-maligned Irish against ethnological attacks by counterarguing that all the races descended from one shared origin and the Celt was now the central type. The article chided, "It is a positive fact that every race thinks itself superior to all others; and the doctor, being a Celt, views all crania through his own peculiar vision. An Anglo-Saxon boasts of his race as superior to all others, while the Celt considers himself the model man. This pride of race is as old as the hills, and just as absurd as it is old." Given the anti-Irish sentiment growing in the United States during this time, however, it is unclear whether *Scientific American* would have dismissed the doctor's argument as absurd had he been presenting the Anglo-Saxon as the pinnacle of human perfection.[110]

Offering a definition of race that diverged from the biological paradigm presented by most white proponents of ethnology, the editors of *Scientific American* pondered, "What is a race? A people speaking a peculiar language and of a certain habitat." Speaking in conventionally gendered language while engaging in a more radical critique of race, they nonetheless offered their readers a surprisingly insightful assessment of the limitations of ethnology. "At the present the German believes the Teutonic to be the model race; the Englishman and American believe the Anglo-Saxon to be the model type; while the French and the Irish boast of the Celt," they pointed out. Hierarchies of men were also constantly evolving, they argued, noting, "The truth is, that virtue, bravery, and industry make a model man and a model race. These are the qualities of character, which in the history of the world, have elevated one race and nation above another." However, they added, "if such qualities were race peculiarities, then the nation first dominant would always have been dominant; the Egyptians would still have been the Prince of Men."[111]

Another noteworthy critic of the American school of ethnology in general and polygenesis specifically hailed not from the United States but Great Britain: Charles Darwin, whose theory of evolution, while initially poorly received in America, would later reshape American racial science. Though not publicly vocal on the subject of ethnology, in a private letter to his friend Charles Lyell, a British geologist, Darwin critiqued Morton's article, "Hybridity in Animals," and concluded, "I do not think Dr. Morton a safe man to quote from."[112] In another letter to Lyell more than a decade later, he again turned a critical eye toward the American school of ethnology, positing that the multiple origin of dogs had no bearing on human beings and "all the races of man are so infinitely closer together than to any ape." Consequently, he could only conclude that "all races of man . . . descended from single parents."[113] His follow-up statement, however, indicates that though he did not share the American school of ethnology's belief in polygenesis, he did share some of its language and belief in racial hierarchy. "I should look at it as probable that the races of man were less numerous and less divergent formerly than now, unless, indeed some lower and more aberrant race, even than the Hottentot has become extinct," Darwin wrote.[114]

Ultimately, though, debates over polygenesis versus monogenesis waned in American ethnology during and after the Civil War. With Emancipation came new ethnological concerns about the status of the freedpeople—and their claims to gendered rights such as citizenship. And later evolution, and its frequent corollary, degeneration, became the dominant paradigm for explaining racial differences. Whereas a quarter of all scientific texts on race published between 1830 and 1859 were focused on the origins question (and many other texts addressed the issue in some way), the number dropped to 2 percent in the 1860s and 1870s and to less than 1/2 percent by the turn of the century. Nott provides an excellent example of this shift. In his 1844 publication, *Two Lectures, on the Natural History of the Caucasian and Negro Races,* discussed earlier, Nott acknowledged his intellectual debt to his predecessors and colleagues as well as his place within a genealogy of American polygenesist thought. He then declared, "The question of the unity of the races is a grave one" and "not a question for mere idle discussion, but one involving others of deep Political, Moral, and Religious import." Twenty years later in 1865, however, he penned a letter to the superintendent of the Freedmen's Bureau in which he referred to "the original unity or diversity of species" as a "mooted question" and turned instead to the more pressing post-Emancipation question of citizenship and the current state of the black race, regardless of how it had originated.[115] Clearly, ethnological defenses of polygenesis declined in the second half of the nineteenth century. But the "pride of race" identified by *Scientific American* certainly did not, nor did ethnologists' tendency to view the body through their "own peculiar vision."

"Like the Relations of the Sexes": Racial Science Confronts Slavery

The consensus among early racial scientists that race was likely more visible in male bodies was not the only reason for the overwhelming focus on men in antebellum ethnology. White ethnological writers—and the African American men who countered their most ignominious claims—were primarily interested in the capacities and destinies of men, for indeed in nineteenth-century America, the social, political, and economic position of a man determined the power and status of his family, his community, his race. White men could not lead if others did not follow; they could not rule if others did not serve. Antebellum ethnology lent scientific credence to existing inequalities, constructing a world in which men competed for power and the fate of the world rested on the "best" men winning the struggle—a struggle in which women had little role.

While not quite as ubiquitous as the debate over the original unity or diversity of the races, slavery was also a source of contestation among early proponents of ethnology, particularly during the 1850s. Even from ethnology's inception as a field of study, some writers had looked to the human body to legitimize slavery's place in the social body. Indeed, much more than aesthetics were at stake in scientific indictments of black physiognomy. Scientists' extensive comparisons between the bodies of African Americans, whites, and occasionally Native Americans did more than show how the races differed anatomically; they also asserted that those bodies were suited, even designed, for specific roles—namely, servitude or civilization.

It would be a mistake, however, to assume that antebellum ethnology presented a uniform front on the issue of slavery. Some scientists, such as Josiah Nott, Samuel Cartwright, and John Van Evrie, vehemently defended slavery as the natural relationship between superior and inferior races. Others tended to avoid offering an opinion on the issue, though to do so was increasingly uncommon by the 1850s. Still other ethnological writers directly opposed the institution of slavery for a multitude of reasons, few of which involved a rejection of the concepts of racial difference and hierarchy. Some of the racial theorists who critiqued slavery conceded that blacks were inferior but argued that the institution itself was either an abuse of power or an all-too-frequent site of human frailty and immoral behavior. Others wanted to rid the United States not just of slavery but of the slaves themselves, who they viewed as either a burden or a moral or physical contagion. And most poignantly, black intellectuals attempted to challenge the central tenets of the field in support of abolition and black equality but often ran into the proverbial problem of trying to dismantle the master's house with the master's tools and employing logic against often illogical claims.[116] Moreover, even when an individual racial scientist did not explicitly support or even address slavery in his work, his claims about anatomical differences between the races could nonetheless be used by politicians or other ethnologists to buttress proslavery arguments.[117] But slavery's ethnological defenders and detractors often shared one common rhetorical strategy: both compared the institution to the relationship between the sexes to argue for either its rightness as a well-ordered "household," with a beneficent patriarch overseeing his inferior charges in the positions nature intended for them, or its wrongness as an abuse of a natural superior against weaker counterparts he should be protecting.

Prior to the 1850s, it was common for white proponents of ethnology to ignore the volatile issue of slavery entirely, in part to appeal to a national audience across growing regional divides as well as to underscore the nascent field's status as a legitimate and politically disinterested *science.* Samuel Morton, the celebrated grandfather of American ethnology, for example, said little about the ethics of the institution itself, simply noting the cultures in which it occurred—though he did refer to slavery as a condition that degraded the faculties of the enslaved by way of explaining that the skull of a Southern slave may not be an accurate representation of the "negro race" as a whole.[118] While his Quaker upbringing may have prevented him from defending slavery outright, Morton's claims about long-standing racial difference and black inferiority fit neatly into such a defense and often mirrored the language of numerous proslavery apologists. "The Negroes are proverbially fond of their amusements, in which they engage with great exuberance of spirit; and a day of toil is with them no bar to a night of revelry," he remarked, adding, "They appear to be fond of warlike enterprises, and are not deficient in personal courage; but, once overcome, they yield to their destiny, and accommodate themselves with amazing facility to every change of circumstances."[119] Morton's reference to "warlike enterprises" and "personal courage," then associated with men, indicates that to him, the race's men and the race overall were synonymous, while his commendation of African Americans' adaptability seemed to suggest that chattel slavery may not be an unbearable hardship for them. Perhaps not surprisingly then, proslavery scientists and politicians alike hailed his work, praise that Morton himself seemed to welcome. Perhaps even more to the point, Morton died before debates over slavery, within ethnology as well as the larger political culture, reached their most fevered pitch, and thus his legacy was malleable for those who invoked his work. Indeed, upon learning of Morton's death in 1851, Josiah Nott lamented the loss of "our leader" in a letter to their mutual friend, fellow ethnologist and diplomat E. G. Squier, who had helped procure skulls for Morton during his travels. Nott's nod to Morton's leadership, however, was likely as much a reference to his advocacy of polygenesis as it was to the fact that his work aligned so well, albeit implicitly, with a defense of America's "peculiar institution."[120]

To be sure, not all ethnological writers prior to the 1850s avoided commenting directly on the issue of slavery. Samuel Stanhope Smith was one early, though exceedingly measured, critic of the institution. Writing in 1810, he foreshadowed Morton's language decades later in characterizing American slaves as a degraded class whose physical nature was hampered by the miserable conditions in which they lived, and thus they compared unfavorably in body and mind to their free counterparts. The "dependent and humiliated race of men" could and would improve as their living and climate conditions improved, a fact he insisted was already born out in the United States where the "most disgusting qualities of the African countenance" were being ameliorated, "especially in the families of wealthy and humane masters" as well as freedmen in the North. Though he described abolition as "an event so desirable to humanity," albeit dangerous if enacted too quickly, he maintained that even "the most oppressed and destitute of [U.S.] slaves, with very few exceptions, are better fed, clothed, and lodged, than their ancestors were in Africa." He then went one step further: harkening back to the several years he spent living in Virginia as a young minister, he noted that while he never owned slaves himself, he was pleasantly surprised to find that most of the slaveholders he encountered were "humane," in contrast to the "very exaggerated descriptions . . . often given of the severities practiced, and the deprivations imposed, on the slaves in the southern states." In other words, he managed to critique the institution of slavery and its deleterious effects without criticizing Southern slaveholders themselves.[121]

Though bitter foes in the racial origins debate, Samuel Stanhope Smith and Charles Caldwell *seemingly* found common ideological ground on the issue of slavery. Caldwell denounced the institution in *Thoughts on the Original Unity of the Human Race* (1830), but it was a critique premised on a belief in racial hierarchy—and a permanent hierarchy at that, as he did not share Smith's optimism about the black race's potential for meaningful improvement under more favorable conditions. "The Caucasians are not justified in either enslaving the Africans or destroying the Indians, merely because their superiority in intellect and war enables them to do so," he remarked. "Such practices are an abuse of power; and where is there privilege that is not liable to abuse?" he added.[122] Caldwell compared this naturalized racial hierarchy to that of gender, noting, "Man protects and cherishes woman because she is feeble, and looks to him for protection. Render her his equal and rival, and he will leave her to protect herself."[123]

Caldwell may have had some kind of pragmatic attitude toward slavery that, while it was "not justified" when the slave trade first began, it was nonetheless the inevitable relationship between a superior and inferior race once living in close proximity, for he himself owned domestic slaves while living in Kentucky and was raised in a slave-holding household in North Carolina.[124] Or perhaps the fact that, like Jefferson before him, he described slavery as an abuse of power while simultaneously owning slaves was more emblematic of attempts by earlier converts to ethnology to divorce the fledgling field from political applications to underscore its scientific legitimacy—a posture that was all but abandoned by the 1850s. But it would hardly be the last time racial science would engender such ambivalence.

In contrast, Charles Brace exhibited no such dissonance between his public writings and his private thoughts and actions in regards to slavery. Diatribes against the institution, which he called a dark "stain" upon the nation, abound in his personal correspondence, but he held especial hatred for the Fugitive Slave Law.[125] In an 1851 letter written to his cousin Jane, wife of Asa Gray (the highly influential botanist who was largely responsible for popularizing the work of his close friend, Charles Darwin, within the United Sates), Brace condemned the controversial law, passed the previous year, as "one of the most unjust, wrong laws ever passed in the history of nations." He further added that he "would never obey it while God preserves [his] reason," for he "consider[s] conscience above all human laws." He railed at the fact that the law made everyone in the country, even those such as himself living on free Northern soil, complicit in the evil of slavery. And in Brace's estimation, the very act of running away showed a fugitive slave to be "more fit for freedom," having exhibited all the spirit and bravery that so many—proslavery scientists included—tried to disassociate from the black race in general and its men in particular.[126]

Though he defended slaves' manhood and manhood rights, Brace was more equivocal when it came to characterizing the white race. Writing in *The Races of the Old World,* he hailed patriarchy, dating back to "antiquity," as one of the hallmarks of Indo-European civilization and superiority, noting, "The relation of husband and wife, the position of the sexes, the absence of caste, and the priestly authority of the Father, were characteristics of our earliest ancestors."[127] In another breath, however, he wondered if some of the features associated with that very patriarchal culture—particularly its emphasis on conquest and disdain for and mistreatment of perceived inferiors—ranked among the worst characteristics of the white race or races across time and space. Sounding remarkably similar to a number of his black intellectual contemporaries, he lamented, "The immeasurable contempt and prejudice against the inferior race, which characterize all branches of the English race, whether the British masters ruling Hindoo servants, or English landlords with Irish Kelts, or Anglo-Americans among Indian tribes, or Southern slaveholders toward slaves, or 'Yankees' toward negroes, is an unfortunate but legitimate inheritance from Teutonic ancestors."[128] Ultimately though, Brace remained more measured in his discussion of slavery in *The Races of the Old World* than he was in his private writings, not for lack of conviction on the subject, but because he explicitly wanted his "Manual of Ethnology" to rescue the field from the overtly political propagandizing with which it had become imbricated by the outbreak of the Civil War. Though he was as ardent an abolitionist as he was a monogenesist, he sought to separate the two issues, for the former was a moral issue and the later was "purely scientific." Slavery was an abuse of power, Brace insisted, but like other antislavery ethnological writers, it was nonetheless an argument premised on innate hierarchy among the races. "The inferiority or superiority of a given race, the questions of justice to the weak, and of Human Brotherhood, have no connection whatever with the scientific problem of Origin," he asserted. "The strong are equally bound to be merciful to the weak: men are equally under obligations to follow the Law of Love, and Slavery is equally wicked and damnable, whether mankind have one parent or twenty parents." Given that he had called white men specifically to task earlier for their aggression and mistreatment of "inferiors," he likely had men as a sex in mind—at least implicitly—when he continued his train of thought in similarly masculine language, adding, "The moral Brotherhood of man does not depend on community of descent, but on a common nature, a similar destiny, and a like relation to their common Father—God."[129]

For black intellectuals, slavery and racial origins were never separate issues. When David Walker proclaimed, "We are MEN," in his *Appeal to the Colored Citizens of the World,* it was simultaneously an assertion of the black race's place in the human family, of the same origin as whites; an assignation of the injustice of enslaving one's fellow man; and a call to black men to stand up as *men,* in the gendered sense of the word, to defend themselves and their families from such tyranny. Similarly, in his cogent critique of American ethnology, Frederick Douglass considered the "manhood of the negro" to be at the crux of both the slavery and origins issues. He pointed out that the *Richmond Examiner*, "a respectable public journal," recently based "its whole defense of the slave system upon a denial of the negro's manhood."[130] And even more directly than Walker, Douglass laid bare the proslavery implications—and political applications—of polygenesis. "Ninety-nine out of every hundred of the advocates of a diverse origin of the human family in this country, are among those who hold it to be the privilege of the Anglo-Saxon to enslave and oppress the African," he remarked, "and slaveholders . . . have admitted, that the whole argument in defense of slavery, becomes utterly worthless the moment the African is proved to be *equally a man* with the Anglo-Saxon."[131] Douglass was not incorrect. By the time Douglass gave his speech in 1854, American ethnology had become dominated by both proslavery and polygenesist factions, and indeed, the two schools of thought were generally intertwined during the 1850s.

As slavery faced new limitations on its expansion west and mounting attacks on its very existence from the thriving abolitionist movement, Southern slaveholders rallied around their "peculiar institution" with unprecedented vigor. Ethnology's adherents largely followed suit. The topic of slavery became central in racial science in the 1850s and remained so well after the Civil War, when scientists juxtaposed the natural order of the antebellum plantation with the disorder of Emancipation. The gendered rhetoric of paternalism, which framed much of the proslavery discourse outside the scientific establishment, often figured into ethnology as well.[132] With its plantation-as-extended-family metaphors and its equation of women and blacks with children incapable of understanding their own best interests, paternalism used gender to defend slavery's racial hierarchy. Likewise, interpretations and representations of human bodies, particularly sexed bodies, were also central to ethnological defenses of slavery. In fact, explicit discussions of women or gender roles in ethnology during this period often centered on the issue of slavery. Gender was thoroughly implicated in both ethnological defenses of slavery and ethnologists' considerations of freedom's perils that questions of suffrage after Emancipation raised.

Ethnological proslavery arguments generally presented at least one of three interrelated claims: (1) that Africans had occupied positions of servitude and slavery dating back to antiquity, demonstrated by Egypt and biblical lore in particular; (2) that the very bodies of African Americans were designed for labor whereas the bodies of (male) Caucasians were more suited for intellectual pursuits; and (3) that the social/political order reflected the natural order in regards to race and to gender. Significantly, the body was central to all three claims. The three most vocal proslavery scientists—Nott, Van Evrie, and Cartwright—each advanced these claims, and all relied on the body to support their arguments. All three drew on paternalist logic and rhetoric, in which gender was central. Additionally, black women often featured prominently in their defense of slavery—in stark contrast to scientists' more general considerations of racial taxonomy in which each race was defined by male prototypes and women appeared nonexistent.

In *Types of Mankind,* Nott argued for a thorough examination of the history of the races and for Egypt's central place in such a history.[133] For Nott, himself the owner of sixteen slaves as well as a cofounder of a slave medical clinic in Mobile, Alabama, Egyptian history held the key to both the separate origins of the races and the inevitability of racial slavery.[134] Both Nott and George Gliddon, in their individual contributions to *Types of Mankind,* spent considerable time interpreting images from Egyptian history that confirmed, they claimed, that the ancient Egyptian was white and the dark-skinned African his slave. For Nott, the issue of slavery was intimately tied to the question of racial origins, for in his estimation, not only did blacks have a separate origin, but they were also from the beginning of history servants and slaves, as if by design.[135]

Falling back on a common paternalistic defense of slavery, Nott argued that if they changed at all, African Americans "became more intelligent and better developed in their physique" under slavery due to their "ceaseless contact with the whites" and "the increased comforts with which they are supplied." Although Africans "became healthier, better developed, and more improved in like manner by domestication" under the benevolent influence of U.S. slavery, it was a difference of degree rather than kind. Nothing could make blacks into something they were not.[136] Much was at stake, then, in ethnological discussions of Egypt and comparisons between the black race in the ancient and contemporary worlds: proving long-standing and likely permanent racial hierarchy, naturalizing slavery as the innate role of blacks, and divorcing African Americans from any claims to past civilization.[137]

With its focus on the past and racial taxonomy, discussions of Egypt more than the other ethnological defenses of slavery tended to focus on men. There were notable exceptions in *Types of Mankind,* however. On a few occasions in his consideration of Egyptian slavery, Nott interpreted the black female body. For Nott, black women's physiognomy also provided evidence of the permanence of racial difference as well as the "negro's" long-standing servile position. For example, Nott juxtaposed an Egyptian image of a "negress" with a Roman's description of his African female servant to demonstrate that her physical type had not changed in the interim (Figure 1.6). Like the earlier Egyptian image, the Roman text illustrated that "the physical characteristics of a 'field,' or agricultural, 'nigger' were understood at Rome 1800 years ago, as thoroughly as by cotton-planters in the State of Alabama, still flourishing in A.D. 1853."[138] Moreover, Nott's inclusion of an image in which the breasts of the "negress" were both

Latin description of a NEGRESS, *written early in the second century after* C.

"Interdum clamat Cybalen; erat unica custos;
Afra genus, tota patriam testante figura;
Torta comam, labroque tumens, et fusca colorem;
Pectore lata, jacens mammis, compressior alvo,
Cruribus exilis, spatiosa prodiga planta;
Continuis rimis calcanea scissa rigebant."

"In the meanwhile he calls Cybale. She was his only [house-] keeper. African by race, her whole face attesting her father-land: with crisped hair, swelling lip, and blackish complexion; broad in chest, with pendant dugs, [and] very contracted paunch; her spindle-shanks [contrasted with her] enormous feet; and her cracked heels were stiffened by perpetual clefts."

Egyptian delineation of a NEGRESS, *cut and painted some* 1600 *years before the Latin description.*

FIG. 177.

Figure 1.6. Male scientists, female bodies. *Though Josiah Nott tended to focus more often on male physiognomy in his discussions of race, black women's bodies provided an especially useful site through which he could interpret the heated controversy in the 1850s over slavery's efficacy and expansion. Nott juxtaposed visual and textual representations of black female bodies from different historical eras and places to underscore that neither their fecund physical form nor their subservient social roles had changed over time. That is, for Nott, the enslavement of black women—and by extension, the race as a whole—was justified because their very biology demanded servitude and the literal reproduction of servitude from the beginning of time itself (Nott,* Types of Mankind*).*

prominent and exaggerated alongside descriptions of black breasts as "pendant dugs" is indicative of ethnologists' growing interest in black female bodies in the context of the slavery question.[139]

Whereas Nott's *Types of Mankind* looked to representations of the body in history to naturalize slavery in the present, Cartwright read the bodies of living slaves to prove they were thriving in the role for which nature had intended them, an argument in which black women's reproduction often took center stage. Echoing Nott, Cartwright, a physician and slave owner, argued that slavery not only benefited the souls of the slaves but also improved their bodies. In this regard, Cartwright's work exemplifies scientific paternalism. That is, he maintained that the black race would die out if freed from slavery and thus Southern slaveholders acted in African Americans' best interests by enslaving them. Cartwright devoted a number of his medical journal articles to proving not only that Africans were more suited to agricultural labor in the South than Caucasians were but that their bodies were literally designed for servitude.

Cartwright made frequent comparisons between black slaves and infants, both of whom needed and benefited from the care afforded them by their superiors.[140] His "On the Diseases and Physical Peculiarities of the Negro Race" was first read before the Medical Society of Louisiana, then published in the *New Orleans Medical Journal* in 1852 and reprinted in the *Georgia Blister and Critic* in 1854. Therein, Cartwright drew a rather strange parallel between the allegedly similar sleeping patterns of infants and blacks "of all ages and sexes," who "instinctively" cover their faces with a blanket or their hands or arms or by laying facedown, so that they reinhale their own warmed breath. Notably, in such comparisons, black women and black men were not only indistinguishable from infants but also indistinguishable from each other. In the nineteenth century, this was no trivial characterization; a man who was indistinguishable from a woman was no man at all and vice versa. For Cartwright, the "universal practice" of face covering during sleep among infants and African Americans of all ages and sexes was a sign of their immature constitutions. But while white infants apparently outgrew the practice after infancy, for blacks "the inevitable effect" of long-term exposure to their own carbon dioxide was permanent intellectual deficiency. One presumes Cartwright must have traveled extensively and suffered considerable insomnia to have had the opportunity to witness enough babies and African Americans in slumber to declare face covering to be universal among both. Indeed, for Cartwright, the shared sleep habit "proves the similarity of organization and physiological laws existing between negroes and infants," but the similarities did not end there. "Negroes . . . resemble children in the activity of the liver," and "they are liable to all the convulsive diseases, cramps, spasms, colics, etc., that children are subject to," Cartwright remarked.[141]

Such claims provided scientific support for paternalist proslavery rhetoric. In the logic of paternalism, slave societies were ideally structured much like a patriarchal family unit, with the benevolent white male as the head of the household/plantation overseeing his various dependents—women, children, and slaves. He would extend protection to his charges and take care of all their material needs; they in turn would offer both labor and loyalty to their kind "master." Cartwright argued that African Americans, like children, feared violence and shied away from physical pain. More important, he asserted, both were "very easily governed by love combined with fear." Cartwright further argued that the slaves' submissive nature was biological and innate, and "like children, they are constrained by unalterable physiological laws, to love those in authority over them, who minister to their wants and immediate necessities, and are not cruel or unmerciful."[142]

In making blacks "of all ages and sexes" alike in body and behavior, Cartwright placed the black race outside the gender roles idealized for whites. This was particularly true for black men under the logic of paternalism, for they could not themselves be patriarchs, only a patriarch's dependent and loyal servant. The ideal—and physically "healthy"—black male in Cartwright's assessment was not a man at all but a submissive "boy," regardless of age, who nonetheless performed the labor of the strongest adult man. Cartwright's simultaneous construction of black men as childlike and brute laborers is one of the many ironies in nineteenth-century racial thought. During this time, childhood was increasingly idealized as a period of leisure rather than labor.[143] However, Cartwright's work was very much in keeping with the gendered language of paternalism, in which African Americans of all ages were compared to children and denied naming practices that denoted respect. For example, another issue of the *Georgia Blister* the same year included a short piece describing the leg amputation of a forty-five-year-old slave "boy" due to recurring infection from a snakebite twelve years prior.[144] By constantly likening black men to children, scientists like Cartwright not only defended slavery as a natural institution that provided care for people unable to care for themselves but also reinforced the equation of manhood and whiteness.

Cartwright argued that if treated kindly with their basic material needs met, slaves would have no desire to run away. His very biology demanded he love his master, just as his biology determined his submissive position and role as laborer. And indeed, I mean "his" literally here, because Cartwright framed this discussion of runaway slaves in terms of men, master and slave alike. This is not surprising. As noted previously, women tended to factor into ethnological discussions of slavery primarily where reproduction was concerned. Also, in reality, male slaves ran away far more frequently than female slaves, since they were freer to travel between plantations and had less responsibility for the care of young children.[145] Lastly, Cartwright's paternalistic argument about preventing slave runaways centered on blacks' "natural" submissiveness, a point that would only have to be explicitly argued in regard to men, for all women were thought to be submissive regardless of race.

Part of the project of racial science was identifying the biological underpinnings of society. Cartwright insisted that if the master played the part of the stern but loving patriarch, "the negro is spell-bound, and cannot run away."[146] Thus he described slaves who did run away from "kind" masters in terms of pathology. Runaway slaves were not an indictment of the system but a sign of poor treatment or poor health in the individual slave. The latter suffered from "drapetomania," a disease that turned a slave "mad or crazy" and drove him to run away from even a kind master. Accordingly, "with the advantages of proper medical advice, strictly followed, this troublesome practice that many negroes have of running away can be almost entirely prevented."[147] Other forms of slave resistance were also pathologized; "Dysesthesia Ethiopis, a disease peculiar to negroes," referred to "rascality" among the slaves. Rebellious slaves manifested physical symptoms that correlated to their pathological behavior. "[Dysesthesia Ethiopis] differs from every other species of mental disease," Cartwright wrote, "as it is accompanied with physical signs of lesions of the body, discoverable to the medical observer, which are always present and sufficient to account for the symptoms."[148] Of course, any "lesions" that covered the bodies of rebellious slaves were more likely than not the marks of the lash employed to return the slaves to their "natural" position of submission. The assertiveness and bravery that ethnologists lauded in white men as demonstrating their superiority was pathologized as a disease when expressed by slave men. Under paternalism, black men did not become more masculine when they acted like white men; white paternalists defined ideal black masculinity in terms of loyal and obedient service. Passing away in 1863, Cartwright himself did not live to see how many former slaves were suddenly stricken with drapetomania upon Emancipation.

While it may be tempting for historians to dismiss Cartwright's views as absurd and extreme even among proslavery thinkers, the wide circulation of his articles suggests his contemporaries thought otherwise. Moreover, the political implication and intent of his work are manifested by his work's frequent appearance in the *Georgia Blister and Critic*. The *Georgia Blister* was ostensibly a medical journal, albeit one with a decidedly political slant, as seen in the journal's subtitle: "A Monthly Journal, devoted to the exposure of quackery, the development of Southern medicine, and the diseases and physical peculiarities of the negro race." As such, it provides an example of science and medicine explicitly dedicated to a proslavery argument. While the Atlanta-based journal included more traditional medical pieces on disease and surgical procedures, each issue also featured articles that defended slavery through scientific paternalism. Consequently, this journal was the perfect venue for Cartwright's work, a fact not lost on its editors, Drs. H. A. Ramsey and W. T. Grant, who declared, "We wish we had more Cartwrights in the Southern profession—had we, the slavery question would, long ago, have been settled."[149]

Central to settling "the slavery question" was the issue of reproduction. For Cartwright and other scientific paternalists, the fertility of slave women was viewed as an indicator of the health and happiness of the slaves and, by implication, the health of the institution as a whole. In the 1850s, when Cartwright was writing, reproduction was also tied up with slavery's expansion—a topic of considerable debate in American politics at the time. With the abolition of the transatlantic slave trade at the beginning of the century and abolitionist challenges to illegal slave trading, the expansion of slavery depended on black women's reproductive labor. Faced with growing challenges to the South's "peculiar institution" during the 1850s, proslavery paternalists pointed to slave women's fertility not only as evidence of their happiness but also to demonstrate that slavery could and must expand into the territories.[150]

Cartwright repeatedly asserted that black women had larger pelvises than white women and that black infants had smaller heads than white infants, an argument that related to several other gender and racial stereotypes: that black women experienced little pain in childbirth; that white women in contrast were weak and frail; and that whites had larger skulls than blacks as adults, which correlated to differences in intellectual capacity.[151] This fortuitous combination of biological factors demonstrated to Cartwright—and many of his peers—that black women were designed for reproduction.

Black male bodies did not escape notice in this particular argument. Their pelvises were also part of a general anatomical structure designed for labor—physical rather than reproductive labor in their case. "Hence, from the obliquity of the head and the pelvis, the negro walks steadier with a weight on his head, as a pail of water for instance, than without it," noted Cartwright, curiously adding that black men were similar to orangutans in that regard. The same could not be said for white men, whose cranial shape was designed for a different purpose.[152] For Cartwright, the white man's head was built to think and the negro's to carry loads. Furthermore, in one sentence, Cartwright managed to position black men as physically similar to both black women and animals while anatomically and intellectually distinct from white men.

In addition to interpreting black anatomy as designed for productive and reproductive labor, scientific paternalists also looked to fertility rates among the slaves to argue that black people thrived in the position of servitude that nature intended. The May 1854 issue of the *Georgia Blister,* for example, contained an article that compared the life expectancy and fertility rates of free blacks in Massachusetts and slaves in Georgia with the explicit purpose of proving slavery to be beneficial. Summarizing the article's results, the editors declared, "The Southern slave is healthier, lives longer, and is more prolific, than the [black] people of the North."[153] High fertility among slaves not only "proved" they were healthy and contented enough to reproduce at high rates; it also showed that the slave system itself was successful, self-sustaining, and economically viable.

Paternalist rhetoric was just one of many approaches the ever-prolific Van Evrie employed in his diatribes against abolitionism and the black race. Often citing the work of Nott and Cartwright, Van Evrie defended "the social subordination, or so-called slavery of the negro" as the only way in which a superior and inferior race could peacefully coexist. He also argued that the sociopolitical order in the United States reflected the natural order in regards to race and gender. Chattel slavery developed not out of economic imperatives but as a product of nature, according to Van Evrie. Paternalistic legislation and practices "protected the 'slave' from the vices or cruelty of the master, while they provided for the welfare of the latter and the general security of this species of property." But above all, such laws recognized the natural relationship between dominant and submissive races.[154]

Van Evrie's writing naturalized a paternalist ideology that was distinctive to—and largely a product of—the antebellum era, an attempt to reform the institution of slavery from within as it faced growing attacks and challenges from without. The sudden boom in prescriptive literature aimed at plantation owners, which extolled the virtues of an efficiently and compassionately run plantation, and religious sermons that preached about mutual obligations between slave and master represented a larger move to ameliorate slavery's worst abuses. The concept of slavery as an organic institution in which everyone inhabited their proper roles, ordained by God and nature alike, permeated political discourse and ethnological writing, particularly as the threat to slavery's existence loomed more imminent.

Gender played a central role in Van Evrie's representation of slavery as part of a naturally ordered society. Applying a gendered framework to the relationship between dominant and subordinate races, Van Evrie argued that slavery was as natural and inevitable as patriarchal families were in structuring the relationship between white men and white women, "for like the relations of the sexes, of parents and children, etc. [slavery] was inherent, pre-existing, and sprung spontaneously from the necessities of human society."[155] Ignoring the distinctions of race was as absurd as ignoring distinctions of sex, age, or species, Van Evrie maintained.[156] Echoing Cartwright, Van Evrie asserted that the black man could not and should not act like a white man any more than a woman could be expected to demonstrate the capabilities of a man or a child of an adult; to do so was an offense against God, nature, and society. "Each one of God's creatures has his specific organization and his specific life, and it is just as reasonable to expect a white man to be an angel as it is to expect a negro to be a white man; that is, to act as a white man, or to work as a white man," he stated matter-of-factly. Just as men and women of the same race could not be placed on equal levels or be expected to perform the same roles in society, neither could men of different races.

Van Evrie buttressed his claims about racial hierarchy by juxtaposing it with the category of gender. He insisted, "It is not necessary to enter into any proof in respect to which would be the greater evil, abolition of marriage in the North or the abolition of 'slavery' in the South; the perversion of the natural relations of the sexes, or the corruption of the natural relations of the races." Emancipating the slaves would lead to social disorder in regards to both race and gender, Van Evrie argued, ominously warning that the inevitable result would be interracial sex.[157]

He was not alone in using marriage as a metaphor in political considerations of slavery. A political cartoon from 1848—the same year that the Seneca Falls Convention brought national attention to women's rights and a short time before ethnologists took up the issue of slavery en masse—represented the political union between the abolitionist Liberty Party and the Free Soil Democrats and Whigs as an interracial marriage between Van Buren and an exaggerated caricature of a slave woman (Figure 1.7).[158] The apparent absurdity of the latter denoted the strange bedfellow union of the former. The slave woman towers over her soon-to-be husband, who cowers in fear and revulsion. Presiding over the ceremony is Benjamin F. Butler, who later served as a Union general during the Civil War and initiated the policy of treating slaves fleeing to Union lines as contraband of war, a policy that proved a vital step toward emancipation. The cartoon hinted at ethnological arguments that emerged in the 1850s and foreshadowed their introduction of gender analysis as emancipation loomed.

A well-known abolitionist text also demonstrates the free flow of racial thought between politics and science in the Civil War era, particularly on the issue of slavery. In the controversial 1857 book *The Impending Crisis of the South: How to Meet It,* Hinton Rowan Helper, a Southern-bred abolitionist, set out to attack slavery from an economic rather than ethnological standpoint. Yet he nonetheless engaged key tenets of racial science. Though not a scientist or physician, Helper encapsulates many of the complexities and contradictions surrounding the slavery issue as well as the pervasiveness of ethnological thought in antebellum pro- and antislavery discourse alike. His animosity toward slaves second only to his hatred of slaveholders, Helper confounds conventional assumptions about the relationship between region and political inclination. As such, he also serves

Figure 1.7. Marriage as metaphor. *As debates over slavery heated up in American politics, growing numbers of ethnologists rallied to defend the institution. Politics and science shared a common set of racialized imagery and this image foreshadowed the mounting attention of both to gender in the context of slavery. Here, the marriage of a stereotypically exaggerated slave woman to a genteel white man is represented to be as absurd as the recent "marriage" of the Free Soil and Liberty parties. After the war, in contrast, interracial marriage would be represented not as a comical absurdity but a looming threat ("Marriage of the Free Soil and Liberty Parties," Library of Congress, Prints and Photographs Division,* American Political Prints, 1766–1876 *[LC-USZ62-10496]).*

as a fascinating parallel to Van Evrie, a northern advocate of slavery whose ethnological pamphlets were unguarded political harangues. And like Van Evrie, Helper demonstrates the fluidity of exchange between scientific and political thought on race. After the war, both men devoted themselves to exposing the threat posed by a vastly enlarged free black population seeking political equality. Moreover, both men used gender in envisioning racial order—or disorder—an aspect of their work rarely discussed in historical scholarship.

Born in North Carolina to yeomen farmers in 1829 and finding little success in the California Gold Rush as a young man, Helper directed his bitterness toward the Southern planter elite in *The Impending Crisis,* published in 1857.[159] Dedicating his book in part to "the non-slaveholding whites of the South generally, whether at home or abroad," Helper posited himself as a friend of the white working man rather than of the slaves. The purpose of *The Impending Crisis* was not "to display any special friendliness or sympathy for the blacks," he noted. This would prove an understatement when he published a pair of virulently antiblack books during Reconstruction that called for ridding the United States of the black race entirely. But even before the Civil War, his vision of abolition entailed the removal of the slaves themselves.[160] His primary concern in *The Impending Crisis,* however, was in overturning the hierarchical class system that he insisted was retarding the economic progress and development of the South, disproportionately harming the white majority of small farmers to which he belonged.

Though he opened his first chapter by distancing himself from an ethnological investigation into the "peculiarities of difference, mental, moral, and physical" between the races, he nonetheless referenced many of the scientists of race by name and engaged their claims at various turns.[161] Rejecting the ubiquitous argument by Cartwright and other ethnologists that blacks were more tolerant of labor in the southern heat, Helper pointed out that poor whites toiling in fields across the South were living evidence to the contrary and used Josiah Nott's own mortality statistics of black Southerners to the same effect.[162] Countering the frequent claim that slavery was simply an extension of the natural order, Helper played on the gender ideals of the era, by which white women were ideologically defined in opposition to work and thought incapable of manual labor. Yet the class relations created by slavery enabled an offense against the natural order. This offense was embodied by white women toiling in the fields to ensure the subsistence of their nonslaveholding families under the plantation class system. "Too hot in the South for white men! It is not too hot for white women," he declared. "Time and again, in different counties in North Carolina, have we seen the poor white wife of the poor white husband, following him in the harvest-field from morning till night," he continued. Helper then added an even more potent image to the mix: two white women residing near his mother who were forced by financial necessity to "hire themselves out" during harvest season, the double meaning of which further drove home his point about the degradation of (white) women under the slave economy.

Whereas many ethnologists argued that slavery upheld the natural order of race, Helper countered that it subverted the natural order of gender. Appealing to the manhood of his white Southern brethren to prevent such an atrocity, he proclaimed, "That any respectable man—any man with a heart or soul in his composition—can look upon these poor toiling white women without feeling indignant at that accursed system of slavery which has entailed upon them the miseries of poverty, ignorance, and degradation, we shall not do ourself [*sic*] the violence to believe." Ironically, he then asserted that white women laboring in Southern fields would be better suited to work in cotton mills and factories—the very labor from which some reformers in the nineteenth and early twentieth centuries were trying to protect women—and that the abolition of slavery would allow such work to become a reality for the economic benefit of their whole family.[163] For Helper, writing in the 1850s, white women and labor served a rhetorical function in both his indictment of the agrarian South under slavery and his vision of an industrialized South after abolition.

During that volatile decade, some legislators were seeking to limit slavery's expansion while abolitionists increasingly pressed for immediate and universal emancipation throughout the United States. As we have seen, in response, growing numbers of ethnological writers rallied to defend slavery from mounting attack. Slave reproduction, which advocates viewed as a sign of the institution's viability after the close of the transatlantic slave trade, was imperative to such defenses and to the contentious issue of its expansion into the territories. Along with the logic of paternalism, prevalent in both political and ethnological discourse on slavery, this attention to reproduction precipitated a new focus on gender among racial scientists. Then as slavery came under more imminent attack and proslavery arguments grew more strident during the Civil War, many racial scientists began to anticipate slavery's demise and tried to envision the impact of abolition on the U.S. racial order. Their predictions were grim: emancipation would result in dangerous social upheaval. Offering implicit counsel against emancipation, white ethnological scientists largely concurred that even if African Americans were no longer slaves, they did not have the mental or moral capacity to be citizens.

After the war, Helper turned his criticism from slavery to the former slaves themselves and found that ethnology offered powerful evidence that the freedmen should never be given social or political equality in America. For Helper and the ethnological experts he cited, white women served a new function after Emancipation as a point of comparison with black men. Arguing that black men and white women shared similarly limited intellectual capacities, both Helper and Reconstruction-era racial scientists insisted that neither deserved the franchise, which should be reserved for white men alone. The project of defining—and redefining—what manhood meant and how it could be measured along racial lines began during the Civil War, however, with a series of large-scale and influential anthropometric studies commissioned by the U.S. Army that provided postbellum scientists an array of data with which to validate their claims about race.

As ethnology developed into a well-known and influential source of expertise on race throughout the first half of the nineteenth century, the hodgepodge group of men at the center of the new scientific field represented a range of disciplinary, political, and social perspectives. Most, but not all, came from a medical background and drew on nature and anatomy, as well as theology and history, to ground their claims about racial difference and, more often than not, hierarchy. Ethnological writers debated whether the races of the world shared the same origin, especially the two races—white and black—they saw as most divergent. Some, most poignantly African American intellectuals who attempted to rebut the field's central claims about innate black inferiority, argued for a common "brotherhood of man," equal in potential and limited only by circumstance. Ultimately, black writers on race did not drive the ethnological debate, however. They lacked the wide audience their white counterparts enjoyed, and while it was clear that white ethnologists read many of the rebuttals from African American men, they proved remarkably adept at adapting their claims toward a foregone conclusion about black inferiority. Some white theorists concurred with black intellectuals' belief in a shared origin but maintained that the races had diverged so dramatically in physicality, intellect, and attainment that racial equality was a near impossibility—an argument they could employ for or against the efficacy of slavery. By the 1850s, though, proponents of polygenesis had come to dominate American ethnology and were also most likely to use the theory that whites alone were the "sons of Adam" explicitly in defense of slavery, maintaining that blacks originated as a separate species designed by nature from the start to be slaves. What was at stake, then, for white ethnological theorists as the nation edged closer to civil war was a far-reaching institution that naturalized their place at the top of America's racial hierarchy, while the stakes for their black counterparts were even higher—defending not only their right to be free but also their place in the human family.

The debate over the origin of the races, so central to the field, was framed primarily in terms of men, masculine characteristics, and male achievements. So too was the debate over slavery that came to permeate ethnology in the politically contentious 1850s. Scientists' considerations of America's "peculiar institution" occasionally referenced women, but only in conjunction with black women's reproduction, particularly its resonance for the expansion of slavery into the western states and territories. The male focus of racial scientists—as well as their detractors—seemed to stem in part from the common ethnological belief that race was more pronounced in the bodies of men. But even more important, in the patriarchal culture in which ethology's adherents lived and worked, it was the pursuits of men that held the most social and political consequence.

In sum, antebellum ethnology was by and about men, but "man" itself was largely an unquestioned and unexamined category. That would soon change, starting during the Civil War with the U.S. Army's scientific investigation into what, and who, constituted the "best man," a question with implications that extended far beyond the military context. Indeed, by the war's conclusion, and for the remainder of the nineteenth century, fitness for citizenship supplanted the issues of racial origins and slavery as the central concern of American ethnologists. Since full citizenship remained contingent on manhood until the 1920s, racial scientists in the late nineteenth century debated over who was "man" enough to deserve it.

· CHAPTER 2 ·

An "Equal Beard" for "Equal Voting"

Gender and Citizenship in the Civil War, Reconstruction, and Redemption

IN HIS 1861 TEXT *NEGROES and Negro "Slavery": The First an Inferior Race, the Latter Its Normal Condition,* the New York proslavery propagandist John Van Evrie continued the ethnological tradition of using the beard to distinguish between the races but invested new meaning in the physical feature as national conflict over slavery finally came to a head and the institution's demise seemed possible. Faced with this possibility, ethnological writers like Van Evrie were forced to consider what would—or should—happen to African Americans if freed, at a time when participation in American civil and political life was constrained by gender. "Like color or any other of the great fundamental facts separating races," he wrote of black and white men, "the beard is sufficient to determine their specific character and their specific relations to each other." Supporting his claims about pervasive and permanent difference between whites and blacks by invoking a bodily marker of sex, he added, "We have only to apply our every day experience as regards this outward symbol of inner manhood to measure the relative inferiority of the negro." His book's frontispiece juxtaposed images of contemporary black and white men with their respective races' ancient counterparts, again using the beard as an enduring measure of racial difference. The image also underscored that such difference did not change over time by showing the men *within* each race to be remarkably similar in appearance across history, while simultaneously very distinct *across* racial lines in both the ancient and modern eras (Figure 2.1). For Van Evrie, both physical manhood and its corollary, sociopolitical manhood, belonged to white men alone. The abolitionist who pressed for "equal manhood of the negro" demanded of society what nature itself could not create. These "friends of humanity" might as well

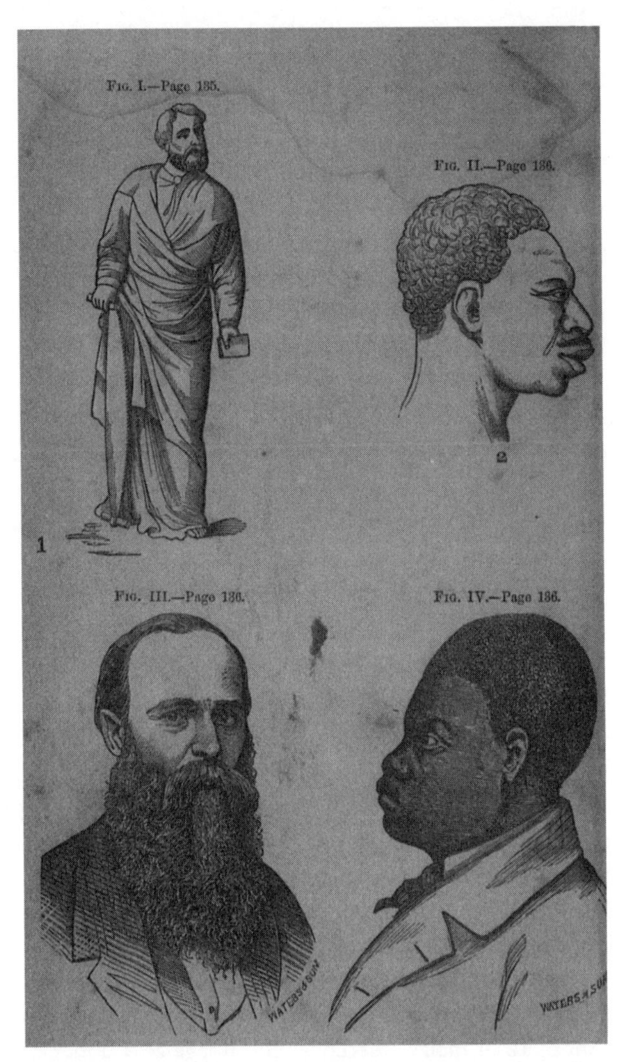

Figure 2.1. The race of manhood and the permanence of race. *With this illustration, John Van Evrie sought to demonstrate that the physical—and, by implication, mental—character of the white and black races remained unchanged over time. As was often the case in ethnology, the black figure was presented in profile to emphasize both his supposedly simian facial angle and his sloping forehead, interpreted as indicative of small intellectual capacity. Moreover, that the ancient and modern Caucasian figures were drawn with beards, in sharp contrast to their smooth-faced black counterparts, served to support Van Evrie's rhetorical connection between beards, manhood, and political capacity (Van Evrie,* Negroes and Negro "Slavery"*).*

ask the black man to grow "the full flowing beard of the Caucasian." An "equal beard"—the "outward symbol of equal manhood"—would have been an "absurdity" on a black man.[1] Black men were not truly men in body or mind and thus better suited to slavery than to any of the rights of men. As the nation stood on the brink of war, what was at stake in ethnological defenses of slavery such as Van Evrie's had intensified and changed; it now included the question of who could fully exercise "manhood rights" if slavery were to disappear.

Van Evrie presented a similar argument about black women, claiming their bodies were not suited for the cultural expectations of womanhood. While black men lacked both the intelligence and the skilled hands necessary for art, science, or anything but the "grosser trades, such as coopers, blacksmiths, etc., which need little more than muscular strength," black women were equally deficient when it came to performing anything but the crudest women's work. "The course, blunt, webbed fingers of the negress," he maintained, could never "produce those delicate fabrics or work those exquisite embroideries which constitute the pursuits or make up the amusements of the Caucasian female."[2] Whereas the bodies of white women were designed for the refinement of domesticity, black women's bodies were better suited to the rigors of domestic service. And since their bodies failed to conform to white definitions of sex, black men and women alike did not merit the rights and protections attached to gender.

Van Evrie's argument is representative of the gendered character of ethnology in the years surrounding the war, a transitional moment in American racial science. During the Civil War and Reconstruction, as the future of American society and the power structures undergirding it were in flux, U.S. physicians and scientists who wrote about race became increasingly preoccupied with the interrelated issues of citizenship and what it meant to be a proper man or a proper woman. Consequently, 22 percent of scientific texts on race published between 1860 and 1879 explicitly focus on issues related to gender, compared to just 6 percent in the 1850s and 0 percent between 1830 and 1849 (Figure 2.2). Whereas 25 percent of antebellum ethnology texts were focused on the debate over racial origins, by the 1860s and 1870s ethnological writers were far more concerned with debating the future of the races, in which gender played a crucial role; just 3 percent of racial science published between 1860 and 1879 focused on the original unity or diversity of the races. Then, in a wave of violent backlash

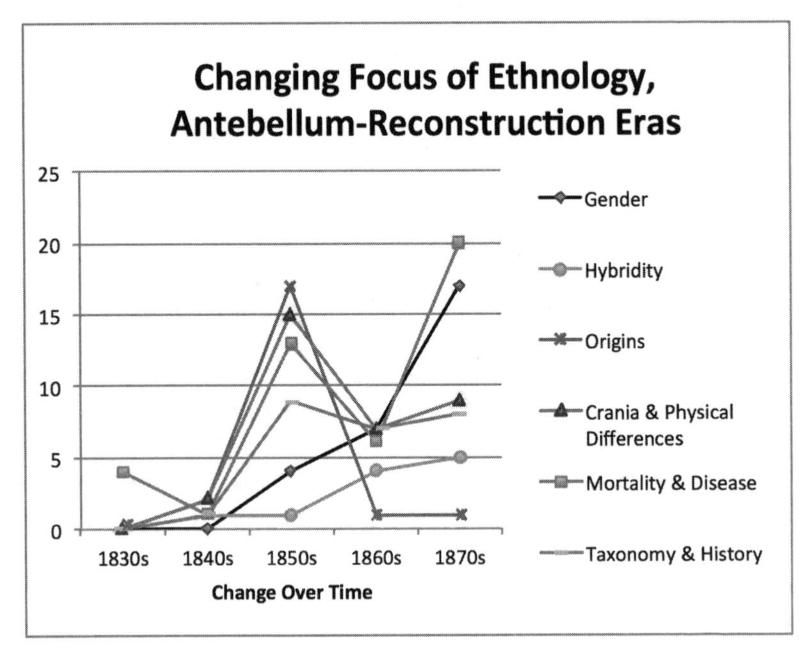

Figure 2.2. Changing focus of racial science texts, 1830–1879. *Ethnologists' interest in the origin of the races declined sharply during and after the Civil War. As their concerns shifted to the future of the races, particularly the political capacity of the freedman, ethnologists engaged in more explicit discussions about gender at a time when citizenship was inextricably tied to manhood (graph by author; data based on the* Index-Catalogue *of the National Library of Medicine).*

to Reconstruction that defined the late nineteenth century, known as the Redemption period, scientists' discussions of racial origins diminished even further, replaced by an overriding interest in evolution and its reverse process, degeneration, and their implications for American race relations (see Figure 2.10). In short, racial scientists' increasingly explicit engagement with gender and sex, which first surfaced in their antebellum discussions of race, reproduction, and slavery, as discussed in the previous chapter, came to a head during the Civil War and shaped the meanings of freedom and citizenship during Reconstruction and beyond.

With his overt use of sex and gender as markers of racial difference, Van Evrie stands in sharp contrast with earlier ethnological writers such as Samuel Morton. Although postbellum racial theorists continued to focus far more often on men than women, ethnological work during this period often interrogated the meanings of manhood as it intersected with race, almost always discussed sex-specific bodies, and made arguments about racial difference in which gender played a central part.

The changing character of ethnology can also be traced through individual scientists whose work bridged the antebellum and Reconstruction periods as well. For example, Josiah Nott's 1844 publication was devoted almost entirely to the origins debate and made no mention of women, white or black, instead comparing the races according to more gender-neutral characteristics like skulls. In *Types of Mankind* (1854), Nott discussed black women, but only in conjunction with a defense of slavery. Finally, his 1866 text, *Instinct of Races,* took up the question of citizenship, questioning black men's capacity for both manhood and the vote.

Several sociopolitical concerns precipitated and sustained ethnological scientists' attention to gender and sex difference during this period. Like Van Evrie, many racial theorists writing during the Civil War continued to defend slavery as the natural relationship between the races, invoking the same gendered logic and language of paternalism common in antebellum proslavery discourse. But with slavery's demise an imminent possibility, they also began to consider the impact of Emancipation on America's social and political power structure. Moreover, after Emancipation, the question of black citizenship would dominate Reconstruction-era ethnology, which often presented black men and (white) women as sharing similarly limited intellectual and political capacities. The Redemption period saw scientists' growing interest in not just gender and bodily sex but also sexuality, which informed their definition of citizenship and who should have it; they rewrote black men's political competition as a sexual threat and miscegenation the inevitable result of political equality. Ethnology's changing targets and concerns reflected larger transformations in American politics and society.

The Civil War itself helped make black manhood an issue for racial theorists. In particular, sparked by these same larger transformations, Union army scientists and physicians conducted extensive anthropometric studies on the living human bodies most readily available to them: the soldiers themselves. Though the studies attempted to quantify differences across the races, they measured only one sex: male, because of the military context in which they worked. Early in 1863, the United States Sanitary Commission, originally chartered two years prior to assist the Union army in improving health conditions and sanitation among the soldiers, embarked on a new project: systematically measuring the "most important physical dimensions and personal characteristics" of the diverse male bodies to whom their work afforded access. The kind of research the Sanitary Commission conducted during the Civil War was scientifically en vogue but also socially and politically pertinent. At a time of national turmoil, the SC's work was deeply invested with concerns about race, immigration, and manhood—itself an explicitly political category in the nineteenth century when full citizenship and "manhood rights" were synonymous but also increasingly contested by suffragists demanding the vote.

Using apparatuses like the andrometer shown in Figure 2.3, which was designed to ensure precise, standardized results among different scientists, they measured body parts and proportions, including "height," "breadth of pelvis," "breadth of shoulders," and "circumference of the waist"—characteristics that not only distinguished the men from each other but also traditionally distinguished men from women. Each investigator was carefully trained in the proper use of the measuring apparatus and instructed, "All measurements are to be noted in inches and tenths, so far as possible," adding that it was better to "omit the answer entirely, than to record an uncertain result."[3] In other words, the scientists took great pains to ensure unbiased results, but what they measured and why, how they interpreted the data, the scope of their study, and the questions they asked were culturally loaded indeed.

The examining scientists recorded their results on a lengthy questionnaire that included questions about not only each soldier's "physical characteristics" but also his "moral and social condition" and, in the context of mounting controversies surrounding immigration in the mid-nineteenth century, his "nativity" as well.[4] Similarly, the SC provided investigators with special instructions for examining African American soldiers. In assessing "muscular development," for example, they were instructed to provide "an estimate of the proportion of black blood, such as Full Black, Mulatto, Quadroon, Octroon," possibly to investigate the common scientific belief that mixed-race men were less vigorous than their fully black counterparts.[5] Indeed, even as they strove to be innovative, the Civil War scientists drew on (and sometimes challenged) existing popular and scientific ideas about race and manhood—categories they thought could be both qualitatively and quantitatively measured.

In the years during and after the Civil War, science and politics intersected on questions of race to reinforce the traditional definition of

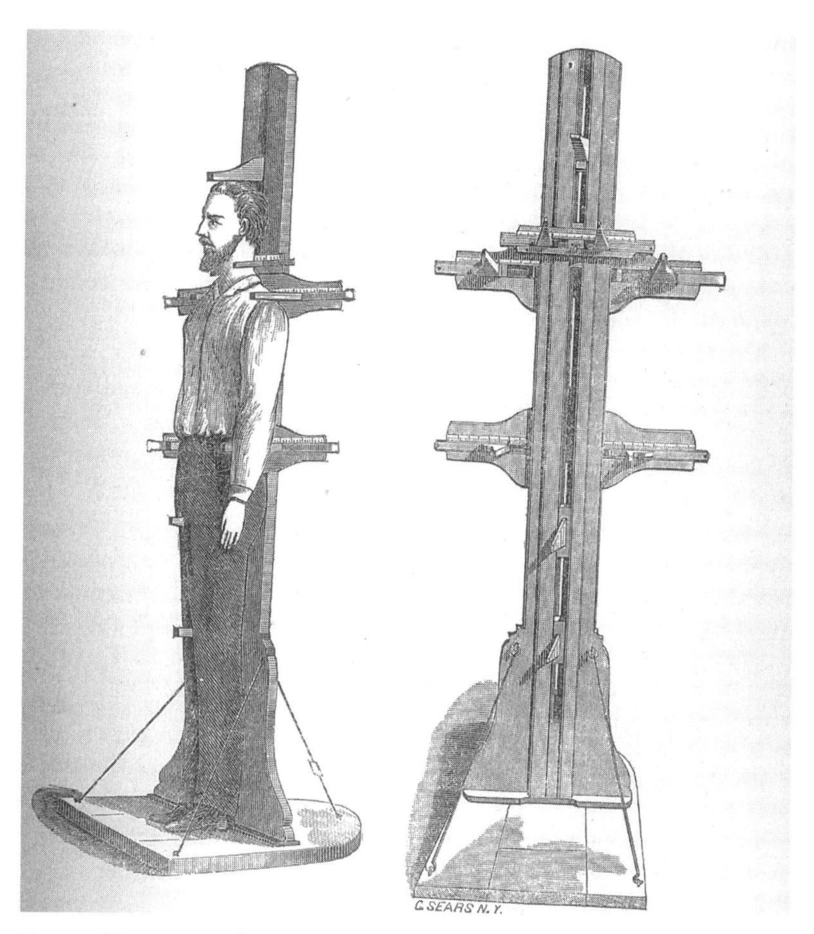

Figure 2.3. The measure of man. *Scientists used andrometers, literally translated the "measure of man," like the one pictured to measure the length and proportions of soldiers' bodies during the Civil War. Sanitary Commission actuary Benjamin Gould included this sketch in his discussion of the history and methodology of the Commission's anthropometric study to demonstrate the scientific precision and, by implication, objectivity with which its scientists approached their investigation (Gould,* United States Sanitary Commission Memoirs, Vol. 2*).*

"citizen" as male and white in the face of the challenges black freedom and the prospect of women's and black suffrage presented. In addition to these shared concerns, racial science and politics also needed each other in more practical ways. Ethnological writers had long publicly affirmed their social relevance by offering scientific expertise on the pressing issues of the day, while politicians used science to bolster their claims about racial difference and naturalize their policies of racial inequality. The line between racial science and politics in nineteenth-century America was thin indeed.

Measuring Manhood in the Civil War

With the possibility of a vastly enlarged free black population imminent during the Civil War, black capacities for both manhood and political agency became major issues. During the war, and in the decade leading up to it, the majority of white ethnological writers were engaged in an explicit and impassioned defense of slavery that also anticipated the potential consequences of slavery's demise. As we have seen, ethnological defenses of slavery in the 1850s and early 1860s occasionally discussed women, usually to invoke slave women's fertility as indicative of the health of the institution overall. But the ethnological vision of American society without slavery was one in which men alone took center stage. Abolition presented new questions about what would or should happen to the former slaves, questions in which gender was central. Newly freed black men, however, would present far greater challenges to U.S. social, economic, and political institutions than black women, who would already be excluded from or rendered largely invisible in these domains on account of their gender. If no longer slaves, did black men have the capacity to be citizens? White scientists debated this question, with most answering a resounding no. And if black men's bodies and minds alike made them fit only for servitude, as many racial scientists themselves had argued, the potential competition for work and wages presented by a newly free race of born laborers would be tremendous. Thus what emerged in racial science during the war was a vision of race relations in the United States as a contest over manhood. Historian Reid Mitchell describes the Civil War as a national rite of passage into manhood, noting that "the very ideas of man, soldier, and citizen were inextricably linked." Indeed, "remaining a civilian was thought unmanly; going to war a proof of manhood," he writes.[6] Prompted by trends in racial science as well as these popular links between service in the Civil War and manliness, army scientists began to critically examine what manhood meant and how it could be measured scientifically and compared along racial lines.

And measure they did, quite literally. Anthropometrics, which measured the dimensions and proportions of living human bodies, expanded in scope and visibility during the Civil War era. In an 1860 article, Joseph Barnard Davis, a British physician who shared Samuel Morton's passion for collecting skulls, argued that racial comparisons relied too often on anecdotal observation and that more precise studies tended to focus on only one measure or body part. The methods and tools for measuring the bodies were often not uniform, which posed additional challenges for racial comparison. Thus Davis called for a "more systematic and comprehensive" study of anthropometry.[7] It was a call that was heeded in earnest during the Civil War, most extensively by the United States Army.

The army entered the scientific debate over racial capacities and manhood under the auspices of the Provost Marshal General's Bureau (PMGB) of the War Department and the Sanitary Commission (SC), which conducted separate anthropometric studies of Union soldiers.[8] Both divisions generated enormous amounts of data on height, limb length, body proportions, head circumference, and facial angle, among other measures. Facial angle was an especially loaded measurement; a perennial marker of racial difference in nineteenth-century ethnology, scientists usually pointed to it to emphasize white men's supposedly prominent forehead, indicative of a large frontal lobe and thus higher-order thinking skills, in contrast to the sloping forehead and "prognathous" (i.e., jutting) jaws of the "lower races," which demonstrated both their alleged intellectual deficiencies and their biological proximity to primates. The anthropometric data was then arranged by race, with white men as the standard to which others were compared. This white normativity was sometimes quite explicit in the Civil War studies. For example, the SC's guidelines instructed physicians to ascertain the "apparent intelligence" of each soldier they examined, with "the ordinary white private solider being taken as the standard of comparison."[9] The paradigm of white male normativity was one of many features the Civil War anthropometric studies shared with the larger endeavor of racial science, despite the former's apparent attempts at scientifically precise, objective measurement.

The army also had more immediate concerns driving its anthropometric research than many racial scientists working outside the military context who were considering the history of the races and what would happen to them in the future should slavery be abolished. For one, the decades immediately prior to the war had seen a steady increase of immigrants to the United States, primarily from the working classes of England, Ireland, and Germany. Faced with growing numbers of new residents and mounting labor competition, American leaders debated over citizenship and to whom, and when, "manhood rights" should be extended. Likewise, with nativism on the rise throughout U.S. culture, military leaders wondered whether native-born American men were better suited physically for military service and more mentally and morally committed to fight for the cause.[10] As a result, anthropometric data was generally arranged not just by race but also by nativity, or place of birth, to develop a physical portrait of the average soldier and, by extension, the ideal soldier and the best *man*.

To be sure, terms like *race, nativity,* and *man* were politically loaded, and the stakes of citizenship were high. Throughout the nineteenth century, U.S. citizenship rights were commonly termed "manhood rights," but full citizenship was not just a gendered category; it was also a racial one and applied to white men only.[11] But steady immigration from Europe during the nineteenth century constantly challenged—and expanded—the parameters of whiteness.[12] An immigrant's passage into full American citizenship depended on whether his "nativity" was considered "white" at that moment in time, which itself was contingent on perceptions of the immigrant group's social worth—the very kind of issues to which the army's studies spoke. Moreover, although American born, "negro males, whether free or slave, were forbidden to exercise 'manhood' rights—forbidden to vote, hold electoral office, serve on juries, or *join the military*."[13] At the same time, women in the United States were also challenging the gendered qualifications of full citizenship, while many politicians in turn were determined to defend the franchise as the right of men alone.[14] So while the Union army was interested in and spoke to these larger issues about who was *manly* enough for citizenship, its decisions during the war itself were ultimately driven most by a pragmatic need for *manpower* that could fight and die for the cause.

Among the military's most immediate concerns, then, was the contentious issue of allowing black men to serve in the Union army. Although abolitionist sentiment had reached new heights in the North by the onset of the Civil War, it was far from universal. Among white soldiers deeply committed to the cause of preserving the Union, ambivalence about—or even support for—slavery was commonplace. Moreover, even antipathy toward the institution of slavery itself rarely translated to a belief in racial equality, and many white military officers and politicians alike were reluctant to arm black men.[15] Banned from service at the beginning of the Civil

War, free blacks lobbied for the right to enlist from 1861 onward, while ever-growing numbers of slaves escaped to Union lines. Faced with a growing shortage of white volunteers as the war progressed and a glut of contraband slaves eager to fight for their own freedom, Union officers and Abraham Lincoln ultimately reversed the ban. In 1862, Congress authorized enlistment of black men in "any military or naval service that they may be found competent."[16] Questions still remained, however, about what kind of soldier the black man would make and whether he had the physical and moral capacity to serve in combat positions.[17]

All these questions and concerns fueled the Union army's anthropometric studies, with the published reports by both agencies referencing contemporaneous debates within the wider culture over the fates of both immigrants and freed blacks in American society. Indeed, the scientists involved in these studies did not see their work as relevant only to the specific needs and concerns of the military but rather as part of a larger ethnological project in the United States. For example, "to give as wide usefulness as possible to these researches in their ethnological relations," the SC distributed the anthropological apparatus it used to measure soldiers to "various institutions of learning in the United States" and supplied "scientific travelers" with copies of the forms and instructions provided to the commission's own scientists.[18] Moreover, the chain of influence was not one-sided. Not only did the anthropometric studies of soldiers during the Civil War influence racial theorists for decades after, but the studies themselves cited prior ethnological work and were shaped by existing scientific ideas about race. For instance, Benjamin Apthorp Gould, a respected Harvard-educated astronomer and surveyor enlisted by the Sanitary Commission to oversee its statistical collection, sought "consultation with friends whose pursuits are of an anthropological or physiological nature," including Swiss-born naturalist Louis Agassiz, an outspoken proponent of polygenesis and key figure in the American school of ethnology.[19] Army scientists, then, assessed the various races' physical, mental, and moral fitness for military service based on prevailing ethnological assumptions about racial difference as well as on their own anthropometric and medical data.

Though numerous scientists and physicians contributed to the Sanitary Commission's two-volume study on the ground by examining, measuring, and interviewing soldiers, the commission charged two men with compiling and analyzing the data for publication: Gould for the volume of anthropometric statistics and surgeon Roberts Bartholow for the medical statistics. Both men were professionally established before their involvement with the SC and would go on to even greater fame—or in Bartholow's case, infamy—in their respective fields after the war.

Gould was born in Boston in 1824 to an old Massachusetts family that traced its lineage to the earliest Pilgrim settlers in the area. His father was the headmaster of the Boston Latin School—where his son studied—and later became a merchant in the East Indian trade. As a student at his father's alma mater, Harvard, Gould developed an interest in physics and mathematics and became a committed Unitarian. After graduation, he taught at the Boston Latin School for a year before deciding to continue his studies of physics and mathematics in Europe, where he forged a formative friendship with famed naturalist Alexander Von Humbolt. After he returned to the United States in 1848—having built lifelong friendships with influential scientists across Europe—Gould started the *Astronomical Journal,* which he largely self-funded for the publication's first decade. He also conducted transatlantic longitudinal studies for the United States Coast Survey, served as the director of several observatories in the United States and South America, and was among the original members of the National Academy of Sciences while traveling extensively and teaching periodically.[20] In the years immediately prior to the Civil War, Gould experienced two events in his personal life that hindered and helped his scientific career. In 1859, his father died and left the junior Gould his by then floundering mercantile business, which he dutifully ran alongside his scientific work until he was able to restore it to its former strength and extricate himself in 1864. And in 1861, he married Mary Apthorp Quincy, an educated Boston woman of means who assisted him in all his astronomy research throughout their marriage. In 1863, President Lincoln named Gould one of fifty men chosen as representative of American science in the newly incorporated National Academy of Sciences. By Gould's own admission, he was surprised by his appointment the following year as actuary to the Sanitary Commission, which put him in charge not of constellations or coasts but statistics relating to the physical and medical character of men, though he approached the work with the same care and vigor he applied to his other pursuits.[21]

Though similarly productive and respected within his profession, Roberts Bartholow did not share Gould's reputation as an affable and generous man with the ability to make—and maintain—professional friendships with everyone he encountered. In contrast, Bartholow's colleagues described him as a chilly, reserved, disrespectful, rude, and even cruel man who "never had an intimate friend or even a close associate," though most acknowledged his contributions to medico-scientific knowledge and methodology, particularly during his highly productive decades teaching and practicing medicine in Cincinnati and Philadelphia after the war.[22] Born in New Winsor, Maryland, in 1831, Bartholow received his bachelor's degree from New Windsor College, where he quickly distinguished himself as highly intelligent and hardworking, and an MD from the University of Maryland in 1852. After working in several hospitals and clinics in Baltimore, in 1855 he enlisted as an assistant surgeon in the U.S. Army, serving on the frontier until the Civil War broke out and he returned east to oversee Union-run military hospitals. In 1862, the young physician married his wife Maria and published *A Manual of Instructions for Enlisting and Discharging Soldiers,* which the Union army adopted as official policy. His work with the Sanitary Commission compiling medical data on soldiers began shortly after.[23]

Unlike Gould and Bartholow, the chief investigator of the PMGB's Civil War studies, Jedediah H. Baxter spent his entire career in medical service to the military, ultimately culminating in his election as the army's surgeon general in 1890. J. H. Baxter, as he was more typically known professionally, was born in Stafford, Vermont, in 1837 and obtained his BS and MD from the University of Vermont in 1859 and 1860, respectively. He briefly worked at Bellevue and Blackwell's Island hospitals in New York City until the onset of the Civil War, when he enlisted in the Union army and was assigned to the Twelfth Massachusetts Volunteers as a surgeon. He rose quickly through the ranks during the war, promoted to major in less than a year, and in 1862, he was sent to Washington, D.C., first to run Campbell General Hospital and then as chief medical officer in the War Department's new PMGB, where he remained through the 1870s. Among his duties was overseeing the collection of medical and anthropometric statistics from more than a million men during the war and later analyzing the data and preparing the PMGB's extensive medico-scientific publication.[24]

For the PMGB, the results were compiled into tables comparing men by race, region, occupation, nativity, and age in *Statistics, Medical and Anthropological, of the Provost-Marshal-General's Bureau, Derived from Records of the Examination for Military Service in the Armies of the United*

States during the Late War of the Rebellion, a two-volume set of medical and anthropometric data the agency collected starting in 1861 under the direction of the secretary of war. The title of the 1875 publication may sound unwieldy and perhaps esoteric to a modern reader, but its editors clearly intended the work to appeal to a wide audience in nineteenth-century America. To a greater degree than the Sanitary Commission, the PMGB made conscious efforts to make their reports comprehensible for the nonspecialist. Toward that end, the editors carefully interpreted the numbers and explained their significance to the reader.

The PMGB's study aimed to speak to the male "masses" that the study sought to define and understand. Their statistical data did not relate to "picked men, in no wise representing the masses—but to the people"—that is, the men who voluntarily enlisted for service. The enlisting masses represented "men engaged in every occupation; the professional man and the man of letters, the trader, the merchant, the clerk, the artisan, and the unskilled laborer; the rich man and the poor man; the robust and the crippled." The army's subjects were "in short . . . the citizens of the United States, both native and foreign-born," who would "illustrate the physical aptitude of the nation for military service."[25] Of course, not all individual men shared the same physical aptitude, and army scientists intended to determine whether patterns emerged among racial types. As a result, comparison was central to the study and to anthropometrics in general. Investigator J. H. Baxter explained, "The tables in which nativity is an element of the comparison show the physical condition of foreign-born citizens of various nativities in relation to each other and in relation to native Americans, both white and colored."[26] In reality, however, the tables that followed broke native-born Americans into three categories: "Colored," "Indian," and "White." Baxter's omission of "Indians" in his introductory remarks is revealing. Though "Indians" are included in the tables themselves as a distinct subcategory of native-born Americans, much of the accompanying discussions of race were literally black and white (See Figure 2.4).

The army's concern with race in their search for the "average man" was further demonstrated by a revealing quotation selected to introduce and frame the first chapter, which reviewed the tables and their results: "If we are to devote our attention, before all things, to what can be measured and weighed, the living man is the first object which demands our investigation. The 'average man' of Europe having been determined by [Belgian

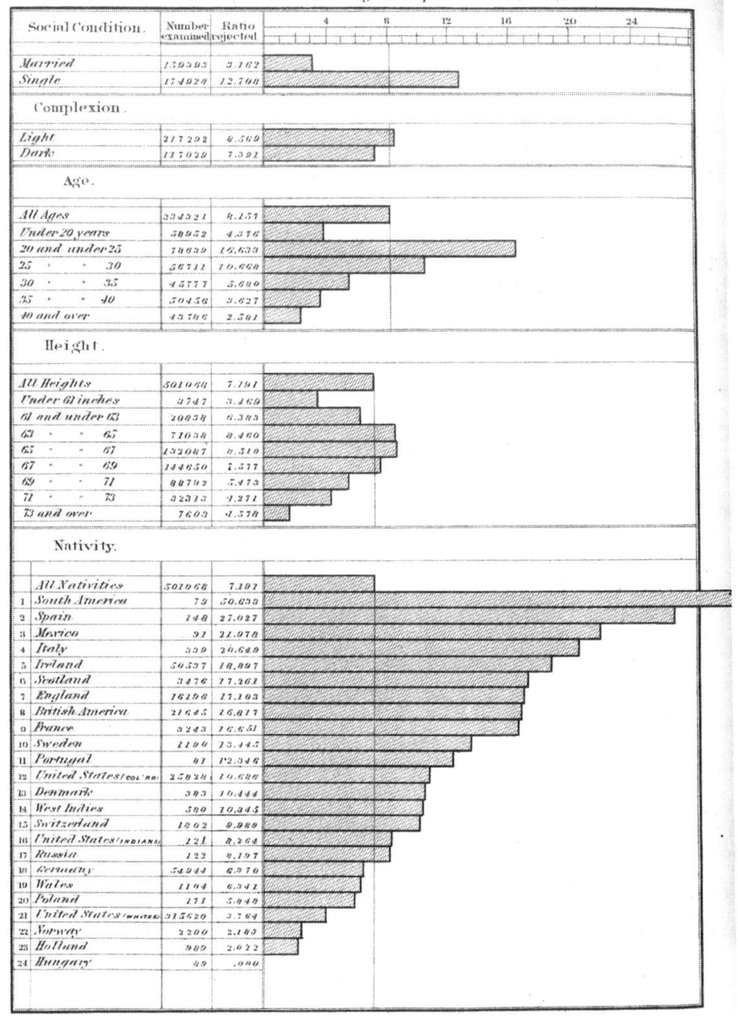

Social Condition.	Number examined	Ratio rejected
Married	159593	3.162
Single	174928	12.708
Complexion.		
Light	217292	8.569
Dark	117029	7.391
Age.		
All Ages	334321	8.157
Under 20 years	58952	4.376
20 and under 25	78839	16.633
25 " " 30	56711	10.668
30 " " 35	45777	5.680
35 " " 40	50456	3.627
40 and over	43786	2.581
Height.		
All Heights	501068	7.191
Under 61 inches	3747	3.469
61 and under 63	20838	6.383
63 " " 65	71038	8.460
65 " " 67	132087	8.510
67 " " 69	144850	7.577
69 " " 71	88792	5.473
71 " " 73	32313	4.271
73 and over	7603	1.578
Nativity.		
All Nativities	501068	7.191
1 South America	79	50.633
2 Spain	148	27.027
3 Mexico	91	21.978
4 Italy	339	20.649
5 Ireland	50537	18.897
6 Scotland	3476	17.261
7 England	16196	17.103
8 British America	21645	16.817
9 France	3243	16.651
10 Sweden	1190	13.445
11 Portugal	81	12.346
12 United States (Col'rd)	25828	10.686
13 Denmark	383	10.444
14 West Indies	580	10.345
15 Switzerland	1802	9.989
16 United States (Indians)	121	8.264
17 Russia	122	8.197
18 Germany	54944	6.370
19 Wales	1104	6.341
20 Poland	171	5.848
21 United States (Whites)	315620	3.764
22 Norway	2200	2.183
23 Holland	989	2.022
24 Hungary	49	.000

Figure 2.4. Race in numbers. *This chart is a typical presentation of medical and anthropometric data collected by the War Department during the Civil War. Arranged by "nativity," the United States was subdivided into three racial groups: "Colored," "Indians," and "Whites." Analysis of the data, however, often focused on comparison between black and white soldiers (Baxter,* Statistics, Medical and Anthropological, of the Provost-Marshal-General's Bureau*).*

statistician Lambert Quetelet], his system is now applied to races."[27] To some extent, however, the army's goals of racial comparison and of defining the "average man" were at odds. Instead, army scientists sought to define average *men,* each representative of a specific race or nativity. Notably, in a discussion of height distributions and mean value among different nativities, Baxter produced separate bell curves for white and black men, even though they shared a common country of birth. Given "the varied origin of the population of the United States," he explained, the army scientists had to arrange their data accordingly, separating the country into its distinct racial parts in the interest of scientific precision.[28] In the past, he noted, attempts to determine average height were "confused" and inaccurate, largely due to their failure to adjust for categories of difference like race and sex.[29] However, Baxter seems to imply that women might best be left out of such investigations entirely rather than simply separating the results by sex. It was the height of man specifically that was of consequence for Baxter.

And indeed, manhood was the scope of the army's concern overall, not just because male bodies were what the circumstances afforded them to study, but because manhood was central to social and political concerns that preoccupied Americans during the Civil War. Scientists who studied race outside the military context during this period, such as Josiah Nott and John Van Evrie, often based their arguments on personal, anecdotal observation rather than systematic measurement of actual human bodies. Yet they too focused predominantly on men, though theoretically, at least, they would have had equal opportunity to observe men and women alike.

Certainly, the PMGB's examining physicians observed their subjects in far more intimate detail than many ethnological writers, though, as we shall see, not without their own racial preconceptions. In volume 1 of its report, the PMGB provided an exhaustive description of the process by which the soldiers were examined. Each soldier was ushered into a bright room and required to strip, in particular to reveal any disqualifying defects a volunteer or substitute might seek to hide while exposing any drafted man who feigned sickness. But it was also deployed so that each soldier would be "thrown off his guard" to provoke truthful responses to the questions asked of him by the examining physician. Nude, each soldier was questioned about his occupation, nativity, health, and family history, which served the added function of allowing the physician to judge "his mental as well as of his physical qualifications."[30] From there, the physician examined each part of his body for medical problems that would interfere with military service and then measured for the purposes of the anthropometric study.

In addition to health problems and general physical defects, the examining physicians were also assessing manhood. Among the list of otherwise physical disqualifying defects were *moral* defects, which many believed would impinge on bodily integrity and civilized manhood. "An impaired constitution, the result of the constant abuse of stimulants, or of indulgence in the habit of masturbation, was an authorized ground for exemption," for example.[31] Medical investigators also looked more literally for manhood among the naked soldiers. One examining physician complained that even "boys with hairless pubes" have the audacity to lie about their age, such that "the surgeon must be constantly upon the alert" that only prime specimens of adult manhood would make it into the military's ranks.[32] More notably, "loss of penis," without further explanation, was listed as a disqualifying factor, as was "hermaphroditism."[33] Indeed, during each soldier's exam, "the penis was examined for epispadia, hypospadia, and venereal disease," as the latter impugned a soldier's conformity to antebellum gender standards of masculine restraint, while the first two conditions, abnormalities of urethral placement that were themselves sometimes linked by scientists to "hermaphroditism," called into question the soldier's biological sex.[34]

Ever vigilant in evaluating proper male genitalia, the same doctor who railed against the fraud perpetuated by hairless boys also reported that "a conscript appeared in the office for examination who came as near being a genuine hermaphrodite as any reported case." "The man was about thirty years of age, five feet four inches in height, with very little beard," Dr. Dixi Crosby observed, "but a luxuriant growth of hair about the pubes." His exhaustive investigation went well beyond what was necessary to determine the disqualifying condition of hermaphrodism, which in this case was seemingly apparent to the naked eye, as the conscript had a body that mixed male and female sex characteristics, including large breasts, broad hips, a small penis, a short "cul de sac" vagina, and testicles located in the labia.

Crosby was equally interested in the conscript's life history as well. He had worn women's clothes until the age of nineteen, Crosby noted, but after that, "his parents, becoming convinced of their mistake, changed his garments and the family residence at the same time." Crosby further explained that he married—and consummated—two marriages with women.[35] The conscript's stated identity and his ability to function as a man in society with "no issue" did not help his efforts to join the army, however, since physicians such as Crosby were looking for soldiers who were normatively male in body as well as behavior. Crosby does not note what became of the potential solider, but he was in all likelihood rejected for military service since "hermaphrodism" was listed as grounds for disqualification. He also does not note the conscript's race, though one suspects that if he had been "colored" or "Indian," Crosby would have included this information. Since white was the default category throughout the reports, it was standard practice to indicate the race of a soldier only when he was *not* white.

The results of the Sanitary Commission's foray into anthropometrics and statistical analysis generally mirrored those of the PMGB studies despite tensions between the two agencies.[36] And like the Bureau, the SC's scientists were keenly interested in the body and mind of the "average man"—an investigation deeply imbricated with existing norms of both sex and race. The SC was measuring only the average man's "external manifestations," using "numerical expressions" to "construct the typical or average man," Benjamin Gould explained. However, such research nonetheless allowed scientists to "discover not merely the outward semblance of this abstract being, but his needs, capacities, intellect, judgment, and tendencies."[37] As indicated by Gould's quotation of Quetelet, widely considered the father of anthropometrics, much was at stake in this average man: "The average man is for a nation what the center of gravity is for a body."[38] Left unsaid by Quetelet but certainly implied throughout the SC's report was that the body was at the center of the man—who he was and what he could do. Moreover, while the SC constructed an average physical and mental type for each race, the races were all compared to a white norm. The "average" man and the ideal man were white; the white man was the center of gravity for the U.S. national body.[39]

That the SC sought to define not just the ideal solider but also the ideal man was revealed in a short but fascinating section titled "Pilosity of Negroes." Here, it becomes quite clear that the SC was interested in physical characteristics well beyond the scope of what ostensibly could be useful on the battlefield. "The question as to the relative amount of pilosity, or general hairiness of the body, in the white and black races is one of some anthropological and ethnological interest," Gould mused. Gould was correct; as we have already seen, secondary sex characteristics like body and facial hair were of keen interest to white scientists during and after the Civil War, most of whom defined manhood along racial lines. Also interesting about this particular measure is that only the bodies of black men were examined. Though the results were ultimately compared to white men, no actual examination of them appeared to have been conducted—no numerical data was given. The nature of the typical white male body was simply assumed here, a norm that needed no measure. It was the manhood of the black race specifically that interested the SC.[40]

Even more interesting than the uncharacteristically limited scope of this particular measure was the methodology employed to obtain it. In fact, Gould described a far different scene than that of soldiers lining up to be inspected by doctors wielding andrometers and other measuring tools. That is, one "Mr. Russell, when accompanying the 25th Army Corps to the Texan boundary, was requested to avail himself of any opportunity which might occur, to observe the colored troopers when unclothed" and record their "pilosity" on a scale from zero to ten, with ten representing "the maximum which he had ever seen or should see in a white man." He did so by "observing the men while bathing."[41] Russell was a busy man indeed and apparently had good eyesight. From his observation post near the bathing men, he recorded the "relative pilosity" of 2,129 black or mixed-race soldiers. One can only imagine what the black soldiers made of the good scientist, spying on them in their nudity, writing furiously in his notebook as they bathed.

Russell provided his results in a table that correlated the number of men to degree of "pilosity." Though none of the black or mulatto men obtained a ten—the height of hairiness for white men—Russell's results conformed more or less to a bell curve. Thus Gould ultimately conceded that with the "excellent distribution" of Russell's results, "the unavoidable inference [is] that there is but little, if any, difference between the white and black races in this respect."[42] Nonetheless, that the SC endeavored to measure a secondary sex characteristic in African Americans alone reveals they were not just interested in black masculinity in relation to military service but also by extension his potential for manhood more generally, with all the sociopolitical rights afforded it—which included the right to vote and hold office, as well as property ownership, paid labor, and, more generally, power and status. The inherent manhood, body and mind, of the white race was a given; the issue of black manhood, however, was up for scientific and political debate at a time when the ideal *man* was also the ideal *citizen.*

The Definition(s) of Manhood: Bodies and the Character of Races and Nations

For historians, the most illuminating aspect of Civil War anthropometric studies is not the data itself but what the scientists were looking for, how they interpreted what they saw, and the ways in which they represented their results. Also far more revealing, and decidedly less dry, are the reports compiled by the PMGB from physicians overseeing the routine examinations conducted on recruits and enlistees for military fitness. These reports offer a fascinating lens into the racial thought of hundreds of white physicians sympathetic to the Union cause, who were not immune to popular anxieties about immigration and ambivalence over the place of the black race in American society once freed from bondage. They invoke many of the same racial tropes pervasive in ethnology, but their responses also reveal how profoundly each physician's own cultural politics and experiences prior to—and during—the war colored their assessment of the soldiers they encountered.

The PMGB sent questionnaires to the head physician of the examining board in each congressional district in every Union state, and nearly every physician returned detailed reports that were reproduced in full in the published volume. Two of the questions are particularly relevant here: "What nationality presents the greatest physical aptitude for military service?" and "What is your experience as to the physical qualifications of the colored race for military service?" These questions prompted many of the responding physicians to expound upon racial categories and give definitions of manhood. Not surprisingly, the overwhelming majority answered native-born Americans to the first question. But it is the way this information was presented that is most telling. Summarizing the results of the questionnaire, editor J. H. Baxter tabulated the responses to the question of the nativity with the greatest physical aptitude for military service in a small table (Figure 2.5). He listed the various nativities in descending order by number of positive responses, except for "colored men," who he placed out of order at the bottom of the list, even when the study's own "empirical" research should put them fourth. His apparent oversight demonstrates in fascinating fashion the teleological nature of racial science and

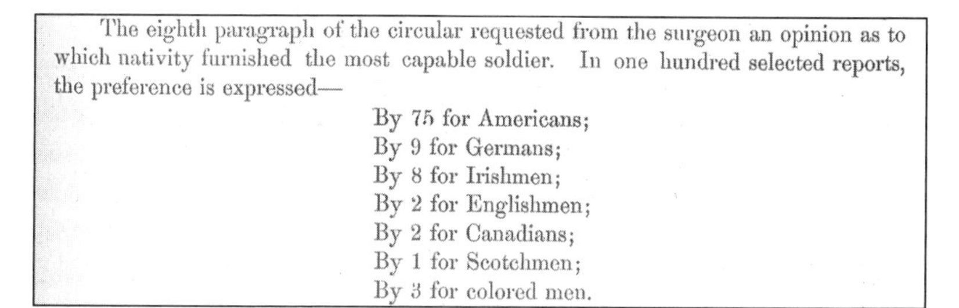
The eighth paragraph of the circular requested from the surgeon an opinion as to which nativity furnished the most capable soldier. In one hundred selected reports, the preference is expressed—

By 75 for Americans;
By 9 for Germans;
By 8 for Irishmen;
By 2 for Englishmen;
By 2 for Canadians;
By 1 for Scotchmen;
By 3 for colored men.

Figure 2.5. Ordering the races. *Even scientists who resolutely expressed their commitment to objective measurement and data collection, like those employed in the army's anthropometric studies, were susceptible to their own preconceptions about racial hierarchy. As part of the army's study, military surgeons were questioned as to what race and nativity produced the best soldiers. Tallying the results, Baxter placed "colored men" at the bottom of the list, even though their number of positive endorsements should have put them fourth, ahead of "Englishmen," "Canadians," and "Scotchmen" in the physical and moral qualities conducive to military service (Baxter,* Statistics, Medical and Anthropological, of the Provost-Marshal-General's Bureau*).*

army scientists' unconscious expectation that black men would fall in last place in the Civil War's contest of manhood.

A large number of the doctors surveyed declined to answer the second question about the Negro's fitness for military service, citing a lack of personal experience examining black troops upon which to base their conclusions. Of the respondents who did address the question, most answered in the affirmative, that the black men they examined were among "the finest specimens of physical development" and good soldiers.[43] However, in so doing, they often created competing—and racially specific—definitions of manhood, which gave blacks credit for a more brute physical manhood deemed conducive to military service rather than the civilized intellectual manhood attributed to whites, who were empowered to make decisions about the war itself. The "complimentary" assessments of the negro's military fitness often involved mixed compliments as a result. For example, Dr. Alex Burbank of Maine believed "the colored race leads off as to physical qualifications for the military service." He explained, "They are generally of a straight, soldier-like appearance; their habits of life have been such as to inure them to exposure; and there is less sensibility and refinement among them than we find in the white race, while there is just animal enough about them to make good soldiers."[44]

Others survey respondents reported that black men made good soldiers because of their "imitative" nature, obedience, and endurance of harsh conditions. Some of these doctors suggested, however, that these qualities were not biologically determined but rather born of the "infamous and cruel history of the race." Facing such pervasive mistreatment in the United States, "it has become his nature to obey," New Hampshire physician Robert B. Carswell noted. "The necessity of this obedience forms the foundation of all the little he was ever taught. It has been forced into him by the branding iron and knotted lash of his inexorable teacher. Yes, the negro has already learned 'the first duty of a soldier.'" Such material and social circumstances resulted in his unparalleled "power of endurance," but Carswell found much to admire in the black soldier's physical form as well. After praising black men's "good ear for music," vital to the mastery of military drills, he further remarked, "I have been struck with admiration at the wonderful display of symmetry, blended with muscular power, in many of these tawny sons of a common parentage." Like Burbank, Carswell asserted that the Negro was not just fit to serve but possessed "the greatest physical aptitude for military service" of all the races and nativities to come before him, for "what constitutes 'the physical qualification' of a man 'for military service,' is mainly muscular development and power of endurance."[45] The ostensibly positive qualities Burbank and Carswell attributed to the black soldier, from his natural propensity for military drill to his physical and psychological endurance, were echoed throughout the War Department's reports.

Army physicians like Burbank and Carswell exhibited a kind of ambivalent respect for black soldiers. The characteristics they attributed to black men were valued in military service, and the scientists were frequently impressed with black men's physical strength. Many physicians admitted black soldiers far exceeded their expectations, while others seemed genuinely moved by the soldiers' deeply personal commitment to the Union cause. Ultimately though, even the scientists who found much to admire reinforced that black men were well suited to follow but not to lead. And scientists' praise of black soldiers was often premised on racial differences rather than similar abilities across racial lines.

Other scientists emphasized black men's capacities less than they praised white men as possessing a birthright to greatness in every respect. One physician from Massachusetts raved that New Englanders "descended from the Anglo-Saxon" were "mentally and physically the most vigorous of modern races." Dr. John Sullivan's praise of Yankee superiority extended well beyond military service; these men had inherited "qualities of mind and body which admirably fit them for the twofold task of developing the resources of an unexplored continent, and of conducting on a grand scale the experiment of popular or democratic government."[46] In Dr. Sullivan's assessment, Anglo-Saxon men were by their very nature citizens, leaders, and conquerors of the land.

Among the few physicians whose assessments of the black soldier were mostly negative, "flat feet" tended to prevail among his deficits, while others pointed to features as varied as a "crooked spine" and "loosely-knit joints."[47] In this regard too, black men were measured against a white prototype; their feet were "almost entirely wanting" of "the ordinary arch which exists in the European." As was generally the case in racial science, the white race represented the "ordinary" and African Americans the deviation. Dr. Lord of Connecticut found black men unsuitable for the infantry, "by reason of the peculiar conformation of their bodies," which included narrow pelvises, slender and overly long limbs, ill-developed calves, and again, flat feet. In Lord's estimation, the black man did however possess "a sound and vigorous body" and was thus "in every way physically adapted for garrison duty, assailing earth-works, as well as for short marches, or charging upon the field of battle."[48] In other words, while black men did not represent ideal soldiers themselves, their bodies were well-suited to labor for soldiers in the army's camps.

Some physicians argued that military service required both brains and brawn and that African Americans fell short on the former. Dr. H. S. Chubbuck of New York, for example, noted that despite black men's praiseworthy musculature, "they do not seem to have the nervous energy or intelligent activity in that ratio that the native-born white possesses."[49] Other physicians drew on common wisdom to maintain that the Northern free blacks they encountered volunteering for service in the Union army were of poorer stock than their Southern counterparts, "for it is a well-known fact that they physically degenerate in this northern clime."[50] Most of the physicians surveyed, however, seemed to agree with New York physician William Roberts that "unless any moral deficiency annuls their physical vigor, they ought to make good soldiers."[51]

In contrast, the physicians surveyed generally gave more negative responses regarding mulatto soldiers. While most physicians lauded the black soldier's physical fitness, many found that "the mulatto, however, is comparatively worthless." The vast majority of the army physicians echoed prominent racial theorists like Josiah Nott in deeming mixed race men as decidedly less hardy men of either race alone. "Subject to scrofula and tuberculosis," they concluded, mulattos were largely unsuited to the rigors of military life.[52] In this regard, the reporting doctors seemed to draw at least as much on the prevailing racial stereotypes of the day as on their own experience examining black and mixed race servicemen. Their assessments rarely challenged prevailing racialist assumptions. Instead, they reinforced them, and even the highest praise for black men focused on the military context alone and racially specific attributes.

Revealing their own ambivalent interest in immigration and the fate of the freedmen, the Sanitary Commission's scientists also found that military examination offered the perfect opportunity for racial comparison—as well as for ranking and characterizing the various groups. "The races composing our volunteer army consisted, chiefly, of American, Celtic, Teutonic, Negro, and the mixed Spanish-American of New Mexico," Dr. Roberts Bartholow, U.S. Army assistant surgeon, explained in the introductory chapter of the first volume of the SC's report; here too, "American" meant *white* Americans specifically.[53] Like many nineteenth-century men of science, Bartholow presented himself as scientifically curious and politically disinterested. However, his interest was not just in "contrasting" the races but also in ranking them. There was nothing arbitrary, then, about his list of the races under the military's examination, which were listed in order of fitness, ostensibly for military service but also for manhood—with native-born white Americans at the top.[54]

Bartholow went on to characterize briefly each of these race's men, though only native-born white men warranted a separate detailed paragraph on their "mental characteristics" in addition to their "physical qualities." In nonwhite and foreign-born men, by contrast, "mental and physical qualities" so overlapped that they were perhaps one and the same. Indeed, the SC largely reinforced the old ethnological adage that the physical man determined the intellectual and moral man, particularly in the "lower races" who could not seem to transcend their biology. His minimal interest in the minds of nonwhite men reflected the common belief that while some races had certain physical attributes like strength or stamina that made them suitable rank and file soldiers, native-born white men alone had the mental acuity necessary to be good military leaders.

Bartholow's assessment of the white race was glowing. Drawing at least as much on cultural stereotypes as biology in the results of his research, he maintained, "The mental characteristics that fit the American for the military service consist of a spirit of enterprise and an intellectual hardihood which render him superior to fatigue." Likewise, white men had "an easy bearing under defeat, and a buoyant self-confidence which misfortunes do not easily depress." Bartholow dismissed charges that this independent spirit and confidence made white men bad soldiers, unable or unwilling to submit to the will of others or to group interests by simply noting that "events have not justified the harsh criticism." Faced with the SC's own data that did not place white men at the top of charts measuring height and muscular development—typical measures of manhood—Bartholow had to spin this information in favor of white men's fitness for military service. He did so in part by looking deeper into the body. "The physical qualities which fit the American for military service consist, not so much in muscular development and height as in the toughness of his muscular fibre and the freedom of his tissue from interstitial fat, whereby active and prolonged movements are much facilitated," he remarked. Even the white man's propensity to succumb to disease on the battlefield was more a product of the delicate digestive system of the civilized than a sign of "imperfect physical development" according to Bartholow.[55]

By comparison, the "Celtic races" shared physical characteristics with white men but had "less tenacity of purpose and mental hardihood," submitted less readily to authority, adapted poorly to hardship, and chronologically complained about their situation. And Germans fell short of native-born white and Celtic soldiers in terms of physical characteristics, suffering as they did from flat feet, varicose veins, and "unusual weakness of the abdominal rings." The German did have "mental and moral qualities" that compensated somewhat for his bodily inadequacies, including "thrift . . . and a love of ease and enjoyment," but his "fondness for good living" rendered him an uncommitted soldier when the rations were not up to snuff.[56]

Examining and characterizing the black soldier in particular was an especially important and high-stakes venture in the Civil War era. A month after being appointed the SC's actuary in 1864, Benjamin Gould appealed to the SC to expand the scale and scope of its own anthropometric investigation. Identifying what he saw as a glaring gap in the research, Gould declared, "No examination of the negro troops seem to have yet been made, and the importance of such inspection needs no comment."[57] Likewise, Bartholow paid special attention to the black race in his introduction to the first volume. He concluded that black men shared the Germans' bad feet, as well as "small, ill-developed calves" and a proneness to pulmonary disease. But Bartholow noted, "The Negro possesses many of the physical qualities pertaining to the highest type of the solider," including height and, again, "considerable power of endurance."[58]

Although his findings supported the army's controversial decision to admit black men into the ranks, Bartholow seemed loathe to imply that black soldiers were as manly as white soldiers. Instead, he constructed an image of the black man as a good rank and file soldier that drew on popular stereotypes about African Americans as strong in body but obedient and childlike in mind. "Having the faculty of imitation highly developed and being fond of the exterior show and parade of military life," he maintained, "[the black man] readily becomes an adept in the mechanical training of the soldier." Bartholow nonetheless concluded that whites were best suited for military service. Seemingly contradicting his own prior praise of blacks' power of endurance, he averred, "The Negro solider is, unquestionably, less enduring than the white soldier; less active, vigilant, and enterprising, and more given to malingering." As demonstrated by the summary reports penned by PMGB's examining doctors, the language and tenor of Bartholow's consideration of black soldiers was typical of Civil War physicians, even if the exact numbers in the anthropometric data occasionally differed between the two studies.

Ironically, despite the SC having collected enough medical data and anthropometric statistics to fill two large volumes, Bartholow ultimately based his judgment of each race's military fitness on characteristics the SC did not measure. For example, he noted the white man's "toughness of . . . muscular fibre" and the black soldier's insufficient activity and vigilance, which was supported nowhere in the body of either volume. Bartholow not only gave weight to traits outside the scope of the medical and anthropometric studies he introduced but also omitted research results that did not support the hierarchical order of race in his summary. The data in the second volume attributed the greatest height and head size to the "red man," who was entirely absent from Bartholow's discussion, for instance. Enormous importance was placed on head size as reflecting intelligence during this period, yet Bartholow did not analyze Indians' seeming superiority in his discussion of that particular data.

He returned to more familiar medical terminology—as well as familiar racial tropes—with his assessment of the mulatto and the mixed Spanish American of New Mexico. Though the mulatto ranked between the white and black races by several measurements in the anthropometric tables in the second volume, Bartholow echoed the war PMGB physicians in placing biracial men below both races in general health and vigor.[59] Likewise, he placed members of the "mixed race of New Mexico," who suffered from "feebleness of constitution, the syphilitic cachexia, impaired vision, deformities of the hands and feet, and diseases of the urinary organs," below the black race. Mixed race peoples had notable moral defects as well. According to Bartholow, "They are cowardly, unreliable, and difficult to control, in consequence of a very mercurial temperament." Drawing far more on popular stereotypes about race mixture as anything contained within the SC's studies, Bartholow painted mixed race men as physically and socially diseased, weak, and destined to die out.

Postwar Legacies

After the war, the three men at the center of the SC's and PMGB's medical and anthropometric studies—Bartholow, Gould, and Baxter—went on to achieve considerable professional success and acclaim, though for Bartholow, and to some extent Baxter, that success also came with some controversy. Bartholow left the military in 1864 after his first child was born to accept a professorship at the Medical College of Ohio in Cincinnati, where he also joined the staff of the Good Samaritan Hospital. Over the next decade, he published a series of influential articles with topics ranging from male sexual dysfunction to sanitary surgical practices to techniques for administering medication by syringe; founded a weekly medical journal, *The Clinic;* and advocated for the application of rigorous scientific methodology to medical practice, particularly knowledge gained from experiments conducted on live animals, on which he drew heavily in his own research and teaching—to the admiration of some and horror of others, who described him as unnecessarily cruel and exhibiting little regard for the animals' suffering.[60]

Nor did his experimentation end with animals. In 1874, Mary Rafferty—a roughly thirty-year-old "feeble-minded" Irish immigrant already heavily scarred from a fall into a fire as an infant and employed as a domestic servant—entered Good Samaritan as a patient suffering from a hole in her skull caused by an infected ulcer that physicians soon diagnosed as cancerous.[61] When their attempts to treat her surgically failed and her condition seemed likely terminal, Bartholow began a series of experiments that involved inserting electrode needles into her exposed brain to conduct electrical currents through her left motor cortex and provoke involuntary motion. According to Bartholow's own report, he continued to increase the strength of the current after she had begun to cry, and soon she lost consciousness and seized violently before slipping into a coma. She was revived twenty minutes later and underwent additional experiments at Bartholow's hands before she had another violent seizure several days later and died—though Bartholow insisted that she ultimately died of the cancer itself rather than from any damage his experiments caused. After he published his findings in the *American Journal of Medical Sciences* the same year, he was met with sharp approbation from physicians in the United States and Great Britain, along with official condemnation from the American Medical Association, which declared his actions in violation of "the spirit of our profession" and "our feelings of humanity." Bartholow wrote a letter of apology to the *British Medical Journal,* which seemed to satisfy some of his critics, but his interest in the possible medical and surgical applications of electrical currents continued, and he went on to publish an additional article on the subject in 1882.

His career ultimately suffered little harm. He served as the dean of the Medical College of Ohio from 1874 to 1879, the height of the controversy over his experiments, and he was elected president of the American Neurological Association in 1881.[62] He maintained a thriving clinical practice in Cincinnati after the Rafferty experiments and continued to publish prolifically for the rest of his career as well, including *Materia Medica and Therapeutics* in 1876, which sold so well and was so well reviewed it underwent twelve editions by 1909. The book's popularity led to Bartholow accepting a chair position at Jefferson Medical College in Philadelphia, where he also served on the staff of Philadelphia Hospital after moving to the city in 1879. The remainder of his career was marked by awards, invited lectures, and more acclaimed publications, and upon his death in 1904, he was memorialized as a hero within the medical profession, with little mention of his experiments on Rafferty.[63]

The minor controversies that swirled around J. H. Baxter paled in comparison. Something of a prodigy in the military, Baxter continued to rise through the ranks after the war ended but was contested at every turn by his competitors who were resentful of his young age and the fact that he had initially entered service as a volunteer and thus did not undergo some of the same extensive testing as his peers. Fortunately for Baxter, he consistently had the support of both his superior officers, who praised his meritorious service and sweeping vision for needed improvements to the Medical Corps, and powerful politicians in Washington, many of whom he treated in the private practice he maintained in addition to his position with the military. (The ever-busy and ambitious Baxter also found time to obtain a law degree during this period as well.) Perhaps not surprisingly, his penchant for courting powerful patients who could—and did—advance his military career, including active presidents and congressmen, earned him additional rancor from his competitors in the military, as well as local Washington physicians. He remained enormously popular within the larger medical profession nationwide, however, and when the army's surgeon general retired in 1890, leaving the highly coveted position vacant, Baxter's candidacy was endorsed by the *New York Medical Journal,* the *Southern Practitioner,* the *New York Times,* the secretary of war, and then president Benjamin Harrison, who Baxter counted as both a patient and a friend. He was appointed surgeon general in August 1890, though his term was cut short only four months later by a stroke, which took his life at fifty-three years of age. His funeral was attended by a virtual who's who of military and political elites, and he was buried in Arlington Cemetery.[64]

Meanwhile, the accomplished and well-liked Benjamin Gould always seemed to court far more friends than enemies throughout his career. After the war, he returned to his work supervising and improving observatories, particularly an ambitious new observatory in Argentina, where he enjoyed considerable support—monetary and otherwise—from the Argentine government and also became involved with the Freemasons, serving as an official intermediary between the North and South American wings. He also saw devastating personal tragedy during this period as well, first with the accidental drowning of his two eldest daughters in Argentina in 1874 followed by the death of his wife while on a brief vacation to Boston in 1883. He leaned on the many friendships he made in both the Sanitary Commission and wider scientific circles for support during that dark decade and in 1885 returned to Boston permanently, where he threw himself into the city's civic life.[65] In November 1896, while celebrating Thanksgiving at the home of one of his three surviving children, he fell down the stairs and passed away shortly after at the age of seventy-two. He was eulogized in the U.S. and international press for both his amiable character and his varied scientific accomplishments.[66]

Just as the career paths of the men who oversaw them went in new directions after the war's conclusion, the specific racial questions and concerns that fueled the collection of anthropometric and medical data during the Civil War would be transformed during the new postbellum sociopolitical contexts in which the reports were published and cited. For one, the changing face of immigration in the late nineteenth century largely did not correspond to the immigrant groups most frequently discussed in the Civil War studies. The War Department's publication, for example, reports that its physicians examined just 171 soldiers native to Poland, 122 from Russia, and 339 from Italy—southern and eastern European nations that represented some of the largest, and most maligned, immigrant groups to the United States in the last decades of the nineteenth century—as compared to 54,944 of German birth and 16,196 English immigrants, with soldiers from Wales, Scandinavia, Scotland, and France each numbering in the thousands as well.[67] These groups most represented immigration in the United States at midcentury, particularly the English and Germans, who received the most sustained attention in the army's reports, including the questionnaires completed by the examining physicians. Consequently, scientists and social commentators in the late nineteenth and early twentieth centuries were far more likely to invoke the Civil War anthropometric and medical studies in discussions of African Americans than of immigrants; when they did discuss the studies in conjunction with immigration, it was usually toward a more general argument that native-born Americans were physically and morally superior to the foreign-born, with little or no reference to specific groups.[68] Moreover, dating back in the United States well before the Civil War, military service was widely understood as a route to expedited citizenship for immigrant men, an opportunity to prove both their individual worth and the social value of the group to which they belonged.[69]

Though they had similar hopes, the legacy of black men's participation in the Union army, as both soldiers and anthropometric test subjects, was far more enduring and, in many ways, tragic. The cultural and political context in which scientists considered the black race—and its future in American society—changed considerably in between the collection of anthropometric and medical data during the Civil War and its publication in Reconstruction. Notably, the physician questionnaires included in the PMGB's report, published in 1875 near the end of Reconstruction, were mostly written and returned in 1865, shortly after the war ended. Thus the ideas about race in general and black men in particular that physicians expressed in these questionnaires were very much a product of the Civil War era and specific to the military context. The 66 percent of respondents who ranked the black soldier as equal or superior to the white (as opposed to the 16 percent who found black soldiers inferior and the 18 percent who declined to answer based on insufficient experience with African American recruits on which to base their conclusions) at least *intended* their endorsement of black men's physical strength (if not their mental acuity) to be praise—even if these assessments were clearly enmeshed with their beliefs in racial difference and black inferiority.[70]

In the new context of Reconstruction, however, such characterizations took on an explicitly negative connotation, divorced from the army's earlier need for manpower and with questions about citizenship and the franchise now dominating discussions of black men in American scientific and political discourse. Though politicians and racial scientists writing during the Civil War itself considered issues related to the possibility of a vastly expanded free black population, such concerns, particularly enfranchisement for black men, gained new urgency once emancipation was a reality. Indeed, one wonders whether Baxter, compiling and editing the PMGB's study in the early 1870s, had these heated debates about black manhood in mind when he (seemingly unconsciously) placed African Americas at the bottom of the list of groups most suited for military service, even though the questionnaires he was tabulating placed them fourth, as discussed previously. In other words, in contrast to the questionnaires themselves, written immediately after the war's conclusion, Baxter's interpretation of the earlier results and his memory of black men as soldiers were likely influenced by the cultural politics of Reconstruction. After all, manhood was a *political* category as well as a physical one in the nineteenth century, and outside the military context, scientists were far more likely to measure the best man by brains rather than brawn. Thus even the most "positive" assessments of black men as soldiers in the army's reports lent support to postbellum ethnological attacks on black men as citizens. "Obedience" and an "imitative nature," coupled with physical strength, could serve black men well on the battlefield, but not in the voting booth.

Ultimately, when the army's reports were published during Reconstruction, it was assessments such as these to which many scientists and politicians would point to argue that black men, though free from chattel slavery, were biologically designed to be laborers, not leaders. Moreover, postbellum racial scientists generally cited the measurements that supported their arguments and ignored those that indicated no difference between the races or placed black men above their white counterparts.

At the onset of the Civil War, antipathy for slavery and, to a lesser extent, sympathy for the slaves themselves were widespread throughout the Union. After the war, however, support for full citizenship and equality for the freedmen was far from universal among whites—in the South or the North. With white male dominance facing mounting challenges from women and black men agitating for political rights, postbellum writers could point to the Civil War anthropometric studies as confirmation that the races were fundamentally different, arguing that citizenship was a biological mandate reserved for white men alone. In other words, read through the lens of Reconstruction-era debates over black enfranchisement, the Civil War reports seemed to support an interminable intermediary status for black men: apparently physically masculine enough to fulfill the most basic obligation of citizenship—potentially giving one's life in battle for the nation—they were nonetheless undeserving of the rights it provided to civilized gentleman, in the racialized and gendered parlance of the late nineteenth century.

Aside from how racial theorists employed this data after the war, however, there is a rather obvious limitation to the data itself that warrants highlighting here. As Benjamin Gould admitted in the Sanitary Commission's reports, "The anthropological results here given are of course restricted in their very nature, pertaining as they do . . . to one sex only."[71] For half a century following the war, scientists bolstered their claims about racial difference with studies conducted solely on men. If much of antebellum ethnology focused implicitly on men, scientists looked literally to male bodies during the Civil War and its aftermath. Though discussions of black women and their reproductive capacities had been increasing in ethnology prior to Emancipation, specifically around the issue of slavery, the "peculiar institution's" demise refocused ethnology on citizenship and the capacities of black men.

The Beard Revisited: Race, Sex, and Citizen Bodies during Reconstruction

Starting immediately after the Civil War ended with the surrender of the Confederate Army in April 1865 and lasting a little over a decade, Reconstruction was a period marked by tremendous social, political, and economic change. The sectional tensions and hostilities certainly did not disappear when the Civil War battles officially ended, though on paper, a unified country existed once more. "You seem to be in a good deal of political excitement at the North and I think the whole country must see much trouble before order is fully instituted," Josiah Nott wrote to his friend, Philadelphia physician and paleontologist Joseph Leidy, in 1866. "The South is whipped into quietude and except for the politicians, the people take little interest in Congressional proceeding," he demurred, and "most of us do not know what the Constitutional amendment is that they are quarreling about."[72] What politicians were "quarrelling about" would ultimately be not one but three acts of legislation led by a Republican Congress and collectively known as the Reconstruction Amendments, which sought to secure rights and protections for the newly expanded free black population. The Thirteenth Amendment, passed in 1865, abolished slavery anywhere within the United States or its jurisdiction. Ratified in 1868, the Fourteenth Amendment granted citizenship to all people born or naturalized in the United States, followed by the Fifteen Amendment, ratified in 1870, which guaranteed that no state could deny its citizens the right to vote "on account of race, color, or previous condition of servitude."[73] Notably, the Fifteenth Amendment did not prevent denying citizens the vote on account of their *gender,* and it was not until the passage of the Nineteenth Amendment in 1920 that women were granted full citizenship and guaranteed the right to vote nationwide.[74]

Reconstruction did not just play out in the halls of Congress, however. On the ground, the Reconstruction era was also marked by black men flocking to the polls to enact their new voting rights as well as entering political office themselves on both the local and national levels. The period also saw the growth of black-owned business throughout the South and the development of a black middle class—though for the majority of Southern blacks, their legal status had changed from enslaved to "free," but their social status and material condition had changed little. Congress created the Freedmen's Bureau in 1865 to ease the former slaves' transition from slavery to freedom, but even though it was run by the War Department, the agency had few practical resources with which to enforce its authority and was met at every turn by angry white Southerners who thwarted many of its efforts to protect the freed people's rights—nor were all local Freedmen's Bureau agents equally committed to such efforts in the first place. The Freedmen's Bureau was most successful in its education work, establishing freedmen's schools throughout the South that blacks of all ages attended, sometimes overcoming great obstacles to do so after literacy had been denied them under slavery. In other words, the South during Reconstruction, with a black population increasingly educated and newly empowered economically, politically, and socially, looked a lot like the worst fears white racial theorists like Nott imagined of a racially disordered society in the absence of slavery.[75]

Although he positioned himself as a politically disinterested member of "the people" who took no special notice of Reconstruction in his letter to his influential Northern friend, Nott's published writings during the period reveal him to be anything but disengaged and apolitical. Just one year prior, for example, Nott penned a fiery missive, "The Negro Race, Its Ethnology and History," in the form of a letter to the superintendent of the newly formed Freedman's Bureau, Major-General O. O. Howard. Ironically, Howard had appropriated the Medical College of Alabama in Mobile where Nott worked to use as a freedmen's school—much to Nott's chagrin. In the letter, Nott argued that black men were physically incapable of citizenship, their bodies designed solely for labor, and that any attempt to educate or enfranchise them ran counter to nature itself.

Whereas Nott was previously concerned with where the races came from, by 1865 he was more concerned with where they were going. Considering the once "grave question" of racial origins now a moot point, he turned his attention instead to the pressing—and gendered—issue of citizenship.[76] For decades leading up to Emancipation, women's rights activists had been profoundly shaping the ways in which citizenship was represented and conceived in American public discourse. In 1852, Paulina Wright Davis, who wrote on and organized around women's suffrage, described women as a political "enigma." "Freeborn and thus logically entitled to their privileges and rights as 'birthright' members of the United States, women nevertheless constituted a politically and legally 'disabled caste,'" historian Nancy Isenberg explains. "The enigma, then, was this: freeborn women had the appearance of citizenship but lacked the basic rights to be real citizens."[77] True citizenship involved far more than place of birth, women continuously argued; it was a complex web of rights and protections routinely denied to large portions of the American populace.

After the Civil War, black Americans found themselves in a similarly complex middle ground as neither slaves nor citizens but rather a free labor force with few tangible rights or protections under the law. Prominent women's rights advocates pointed out, though, that freedmen were already one step ahead of women since married women lacked rights to contract, and the fruits of their labor legally belonged to their husbands—a situation they deemed as akin to slavery. In other words, racial inequality provided women a language in which to frame their own protests about gender inequities.[78] In using politically charged analogies to race, slavery, and freedom, these early feminists also highlighted the limits of birthright citizenship versus the full citizenship rights extended to white men. When Reconstruction-era racial theorists then discussed the capacity of the black race for citizenship, it is this broader definition to which they referred—one that included full participation in American political life and, mostly notably, the franchise.

With the end of the Civil War, emancipation became an unavoidable reality and white thinkers struggled to make sense of a racial order unbounded by chattel slavery. For Nott, "the first question, then, to be settled is, the capacity of the negro for self-government." He wondered, "Is he capable of taking any part in the march of civilization beyond that of a mere 'hewer of wood and drawer of water?' Does his history afford proof that his intellect is susceptible of any really useful development?" Nott's answer was an emphatic no.[79]

Nott's letter to the Freedman's Bureau, which was reprinted for a wide audience in 1866 in the *Mobile Daily Times,* was typical of Reconstruction-era white racial thought in several important regards. Employing science to support explicitly political ends, the letter rooted capacity for citizenship in the physical body. Moreover, the citizen body it constructed was both male and white. His letter also demonstrates that although slavery had been abolished in the United States, it did not disappear in American ethnology. On the contrary, racial science published after the Civil War still often juxtaposed citizenship and slavery as the natural roles of men on opposite ends of the racial hierarchy and presented the work of Reconstruction as at best uninformed by scientific fact and at worst dangerous. Forced to integrate the reality of slavery's demise into their arguments about racial hierarchy, postbellum racial theorists typically argued that although the structure of society had changed, the black man's biology had not, regardless of any legislative or philanthropic attempts to change his position. Notably, with questions of citizenship driving Reconstruction-era ethnology, scientists like Nott were now primarily concerned with the bodies and pursuits of men.

In his "Instinct of Races," published in the *New Orleans Medical and Surgical Journal* in 1866, Nott reasserted his belief in race as a permanent, physiological force that resisted human intervention.[80] Throughout the text, Nott positioned his expertise on black men's physiological and intellectual nature as vitally important to politicians, for "governments can never legislate wisely, without due consideration of these important facts."[81] For Nott, man's natural history also determined his civil future. He asserted that his professional efforts to "calmly" enquire into the "future of the white and black races" in America were not just relevant to scientists and political leaders; they were vital for all concerned citizens—and thus his "duty."[82]

Josiah Nott's demeanor regarding this question seemed anything but calm, however. He had spent the Civil War as military surgeon and hospital inspector for the Confederate cause, while two of his surviving sons (Nott had lost *four* of his eight children in one week in the yellow fever epidemic that decimated Mobile in 1853) served, and eventually died, on the front lines. At the end of the war, Nott's worst fears had come true. Slavery had been abolished, his city was occupied by Yankees, and his own family was in shambles.[83] Nott furiously denounced Reconstruction-era policies that attempted to legislate the Negro into becoming something his own biology made impossible and endangered whites who must live among him without the structure and protection of chattel slavery—although in reality, it was Mobile's black population who were subjected to waves of violence at the hands of defeated and enraged whites in the immediate aftermath of the war.[84] Nott's bitterness reached new heights during this period, and as Nott biographer Reginald Horsman explains, Nott "was absolutely unwilling to say, 'slavery has ended, let's make the best of it,' because he had raised his innate prejudices to the level of scientific truth.

Nott believed that natural law dictated that free blacks in the presence of whites were destined to permanent inferiority and eventual extinction."[85]

Nott was also unwilling to wait for nature to eventually take its course and return the South to the white man, especially since Congress refused to do so in the meantime, and a year after he published "Instinct of Races," which heavily cited John Van Evrie, he left Mobile in disgust, settling in Van Evrie's city of New York after a brief time living in Baltimore. Indeed, the 1866 text would be his last publication addressing race *or* Reconstruction, and he devoted the rest of his career to building his medical reputation in his new Northern world, where he seemed to be welcomed with open arms by the city's medical and scientific elites. While Nott maintained contact with old peers in ethnology and was a member of the New York Ethnological Society, he insisted that the time he needed to devote to his medical practice to keep up with the city's high cost of living left little for ethnological work. Horsman, however, suggests that Nott's shifting energies were also due to the fact that Nott "was now living in a mostly white society and no longer felt the compelling need to warn of supposed racial dangers."[86]

Of course, when Nott moved to New York in 1867, he did not enter a city without its own racial tensions or conflicts. Indeed, just four years earlier, the city was racked by draft riots brought on by local Democrats enflaming existing animosity toward blacks among the white working class, particularly Irish immigrants, by suggesting that two recent developments—the Emancipation Proclamation and a military conscription law—would result in former slaves flooding New York to compete for their jobs *and* their women. Most of the military stationed locally had been sent to Pennsylvania to help fend off the encroaching Confederate troops there, leaving New York with vastly diminished defenses from the surging mob. For four days, the mob destroyed property, burned homes and businesses to the ground—especially those owned by African Americans and Republicans—and lynched at least eleven black men while attacking countless other African Americans of all ages and genders. Though with the Confederate defeat at Gettysburg, the Union army was able to send troops back to New York to quell the fighting, by the time they were able to do so on July 16, 1863, 119 New Yorkers lay dead and many African Americans fled the city permanently, their numbers dropping to 9,945 in 1865—a staggering 20 percent drop. The city also remained bitterly divided for years between support for the Democratic and Republican platforms, particularly on the issue of black enfranchisement.[87] Radical Republicans in New York raised the issue of enfranchising black citizens for the first time at the state constitutional convention in 1867, the same year that Nott moved to the city. Local democrats, particularly state senator and former mayor of Brooklyn Henry C. Murphy, directly referenced craniology and ethnological arguments about black inferiority to urge white voters to strike down any effort to place black men in a position of political equality dissonant with natural law. And they succeeded. In 1869, New York's citizens voted down a proposal to amend the state constitution to enfranchise black men by more than 70 percent.[88] Horsman's assessment that Nott no longer felt "the compelling need to warn of supposed racial dangers" after his move to New York seems apt, then, considering powerful politicians in Nott's adopted city were apparently heeding the lessons of his life work, drawing on ethnology to successfully deny black men the vote while white men on the street used violence to drive many African American citizens from its borders entirely.

While racial theorists like Nott were addressing their tracts explicitly to the Freedmen's Bureau and the Republican Party in the years immediately following the end of the Civil War, their work clearly had other audiences as well. As Nott predicted, the future of the white and black races was of great interest to politicians and concerned citizens, as evidenced by Henry C. Murphy's ethnologically informed campaign against black enfranchisement in New York. Similarly, this poster from the 1866 Pennsylvania gubernatorial race, promoting Democratic candidate Hiester Clymer, who ran on a white supremacy platform, looks remarkably similar to the kind of illustration one might find in an ethnology text (Figure 2.6). Scientific illustrations in the nineteenth century often juxtaposed black and white heads or profiles to denote physical—and by implication mental—difference and hierarchy. Furthermore, the prominent forehead of the white man in the poster was likely no accident either, nor was the "less capacious . . . frontal region" and "retreating forehead" of the black man.[89] Scientific studies of crania, including phrenology, the most popular of these disciplines among lay readers, abounded with comparative descriptions and images of black and white foreheads as evidence of the former's small frontal lobe, which indicated incapacity for higher-order thinking skills.[90] Moreover, the poster mirrors ethnology's presentation of Reconstruction as a contest between men. The poster reads that Clymer is "for the white man," represented by the idealized head we see on the left,

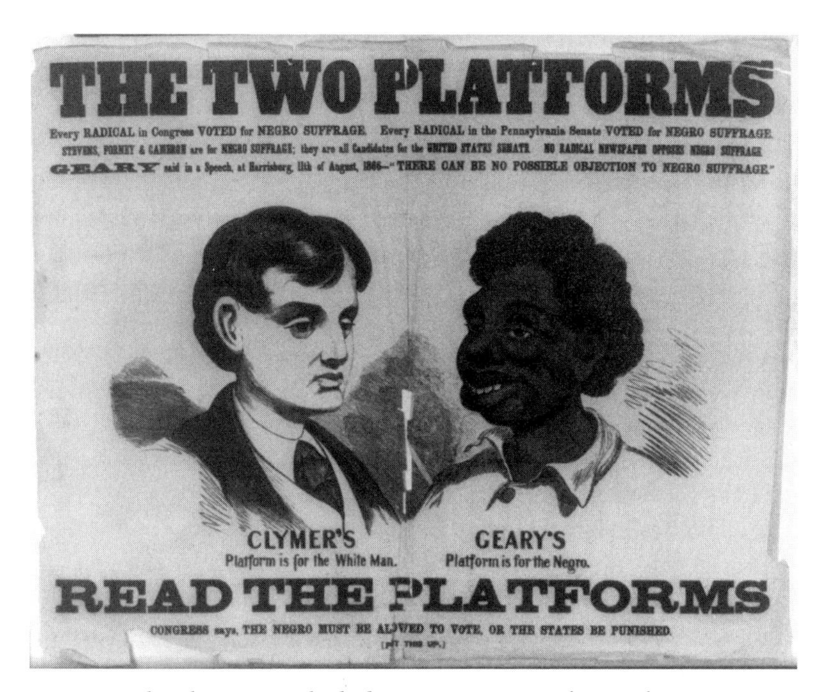

Figure 2.6. Political capacity embodied. *Here, again, we see politicians drawing on ethnological imagery of racial difference. The familiar juxtaposed profiles highlight the sloping forehead of the black man and, by extension, his lower-order thinking and incapacity for informed citizenship. As in ethnology, this political poster represents Reconstruction as a contest of manhood, in which supporters of enfranchisement attempted to push black men into a position nature had never intended (Library of Congress, Prints and Photographs Division,* American Political Prints, 1766–1876 *[LC-USZ62-32498]).*

contrasted with this stereotypically exaggerated black head that represents his opponent's platform as "for the Negro." The exaggerated countenance of the black man here, which stands in for the Republican Party platform, gives visual form to postbellum ethnological writers' critique of black male suffrage as an affront to nature and physiological fact.

This campaign poster and Murphy's invocation of craniology to counter efforts to enfranchise black Americans as well as Nott's *Instinct of Races* and letter to the Freedmen's Bureau demonstrate how much politics and racial science were in dialogue during Reconstruction. The line between them was constantly crossed, and they drew upon each other for support and legitimacy. Ethnology provided anti-Reconstruction politicians with a familiar set of arguments and imagery of black inferiority. Moreover, during Reconstruction, scientific discourse on race and general racist literature and imagery were often nearly indistinguishable. Ethnology's adherents engaged with political debates and used their scientific authority to instruct society on civil affairs while political writers like Hinton Rowan Helper drew on ethnology to support their vision of a white male polity.

After the Civil War, Helper immediately turned his attention from ridding the United States of slavery to ridding the country of the former slaves. Like Nott, he saw racial science and contemporary political debates as interdependent. In considering the removal of the black race from the United States, if not the planet (although he argued that nature would eventually do just that), he cited—and praised—the work of "the most learned naturalists," including Nott, Samuel Morton, and Thomas Jefferson. The scientifically informed "are those whose voices, above all others, should be most attentively heard and heeded" in regards to race, Helper lauded in his 1867 text, *Nojoque: A Question for a Continent.*[91]

Seamlessly blending white supremacy, nativism, and male dominance, Helper envisioned a postbellum United States reserved for American-born white men and drew heavily on ethnology to support his claims. He dedicated *Nojoque* to "that most enlightened and progressive portion of the people of the New World, who have the far-reading foresight and the manly patriotism, to determine irrevocably, by their votes, in 1868 . . . no slave nor would be slave, no negro nor mulatto, no Chinaman nor unnative Indian, no black or bi-colored individual of whatever name or nationality, shall ever find domicile anywhere within the boundaries of the United States of America."[92] Helper thus rendered voting as a manly act and one that was inherently white. In the text itself, Helper contrasted these manly patriots of white America with the "decrepit and effete races," "whose colors are black or brown . . . and whose mental and moral characteristics are no less impure and revolting than their swarthy complexions."[93] By continually feminizing African Americans, he distanced them from the possibility of political and social power. Men alone could be citizens and leaders, and there were no men among the "effete races."

Though his dedication seems to imply that Helper held equal animus for all nonwhites, the chapters themselves starkly revealed that the "question for a continent" revolved around black people specifically. One chapter was titled "The Servile Baseness and Beggary of the Blacks," while other parts juxtaposed the black and white races as polar opposites in every respect: chapter 4 examined "White Celebrities, Black Nobodies," and chapter 2, "Black: A Thing of Ugliness, Disease, and Death," was followed by chapter 3, "White: A Thing of Life, Health, and Beauty." Moreover, he began the book by querying, "What is the best and only true remedy for the present and prospective troubles now brewing in the United States, between the White people and the Negroes?"—to which he answered, "An absolute and eternal separation of the two races." Separation was only the first, immediate step, however. Helper argued African Americans should ultimately be removed from the country entirely, setting July 4, 1976, as a target date for U.S. independence "from the negroes."[94] Likewise, most ethnological writing from this period continued to focus on comparing the black and white races.

The black/white juxtapositions that structured the book also reflected Helper's gendered hierarchy of race. Chapter 6 considered "A Score of Bible Lessons in the Arts of Annihilating Effete Races," while chapter 7 described "The United States of America: A White Man Power." Also noteworthy is that Helper, who eschewed the scientific study of race in his work prior to the Civil War (yet nonetheless frequently engaged with its core tenets and cited its most prominent players), included in *Nojoque* a lengthy chapter on "The Negro, Anthropologically Considered," which sought to prove "that this difference of manhood, this despicable inferiority of the negro, is natural, conspicuous and permanent."[95] Scrutinizing the entire male body part by part, the chapter cited the most prominent white American and European scientists.[96] A new convert to ethnology, Helper alternated between a voice of scientific matter-of-factness and unveiled vitriol, titling one section, for example, "The Negro's Vile and Vomit-Provoking Stench."

While differences of manhood were indeed his primary concern, black women did not escape Helper's scrutiny. In a section on "The Breasts," he included excerpts from European travel narratives of Africa that recounted women whose breasts were so long that they could suckle children over their shoulder, breasts that are "in the eyes of a European, a real object of horror."[97] Questioning black women's femaleness, he also included an excerpt from Sir John Barrow's "Travels in the Interior of Southern Africa" describing in lurid detail the elongated inner labia of Hottentot women he encountered, which "leave the spectator in doubt as to what sex they belong." For Helper, the Hottentot apparently stood in for all black women, and their bodies excluded them from the category of true woman.[98]

Ethnology and gender also figured prominently in Helper's second antiblack diatribe, *The Negroes in Negroland, the Negroes in America, and Negroes Generally: Also, the Several Races of White Men, Considered as the Involuntary and Predestined Supplanters of the Black Race.* Again excerpting ethnological theorists throughout, Helper positioned himself as "a rational Republican" exposing the dangerous folly of Reconstruction endorsed by Radical Republicans.[99] His vision of a post-Reconstruction future was both politically and sexually charged: the white majority would be sexually enslaved by a black minority. Thus he concluded that if successful, the policies of Radical Republicans would force "political, religious, civil, and social equality" between whites and blacks, a horrific fate he hoped to prevent by reminding his readers that nature itself had made the races different and unequal.[100] At the same time, "social equality" often represented a veiled reference to miscegenation, the specter of which ethnological writers would use to warn of the dangers of *political* equality for decades to come.

As a counterargument to such equality, Helper insisted that black men failed to meet the parameters of the male sex. Among the features to which he pointed were the black man's "strange, Eunuch-toned voice" and "the scantiness of beard on his face."[101] Helper looked not only to sex—black men's physical lack of manhood—but also to gender in his indictment of the black race. He examined at length examples of Africans and African Americans failing to act in accordance with white American gender norms, devoting whole chapters to topics like "Nakedness, Shamelessness, and Prostitution in Negroland" and "Courtship, Marriage, and Concubinage in Negroland." Throughout, he expounded on black women's aversion to the trappings and responsibilities of domesticity and painted black men as cowardly, amoral, intellectually deficient, and utterly opposed to work. No idle insults, Helper's bleak assessment of African Americans' inability to adhere to white gender norms spoke implicitly to the sociopolitical rights and protections that should be afforded or denied them at a time when suffrage—for women as well as black men—was subject to constant debate.

Scientific experts on race, for their part, were deeply invested in these public contestations over the meanings and parameters of citizenship and sought to bring their expertise to bear on American politics. Historian Nancy Isenberg notes, "Granting the ballot to 'every man who fights and pays,' in the words of Thomas Jefferson," had long "provided a republican theory of universal white male enfranchisement."[102] But after Emancipation, this standard's link to whiteness could no longer be taken for granted. Black soldiers had fought bravely in the Civil War, even by the military physicians' own accounts. And blacks were not exempt from taxes either. During Reconstruction, prominent scientists sought to reinforce a racialized and gendered definition of citizenship that would ensure that though African Americans were no longer slaves, they could not be full citizens.

White scientific thinkers during Reconstruction presented a largely united front on the question of citizenship for black men. Even writers who had criticized slavery as an abuse of power prior to Emancipation argued for pervasive racial difference and a white standard of citizenship during Reconstruction. Their attack on black (male) citizenship tended to emphasize three major points: (1) the Constitution was never intended by its authors to include African Americans or women as part of the electorate and was written by and about white males; (2) black men were not "male" by bodily standards (and citizenship was legally tied to manhood); and (3) black men did not and could never have the intellectual capacity required of informed citizenship. The black man, now free from the plantation, was still a slave to his biology.

Most of these experts presented the first of these arguments as an uncontested statement of fact—noting the Constitution was only intended to ensure the rights and protections afforded to white men. Demonstrating that black men were not really men and thus undeserving of the political rights attached to manhood, however, was more of a challenge.

During Reconstruction, scientific experts on race found themselves in a rhetorical quandary. They had long represented slavery and citizenship as oppositional political states with no real middle ground. Women were of course not slaves, but they did not vote. The full rights of nineteenth-century U.S. citizenship were contingent on manhood. By this logic, millions of black men would have to be citizens if they were no longer slaves—unless they were not actually men. While Reconstruction-era racial scientists still juxtaposed slavery and citizenship as the natural positions of inferior and dominant races, Emancipation had disrupted that natural order. These scientists were left with the task of providing scientific support for denying black men the franchise so that the races would not be on equal political and social planes, which would obliterate the natural order entirely.

As postbellum-era white racial theorists struggled to place black men in a political middle ground between slavery and citizenship, they did so largely by positioning them in a physiological middle ground regarding sex. In an 1866 pamphlet, "The Six Species of Men," Van Evrie both revived and revised his argument about the black man's supposed paucity of beard as symbolic of his racial inferiority. Whereas earlier he used the beard to impugn black manhood in general, in "The Six Species of Men," he linked the beard specifically to intellectual and political capacity. In 1861, he used this physical lack of manhood to counter abolitionists' characterization of slavery as an institution in which man held his fellow man in bondage. During Reconstruction, Van Evrie employed insufficient manhood to attack black enfranchisement. The Freedmen's Bureau, aid societies, and northern volunteers took the place of the abolitionists as the new object of Van Evrie's ridicule, deluded idealists blind to the realities of nature. In "The Six Species of Men," he discussed the numerous biological differences he perceived among the races, differences that ostensibly determined each race's place in the natural and social order. Among these differences, "the Caucasian is really the only bearded race, and this is the most striking mark of its supremacy over all others." Again, white men were the standard by which others were measured: "All other races approximate to [white beards] in this respect, but the typical, woolly-headed negro, except a little tuft on the chin, and sometimes on the upper lip, has nothing that can be confounded with a beard." The pamphlet's illustrations supported his claims correlating race with facial hair and underscored his focus on men. Small hand-drawn facial portraits were used to represent each of the six races—illustrations that were enlarged to full body length with elaborate backgrounds—in his 1868 text, *White Supremacy and Negro Subordination,* and in both cases, the Caucasian is the only man shown with a bushy, flowing beard (Figure 2.7).

Van Evrie's textual and visual attention to beards was no mere aesthetic preference. In "The Six Species of Men," Van Evrie made an explicit connection between a secondary sex characteristic (the beard) and capacity for citizenship. "If [Freedmen's Bureau agents] expect to make something of Sambo, they must strike for 'equal beard' for him as well as for 'equal education,' or 'equal voting,'" he proclaimed.[103]

(continued)

Figure 2.7. Citizenship and the bearded Caucasian. *In the nineteenth century, citizenship was tied to manhood. In John Van Evrie's view, only white men had "bushy, flowing beards," and therefore they alone were truly men. This illustration, from his "Six Species of Men" (1866), offered visual support for his claim; while the men of other races are drawn outdoors with sparse or no facial hair, the bearded "Caucasian" is surrounded by the trappings of civilization. Accordingly, Van Evrie argued, the Freedmen's Bureau would need to change black men's innate biology—an impossible task—in order to justify extending them "manhood rights" such as voting (Van Evrie,* The Six Species of Men*).*

Figure 2.7. (continued)

Van Evrie, like other ethnological writers in the period, thought intellectual and political capacity was biologically determined. In an era when citizenship and manhood were inseparable categories, these scathing indictments of black masculinity and political capacity were not simply a critique of the performance of gender but rather envisioned a sexed body contingent on race. That is, rather than accusing black men of engaging in unmanly behavior by white nineteenth-century standards, Van Evrie grounded his claims in the body itself by asserting that black men lacked a crucial physical marker of manhood. Ironically, Van Evrie's claims came as growing numbers of surgeons and physicians—Van Evrie's peers—were renouncing long beards despite their long association with distinguished masculinity in the late nineteenth century due to the ascendency of germ theory and its emphasis on contagion and hygiene.[104] John Van Evrie, a hirsute man who seems to have maintained his own thick beard, used the beard not just to differentiate the varieties of men but also to argue that there was no sharp line between men and women among African Americans. The illustrations representing the "Caucasian" type in Van Evrie's *Negroes and Negro "Slavery"* (1861) and *White Supremacy and Negro Subordination* (1868) appear to be of the same individual man (see Figures 2.1 and 2.7); that this illustration in the latter is placed where the author portrait typically would be further suggests that it is likely Van Evrie himself. So while Josiah Nott also used a flowing beard—a physical feature that, like Van Evrie, he personally possessed—as a marker of racial supremacy, Van Evrie seems to have taken it one step further by literally using his own visage and form to represent the pinnacle of human perfection.

In the work of Van Evrie, Nott, and other ethnological writers, the very bodies of the freedmen were deemed incapable of intellectual pursuits or political agency. Political capacity should be contingent on intellectual capacity, they argued, and as a group, white men alone possessed such mental fortitude. Ethnological texts such as "The Six Species of Men" also underscore an important point in regards to sex, gender, and scientific racism. The language of biological difference used to categorize human beings by race in the nineteenth century was often identical to the "biology is destiny" arguments made by doctors, anatomists, and naturalists in regards to the "woman question" even earlier. In other words, a familiar scientific lexicon of sex difference provided nineteenth-century scientists an ideological framework through which to conceptualize race. For many doctors and scientists in the eighteenth and nineteenth centuries, any argument for social and political equality between the sexes ignored the "fact" of immutable physical and mental differences between men and women. Nearly every part of the body was examined and discussed as to the evidence it provided for women's "natural" domestic roles and inability for political involvement. The body then held the key to both race and sex.

The issue of intellectual capacity and efficacy of formal education is perhaps where we see the most similarity and influence between scientific claims about sex difference that began as early as the seventeenth century and the racial sciences that followed in the nineteenth century. Some scientists believed that women's head shape and delicate necks proved incapable of supporting the large cerebellum necessary for the higher-order thinking and rationality characteristic of men. Women's menstrual cycles, too, were thought to render them unfit for higher education and activity outside the domestic sphere and motherhood. In the late nineteenth century and into the twentieth, doctors were concerned with the lower birthrates among educated women, which they saw as evidence not only that higher education decreased women's interest in and commitment to motherhood and domesticity but that it had biological consequences as well. The energy necessary for menstruation and ovulation was being diverted to women's brains, thereby decreasing their fertility—the very thing that defined them as women. Later in the nineteenth century, for example, G. Stanley Hall, a prominent psychologist and outspoken critic of "race suicide" among whites, worried that educated women would become "functionally castrated . . . deplore the necessity of childbearing . . . and abhor the limitations of married life." A gynecologist contemporary of Hall added that educated women were "sexual incompetents."[105] Hall's choice of words is interesting, for a woman may become less like a woman by being overeducated, but being "functionally castrated" through education would not in turn make her like a man. Women then were not only anatomically ill-suited for education, but attempts to educate them would render them unfit for what they were designed: reproduction.

Similarly, nineteenth-century white racial scientists typically argued that the bodies of African peoples could not support education beyond simple vocational instruction. Van Evrie went so far as to suggest that the conformation of black men's skeletal structure could not support the same pronounced forehead and frontal lobe found in white men because it would throw off their center of gravity. Thus he claimed that educating a black man would not help him, it would literally cause him to fall over: "It is obvious, therefore, if Gen. Howard [superintendent of the Freedmen's Bureau] and the Yankee school marms could 'educate' Sambo into intellectual equality with the white man, their protégé would be as incapable of standing on his feet as he would if they had cut his head entirely off."[106] Van Evrie applied similar logic to race as Hall did to the "woman question." Where education would render women biologically unfit for their true purpose ordained by nature—reproduction—education would render African Americans unfit for their purpose—labor—as reflected in their allegedly stooped skeletons naturally designed for toiling in the field. If black men and women of all races could not be sufficiently educated by nature of their bodies, they could not be expected to perform the duties of citizenship or make an informed vote.

Indeed, the sciences of sex difference and the sciences of racial difference were so closely linked that ethnological writers often made analogies between (white) women and African Americans.[107] In "Race and Gender: The Role of Analogy in Science," Nancy Stepan notes, "Gender was found to be remarkably analogous to race, such that the scientist could use racial difference to explain gender difference, and vice versa." Medical scientists determined women and the "lower races" to be similarly deficient in brain weight and structure and corresponding intellectual inferiority.[108]

For Van Evrie, then, a politically empowered black man and a politically empowered white woman were equally absurd and unnatural. Published the year after President Abraham Lincoln's Emancipation Proclamation, Van Evrie's 1864 text, *Subgenation: The Theory of the Normal Relation of the Races, an Answer to Miscegenation,* shows Van Evrie continuing to defend both black enslavement and the theory of polygenesis but also already retooling his argument for the increasingly likely possibility of slavery's defeat. Therein, he defined *subgenation,* a word he fashioned from Latin, as "a race born or created lower than another; hence, the natural or normal relation of an inferior race," while he used *subgen* "to describe the persons of the inferior races thus placed in their natural positions."[109] In a rather convoluted argument even for Van Evrie, he insisted that the very term *slavery* was a misnomer or a myth and that blacks laboring for the benefit of whites in the South simply represented the enactment of natural law, whereas freedom for Northern blacks was an act of policy, and a misguided one at that, under which free African Americans were rapidly perishing and causing social disorder through crime, indolence, and licentiousness—and thus "every negro in the North ought to be placed in a position of subgenation—that is, *enslaved,* as it has been called in our ignorance of the laws of races."[110] Van Evrie also argued that miscegenation would inevitably follow abolition, and "it would only be a question of time when the whites would destroy the negroes, or the negroes the whites!"[111] However, he predicted that if abolition became a reality in the Southern states because of an unjust war—which he claimed resulted from ignorance of the true relationship between the races—the new racial (dis)order would be short lived, for soon whites across the nation would be forced to heed the harsh lessons already born out in the urban North to ensure their survival.

Van Evrie predicted a Utopian future predicated on the return of the lower races—and women—to their proper place in nature nationwide. This "normal order of creation," "subgenation," would see white men "*emancipated* from their burdens to enjoy the full fruition of their faculties," he maintained, co-opting the language of abolitionism for white supremacy.[112] He revived his old argument that neither black men nor black women were physically suited to the gendered refinements of their respective sexes—intellectual pursuits for men and genteel domesticity for women—but took it a step further to propose that subgenation could free the white race from labor entirely so that it could reach its true potential. He explained that the "Creator" endowed the white race with greater intelligence to employ in the "nobler and more intellectual duties of life," while "inferior races" had likewise been adapted for "inferior labor." From this he concluded that God intended for "the race created in his own image" to eventually be freed from "all employment, except that which will develop the intellectual, moral, and spiritual natures."[113] Consequently, "subgenation" would eliminate poverty—for whites—and distribute wealth equitably (again, among whites) and finally allow for true democracy. Again subverting abolitionist rhetoric, he wrote, "The equality of all whom God has created equal (white men), and the inequality of those He has made unequal (negroes and other inferior races), is the corner-stone of American democracy, and the vital principle of American civilization and of human progress."[114]

His emphasis on white men specifically was not accidental. He applied his concept of "subgenation" to the "proper relation of the sexes" as well, devoting an entire chapter to the subject. "The physical, mental, and moral nature of woman was different from man's, and we adapted our laws and institutions to that difference," he noted. He warned, though, that the North—again an instructive site of social disorder—was following the lead of Europe in giving women too much liberty and power, to the peril of the women themselves and the white race as a whole. Abolitionists shouldered some of the blame in Van Evrie's estimation, since they largely favored "Female Reform," which included women's suffrage, alongside their efforts toward racial anarchy. As a result, Van Evrie scoffed, women were so degraded in the North that prostitution and polygamy ran rampant, and members of the fragile sex, with "a more delicate physical structure," were even forced to stand in streetcars while "'the lords of creation,' with their strong sinews and Samson-like muscles" sat lazily in the seats! "As nations become effete, they seem to work wickedness with greediness," he observed, continuing his assignation of Northern manhood before characterizing the Roman Empire as a fallen woman destroyed by her own debauchery.[115] Perhaps most ironically, he also accused the lower races of not appreciating and elevating their women, as the best men among whites did, after he himself had argued that women of inferior races were not physically designed for feminine refinements, and that they, like black men, were mere brutes.[116]

Neither women nor African Americans had the intellectual capacity for the duties of citizenship, and their bodies were designed by nature for other purposes, Van Evrie again insisted in *White Supremacy and Negro Subordination,* published in 1868, the same year that the Fourteenth Amendment—which granted citizenship to all people born or naturalized in the United States, including the former slaves—was ratified. Bitterly noting the frequent critiques of American slavery and race relations lodged by the British, he fired back that they could not possibly understand the natural order of race demonstrated in the United States when England was a country whose natural order of gender had been turned upside down by having a woman as their leader. Appalled, he protested, "A woman is the chief of the nation, whose husband is her subject—thus violating the relations of the sexes—of husband and wife—and thrusting her from the normal position of woman as well as contradicting the relations and duties of citizenship." His characterization of Queen Victoria echoed both scientific discourse on the "woman question" and the racial paternalism he demonstrated throughout his writings:

> God created her, adapted her, and designed her, for a wife and a mother, a help-mate to her husband and the teacher and guide of her children; He endowed her with corresponding instincts to love, venerate, and obey her husband and devote her life to the happiness and welfare of her offspring, and to trample on His laws—to smother these instincts and force this woman to be a queen, a chief of state, the ruler over millions of men, is as sinful as it is irrational, as great an outrage on herself—her womanhood—as it is on the people who suffer from it.

Van Evrie warned against the folly of any attempt to "reverse the natural relations of the sexes" through "human law."[117] The United States had best take heed of the example Queen Victoria represented, Van Evrie argued, and preserve the natural relations of sex as well as the natural relations of race. Nature itself mandated white male leaders and voters alike.

With the same incisive verve he devoted to challenging slavery prior to Emancipation, Frederick Douglass was one of the most prolific black critics of scientifically legitimized rationales for denying African Americans, as well as women, the vote during Reconstruction. In an address delivered in Saint Louis in February 1867—various versions of which he had been giving across the country for months—he maintained, "The fathers of this republic did not learn to insert the word white, or to determine men's rights by their color. They did not base their legislation upon the differences among men in the length of their noses or the twist of their hair, but upon the broad fact of a common human nature. . . . They were above going down, as certain men—Caucasian and Teutonic ethnologists—have recently done, on their knees and measuring the human heel to ascertain the amount of intelligence he should have. They were above that." The scientific search for difference, and scientists' overdetermination of biology as an indicator of intellect or social role, was a "modern invention," not a value intrinsic to American democracy, Douglass insisted.[118]

Similarly, in September 1866, Douglass spoke before a convention of radical and moderate Republicans gathered in Philadelphia to oppose the Reconstruction plan of Andrew Johnson, who had assumed the presidency after Lincoln's assassination the previous year. Johnson, a North Carolina–born Democrat, had served as Lincoln's vice president on the National Union ticket and was eager to bring the former Confederate states back into the Union as quickly as possible, without any planned protections for the freedmen. Although initially ignored at the Philadelphia gathering, along with other black delegates in attendance, Douglass was finally given the opportunity to speak, and his remarks were among the most well received by the audience and the press alike.[119] Subverting white scientists' invocation of "nature" to legitimize their claims about black incapacity for enfranchisement, Douglass intoned, "The great error of the American people [has been] to limit what in its very nature is illimitable; to circumscribe principles intended by the great Creator of the universe for the harmony of the universe, to be equally applicable to all the people of the country." The Declaration of Independence "sets out with the doctrine that 'all men,' not a part of men, 'all men'—not white men, 'all men'—not the Englishman; not all men of the Teutonic or of the Caucasian race; but 'all men,' 'all men are created equal,'" Douglass added, referencing racial categories that were conventional in ethnology. Just as Josiah Nott was committed to the "practical fact" of black inferiority, Douglass reminded the audience before him that they had the opportunity to make equal citizenship for all men "a practical fact for the whole country."[120]

Astute at knowing his audience, Douglass continuously framed his arguments in the Philadelphia speech, given amid a mostly white and all-male crowd, in terms of manhood rights and challenging claims, ethnological and otherwise, that black men were not truly men. He frequently pointed to African Americans' contributions to the Civil War, at a time when military service was a common cultural marker of manhood, to underscore black men as both manly in action and deserving in character. Through their vital labor and their honorable service in the war, they had more than *earned* full citizenship, he insisted, but more simply, he asked the right to "the witness-box, the jury-box, and the ballot-box" because "the negro is a man." "Manhood suffrage" was a "great political revolution" sweeping the United States, and no man could or should be reasonably excluded.[121]

His emphasis on "manhood rights" in his speech in Philadelphia is somewhat misleading, however, and largely strategic given his particular audience that day, for Frederick Douglass was also committed to the cause of women's suffrage. At a time when black male enfranchisement was under attack by many ethnological writers, Democrats, and, increasingly, a faction of women's suffragists—many of whom had been Douglass's former allies in the antislavery movement and who were now angry and indignant that it appeared likely that black men, their racial "inferiors," would get the vote before they did—Douglass remained a conciliatory voice insisting that the franchise should be limited by neither race *nor* gender.[122] For example, in Albany in 1866, Douglass gave a speech titled "Let No One Be Excluded from the Ballot Box" before a meeting called to lobby New York State to revise its constitution to specifically include African Americans and women in the franchise and attended by numerous male and female luminaries from the antislavery and women's suffrage movements. Therein, he appealed to women's moral natures, in common nineteenth-century understandings of gender, to "lift the negro with her" in the fight for the franchise. At the same time, he negated scientific rationales often aimed at black men and women alike that both innately lacked the intellectual capacity for informed participation in the political process.[123] He similarly drew on and, in equal measure, challenged gender norms for women in an 1868 speech at a women's rights convention in Boston. The female vote was "necessary to the welfare and purity of society. She comes to us not so much to assist herself as to assist the whole family," Douglass remarked, again invoking cultural tropes of women as more moral as well as maternal. But he also critiqued the practice of denying education—and by extension, full citizenship—based on quasi-scientific claims about nature designing the negro to be a laborer and no more and women to be domestic and no more. "I want to vote because I am a rational being," he declared, and the same was true of women.[124] As he did in so many other speeches and writings during Reconstruction, he demanded that the contemporary laws of the nation be brought into accordance with the spirit of its founding: "Let us make this government consistent with itself. Let no man be excluded from the ballot-box because of his color, and no woman because of her sex."[125]

Ultimately, though, while Douglass remained an enormously popular speaker and writer throughout the Reconstruction period and beyond, the kind of racist sentiments John Van Evrie espoused fueled the disenfranchisement of black men throughout the South just as they were beginning to enact their newly granted right to vote, and women would not achieve the vote nationwide until 1920. Although the Fifteenth Amendment was ratified in 1870 to prevent "the right of citizens of the United States to vote" from being "denied or abridged by the United States or by any State on account of race, color, or previous condition of servitude," even before Reconstruction formally ended in 1877, African Americans were facing intimidation at the polls. The Ku Klux Klan was founded in 1865 and within several years had spread its reach to most of the South, rallying embittered whites to resist Congress's Reconstruction policies through violence against African Americans as well as Republican politicians of both races. In addition to violence and intimidation, by the late 1870s, states throughout the South had also passed poll taxes and literacy tests to all but eliminate the black vote, exploiting a loophole in the Fifteenth Amendment that allowed states to pass voter restrictions as long as they were applied to both races—in theory if not in effect. It was not until the Voting Rights Act of 1965 that the majority of Southern blacks were actually *able* to vote in accordance with the Fifteenth Amendment passed nearly a century prior.[126]

John Van Evrie's beliefs about the folly of black male enfranchisement were shared by his scientific contemporary and, by the late 1860s, fellow New Yorker Josiah Nott. Although they cited each other's work liberally and inhabited the same city for several years, it is not clear if the two men knew each other personally, but it would hardly be surprising if, at the very least, Van Evrie also attended meetings of the New York Ethnological Society, with which Nott remained halfheartedly involved throughout his time there. While he was no longer publishing his theories on race, Nott thrived in New York, as did the private practice he established there. What happened to Van Evrie in this period is less known. *White Supremacy and Negro Subordination* (1868) appears to have been *his* last ethnological publication, or publication of any kind for that matter, and little biographical information about his later years exists to explain why he stopped his work on race or what filled his time prior to his death in Brooklyn in 1896, the cause of which was listed as "senility" in cemetery records.

What we *do* know hints at some rather ironic notes about Van Evrie's life, death, and legacy. Most notably, Van Evrie, who in life railed against both black and women's suffrage, is buried in death in the same Rochester, New York, cemetery as Frederick Douglass and Susan B. Anthony, two of the most prominent suffrage activists of the nineteenth century—and also a black man and white woman whose careers eclipsed Van Evrie's own during his lifetime and today. Van Evrie is buried in a section of Mount Hope Cemetery reserved for Rochester's most important citizens, including Anson Colman, one of the first physicians in Rochester and a founder of the Monroe County (New York) Medical Society, as well as the city's founder, Nathaniel Rochester, whose great-granddaughter, Sophia Elizabeth Colman, was married to Van Evrie in 1842 in a local church.[127] The young couple had one daughter, Catherine Rochester Van Evrie, the following year, but in 1845, Van Evrie's wife died in childbirth with their second child, who also perished.[128] It is unclear how soon after Van Evrie relocated to New York City, or if his daughter accompanied him, but the fact that he was buried next to his wife, who preceded him in death more than fifty years prior, suggests that he probably never remarried. His prestigious placement within Mount Hope also suggests that, in life, Van Evrie was at least acquainted with important and influential people. But in perhaps the most bitter irony for Van Evrie, his grave site now lies in a state of disrepair, the headstone broken and, like Van Evrie himself, largely forgotten to the ages, while elsewhere in Mount Hope, the final resting places for Anthony and Douglass are immaculately maintained and often adorned with fresh flowers, still visited frequently by those wishing to pay tribute to two figures central to the ultimately successful battle to achieve the vote for women and African Americans (see Figures 2.8 and 2.9). One can imagine Van Evrie rolling in his grave.

Nott, in contrast, did not fade from public view in the final years of his life in New York, though he did refocus his attention away from race to two other areas of medicine in particular. Yellow fever had long been

Figure 2.8. Van Evrie's fractured legacy. *After dying of "senility" in 1896, Van Evrie was buried in a prestigious section of Mount Hope Cemetery, in close proximity to some of Rochester, New York's most prominent citizens, including the city's founder, whose great-granddaughter Van Evrie married as a young man. In life, he was at least acquainted with important people, and his work on race was widely cited by his contemporaries. His inflammatory racial theories, the only reason he is known to historians today, are not mentioned in his simple headstone, which memorializes him only as an MD. In fact, his name has largely been forgotten (even the Friends of Mount Hope Cemetery organization does not list him among the* notable *people buried there) or perhaps dis-remembered as part of a history of scientific racism many would prefer to bury with its most outspoken proponents, even if the legacy of their ideas continues to haunt American race relations today (photos courtesy of Rebecca Scales).*

Figure 2.9. Ethnology's foes remembered. *In contrast to Van Evrie's broken and forgotten grave, the gravesites of Susan B. Anthony and Frederick Douglass remain two of the most visited memorials in Mount Hope Cemetery. Leaders in the fight for women's and black suffrage—both of which Van Evrie adamantly opposed—their fame eclipsed Van Evrie's in life and in death. Douglass, who was also an astute critic of scientific racism, died a year earlier than Van Evrie, but unlike his ideological foe, Douglass did not fade into obscurity before his death, remaining professionally and politically active until the end, and succumbed to a fatal heart attack shortly after speaking at a women's suffrage meeting. Likewise, though Anthony formally retired her leadership of the National Women's Suffrage Association in 1900, she remained widely respected before and after her death in 1906 (Anthony photo courtesy of Rebecca Scales; Douglass photo by Tom Flynn/The Freethought Trail, a project of the Council for Secular Humanism).*

a professional interest to him and, after his family was decimated by it in 1853, a personal interest as well; he returned to the subject often in the last years of his life. His new attention to—and rapidly growing prestige in—gynecology seems more surprising, though, given his prior focus on men in his ethnological publications on race. In short order, he was president of the New York Obstetrical Society and counted celebrated gynecologist J. Marion Sims among his closest friends. Like Nott, Sims was born in South Carolina, received his medical degree in Philadelphia, and later moved to Alabama, where in the 1840s he built his reputation by developing surgical procedures for the treatment of vesicovaginal fistulas.[129] Perhaps his most well-known legacy to historians, he did so by purchasing and experimenting on several slave women with the condition, all without the benefit of newly available anesthesia—possibly working from the common racist trope that black people felt less pain than whites, though surely the reactions of the women under his "care" would have suggested otherwise. Also like Nott, his connections to slavery did little to dampen his career prospects in New York, where he moved in 1853 and founded the first women's hospital in the country, continued to perfect his techniques on indigent women, and eventually went on to serve as the president of the American Medical Association. A statue erected in his honor stands in Central Park to this day, hailing his "services in the cause of science and *mankind*" and making no mention of his experimentation on slave and poor women. And like Civil War surgeon and neurologist Roberts Bartholow, his experimentation on human subjects, shocking to the modern reader, earned him some sharp critics during his lifetime as well, long before the concept of "informed consent" was codified in medical practice, but such criticism ultimately did not impede his career or tarnish his legacy upon his death in 1883.[130]

The friendship between Nott and Sims was ultimately short-lived, as Nott fell ill with symptoms of tuberculosis in 1872 and by the end of the year returned to Mobile to prepare for what he saw as the inevitable. Following his death on March 31, 1873—his sixty-ninth birthday—Mobile shut down to honor one of its most esteemed citizens, who was buried next to his children.[131] Though Republicans were already diminishing their efforts to protect the newly freed black population of the South by Nott's death in 1873, had Nott lived another four years, he would have witnessed the withdrawal of all federal troops from former Confederate territories, marking the formal end of Reconstruction. He also would have witnessed Southerners quickly instituting the kind of de jure and de facto white supremacy that Nott himself had insisted was the natural—and necessary—racial order when he first denounced the Freedmen's Bureau in 1865.[132]

The changes brought by the Civil War, Reconstruction, and Redemption—the period following Reconstruction's demise when the region's racial hierarchy was "redeemed" by white Democrats—were welcome or devastating, depending on who was experiencing or interpreting them. For white Southerners, for instance, the Civil War and Reconstruction represented an enormous disruption in their "way of life" and the hierarchical society they had created; Redemption for them was an attempt to return their world to what it had once been. For Southern blacks, on the other hand, Reconstruction represented a moment of hope and possibility, followed by the brutal backlash and reassertion of "white supremacy" during Redemption. The North emerged from the Civil War victorious but in the process had lost many of its men, with many others returning to their homes maimed and psychologically scarred. At the same time, some Northern ethnological writers like Van Evrie looked to the South during Reconstruction as a racial laboratory that would prove that a superior and inferior race could not coexist outside the confines of slavery and that African Americans were not biologically suited to enact the "manhood rights" of full citizenship.[133]

The Descent of Man: Continuity and Change at the Dawn of a New Century

In the late nineteenth century, an even more sinister trope of African American masculinity—that of the hypersexual "black beast"—emerged in both scientific and popular racial thought that further served to justify black men's exclusion from the rights of citizenship. In lending legitimacy to this new racist trope, racial scientists also constructed the fate of the United States as a contest of manhood between brute, unrestrained, and sexually deviant black men who threatened it through violence and tainted America's "stock" with their inferior blood and the civilized, honorable, and chivalrous white men who endeavored to save it.

The emergence of the "black beast" and racial scientists' growing interest in sexuality more generally in the late nineteenth century was precipitated by several factors in American politics, culture, and science. In the last decades of the nineteenth century, white Southerners struggled, ultimately successfully, to "redeem" themselves from Reconstruction's experiment in racial democracy. A period of often violent backlash against the political and economic advancements African Americans made during Reconstruction, Redemption saw the recently achieved black vote dismantled and large portions of the race returned to positions of social subordination that differed from slavery in name alone. In addition, the publication of English naturalist Charles Darwin's enormously influential 1859 text detailing his theory of evolution, *On the Origin of Species by Means of Natural Selection, or the Preservation of Favoured Races in the Struggle for Life,* as well as *The Descent of Man, and Selection in Relation to Sex* in 1871, proved to have a monumental impact on racial thought in America in the late nineteenth and early twentieth centuries as well. Racial scientists' incorporation of evolutionary theory into their understandings of race prompted their growing emphasis on sex and sexuality as well as their new representation of black masculinity in two primary regards: (1) many racial scientists argued that black men were instinctively driven to fight their inferior race's inevitable extinction by becoming sexually aggressive, particularly toward white women, and (2) the degree of difference between males and females within a species or race corresponded to its degree of evolutionary progress. Asexual organisms represented the lowest rung of the evolutionary ladder, while the top was represented by species in which the males and females were highly distinct and specialized in both their physiology and their roles.

This concept was repeatedly applied to human beings and to races specifically. In that context, the white middle-class *gender* ideal of separate spheres for men and women—who were vastly different in body and mind and designed by nature for different roles in society, many physicians and politicians alike argued—was imbued with racial meaning. Specifically, it served as a marker of how evolved white civilization had become and its superiority over the "lower races" in which muscled and masculine women worked alongside men in fields and factory floors and the men were not quite men.[134] For example, in his 1890 *Races and Peoples,* Philadelphia ethnology professor Daniel Brinton insisted that the differences between the races were more pronounced than the differences between the sexes, particularly among the "lower races"; race trumped sex as a meaningful and visible measure of human variation. "It is a common statement that to our eyes all Chinamen look alike, or that one cannot distinguish an Indian 'buck' from a 'squaw.' Yet you recognize very well the one as a Chinaman, and the other as an Indian. The traits of the race thus overslaugh the variable characters of the family, the sex or the individual, and maintain themselves uniform and unalterable in the pure blood of the stock through all experience," he explained.[135] He believed that ethnologists put too much stock in craniology and craniometry, focusing on skulls "to the exclusion of other elements"—physical features that for Brinton and many other racial theorists during this period included sexual characteristics.[136] Instead, he drew the reader's attention to "the relative position of the sexes in different peoples." Like other turn-of-the-century writers on race, he insisted that "the expression of sex" was least pronounced in the lower races, indicating their position at the bottom of the evolutionary ladder, while "the muscular force of the sexes presents the greatest contrast in nations of the highest culture."[137]

Evolutionary theory and its application to human beings specifically, commonly termed social Darwinism—phrases I use more or less interchangeably here—met some initial resistance in the United States when Darwin's first text was published. As with the ethnological theory of polygenesis earlier in the nineteenth century, some Americans resisted the theory of evolution on religious grounds, while some racial scientists themselves resisted at first because of their persistent belief in the separate origins of the races and their *lack* of change over time. Ethnological writers soon found, however, that social Darwinism was just as compatible with the "practical fact of negro inferiority," in the immortal words of Josiah Nott. As discussions of evolution and its reverse processes, degeneration and extinction, in conjunction with race rapidly increased, the "origins debate" between monogenesists and polygenesists all but disappeared by the early twentieth century (see Figure 2.10).[138]

Historian Richard Hofstadter famously characterized the United States at the turn of the century as "*the* Darwinian country." As he noted, "England gave Darwin to the world, but the United States gave to Darwinism an unusually quick and sympathetic reception. . . . American scientists were prompt not only to accept the principle of natural selection but also to make important contributions to evolutionary science." This was especially true of America's scientists of race. At the same time, the influence of social Darwinism reached well beyond the scientific establishment, and bourgeois Americans eagerly consumed and discussed writings that framed social inequalities in evolutionary terms.[139] By the end of the

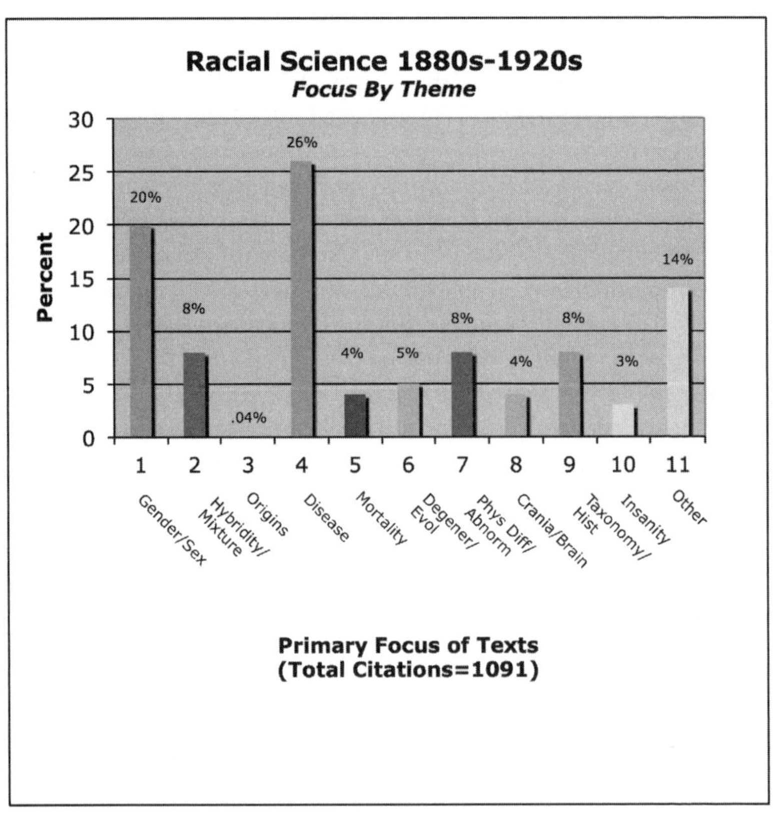

Figure 2.10. Racial science in the age of evolution and extinction. *At the turn of the century, racial scientists were far more interested in where the races were going than from where they came; consequently, texts focused on the "origins debate" dropped from 25 percent in the antebellum period to just .04 percent of ethnological work published between 1880 and 1920. Instead, scientists focused on evolution as it related to race in general or on specific indicators of degeneracy (evolution's reverse process usually attributed to nonwhite races) and the related black extinction thesis most famously articulated by statistician Frederick Hoffman: disease, mortality, and insanity. At the same time, texts focused on gender and sex rose from 5 percent to 20 percent between the two periods as ethnological writers began to frame the future of American race relations as a contest of both manhood and numbers and thus looked increasingly at sexual instincts and sex characteristics as measures of racial difference (graph by author; data based on the* Index-Catalogue *of the National Library of Medicine).*

nineteenth century, evolutionary theory was nearly ubiquitous in American cultural, political, and scientific discourse.

A particular strand within evolutionary theory—Frederick Hoffman's thesis about the inevitability of Negro extinction, itself a take on Darwin's "survival of the fittest" concept—framed many of the scientific discussions of race at the turn of the century and beyond. Though he had no formal training in science or medicine, Hoffman was an enormously influential figure in ethnological thought as well as public health during this period and was a member of the Committee on Anthropology and the chairman of the Sub-Committee on Race in Relation to Disease of the National Research Council, among other scientific affiliations. Born in Germany in 1865, Hoffman emigrated to the United States when he was nineteen to work as a shopkeeper's assistant in Cleveland. Dissatisfied with manual labor, he left the position after a short time to live a transitory life, moving from place to place, mostly in the South, before he got into the insurance business, eventually moving to Norfolk to work for the Life Insurance Company of Virginia, where he became interested in mortality rates among African Americans. The article he published in the journal *Arena* in 1892 based on his research into the subject led to a job two years later in the statistical department of the Newark, New Jersey–based Prudential Life Insurance Company, where he worked for the next four decades.[140] He published his most famous work, *Race Traits and Tendencies of the American Negro,* in 1896, the same year that the Supreme Court's *Plessy v. Ferguson* decision codified Jim Crow segregation into law by declaring "separate but equal" accommodations for the black and white races to be constitutional. Like so many of the racial theorists who came before him, Hoffman emphasized his unbiased, *scientific* approach, maintaining that he dealt in statistics—and, echoing the scientists involved in the Civil War anthropometric studies three decades prior, which he himself cited in his work, Hoffman insisted that numbers do not lie.[141]

In the enormously influential book, Hoffman pointed to disease rates among black soldiers during the Civil War—largely without reference to the poor conditions and health care disparities they faced—alongside more recent data to argue that the black race was rapidly deteriorating outside of slavery and ultimately headed toward extinction.[142] As a statistician, however, Hoffman was also deeply interested in the anthropometric data generated during the Civil War. Citing both agencies' reports, he hailed the unique opportunity the military offered for understanding human anatomy and variation. In a detailed 1917 paper on military anthropometry, he described the strengths and failings of the U.S. Army's continued work in the field in the years following the Civil War. Ironically, he charged military scientists with not paying nearly enough systematic attention to race, and, after all, "*the racial factor* . . . is invariably of paramount importance."[143]

Despite Hoffman's concerns, with the publication of the SC's and PMGB's respective reports, postbellum ethnological theorists were armed with vast anthropometric and medical data on thousands of living men to support ongoing arguments about pervasive racial difference. Just as a long history of racial science influenced *what* the army measured, so too did the specific measurements have a long life in American ethnology after the Civil War. Antebellum experts on race had often claimed that black men had longer forearms and more protruding jaws than white men, which supposedly signaled the former group's closer proximity to animals and by extension, for some proslavery scientists, an indicator that they were designed by nature to be "beasts of burden." In the late nineteenth century, these measures coupled with the physicians' characterization of African American men as physically strong if intellectually lacking fit seamlessly into the new trope of black masculinity, the "hypersexual beast" that dominated at the turn of the century.

Hoffman, himself a father of seven, pointed to statistics on birth rates, mortality, insanity, and disease among African Americans to make the evolutionary claim that nature's law of "survival of the fittest" was solving the country's "negro problem," as it was often termed by scientists and politicians alike at the turn of the century—many of whom were influenced by Hoffman's work. For Hoffman, high disease and mortality rates among African Americans were not indicative of the poor conditions in which they lived and worked in a racist society; it indicated their inherent inferiority. He maintained that the black race—which he deemed a "lower race" from the start, and now a profoundly diseased one as well—was degenerating since Emancipation and would ultimately go extinct but posed a grave social and health threat to the nation in the meantime. "Race deterioration once in progress is very difficult to check," he intoned, and throughout history deteriorating races "have invariably become useless if not *dangerous* factors in the social as well as political economy of nations."[144] It was this claim about danger in particular that resonated for many ethnological writers, who provided their own assessments of the extent of the "Negro problem" and what should be done to address it.[145]

For many of these writers, the danger posed by black degeneracy was a sexual one. Not only were African American men reverting to their innate savagery—exemplified by sexual excess—outside the confines of slavery, but they were compulsively driven to sexually pursue women, particularly white women, at all costs in a vain attempt to stave off their own extinction, a number of white racial theorists argued.

Retired army surgeon and naturalist R. W. Shufeldt, who cited Hoffman's work often, was one such theorist. Shufeldt, a native New Yorker, pointed to the American South, where the "dangers" posed by the black race were "vastly intensified and increased," to illustrate the racial threat facing the nation overall if the problem was not addressed. "The bulk of the white population throughout those states are slowly awakening to the horrors of the situation that now confronts them," he began ominously. "Negroes are massed everywhere, and being no longer under control, menace all that is decent in a developing nation—menace morals; menace progress and development; menace legal and political stability; and threaten, in no uncertain manner, the very existence and purity of the American race and its career." Like the scientists involved in the Civil War anthropometric studies, Shufeldt used "American race" to refer to whites specifically, while blacks constituted a foreign element that imperiled the nation's security and progress.

At the same time, as demonstrated throughout his lengthy 1907 invective, *The Negro: A Menace to American Civilization,* "purity" held multiple meanings for Shufeldt. He warned that the black race was a moral "menace" that would negatively influence civilized whites in close proximity. He also warned that black men were a threat to white women's *sexual* purity through their supposedly innate propensity toward rape. And relatedly, they threatened the racial purity of the nation's white stock through miscegenation, often through force. The situation was grave, and it called for a scientific expert like himself. He added, "Ask any intelligent Southern man or woman and he or she will tell you something of the state of affairs there existing, even if such people do not grasp the danger in its fulness [*sic*] as the far-seeing and philosophic biologist and anthropologist does."[146]

Indeed, R. W. Shufeldt had the scientific and medical training that Hoffman himself did not, though Shufeldt also far exceeded the comparatively measured statistician in unveiled racial vitriol and, on a personal level, seemed equally adept at courting important friends *and* enemies. Born in 1850 in New York, he was the son of a Navy admiral (also named Robert

Wilson) with whom he often traveled, obtaining his childhood education in the United States and Cuba. The adolescent Shufeldt served in the last year of the Civil War as a clerk and signal officer aboard a gunboat under his father's command—which would not be the last experience he had with military service, though unlike his father, his involvement was primarily through medical and scientific work. After receiving his medical degree from what is now George Washington University in Washington, D.C., in 1876, he was appointed to the Army Medical Department and first was stationed as an assistant surgeon at Fort McHenry in Baltimore and then saw service in the West during the military's skirmishes with Native Americans, again as a base surgeon.[147] Shufeldt also developed a lifelong passion for ornithology and paleontology, collecting bird and animal skeletons as well as human skulls during his travels with the army—not unlike the friends of Samuel Morton decades earlier who also dug up Native American burial mounds to procure specimens during their military service in the Western frontier—which he recounted in a biographical essay, "Personal Adventures of a Human Skull Collector."[148] Perhaps because of his penchant for collecting such specimens, Shufeldt was hired as a curator in the Army Medical Museum in 1882.

He retired from his army position in 1891 because of a heart condition, but the avid ornithologist was then made an honorary curator of comparative morphology at the Smithsonian Institute, a position from which he was dismissed just six years later amid personal scandal involving his second wife, Florence Audubon, the granddaughter of famous naturalist and painter John James Audubon. Historian of science Mark V. Barrow describes Shufeldt as a man whose misogyny rivaled his racism but also a widely respected ornithologist when the scandal broke.[149] Shufeldt married Florence, with whom he had corresponded about an article he published related to her grandfather's work, just a month after they met—and three years after his first wife committed suicide in an insane asylum in 1892. The marriage fell apart almost instantly. When she accused him of adultery with their maid, a Norwegian immigrant he later married, and filed for divorce, Shufeldt responded by writing an article on female sexual dysfunction that was obviously about her. As Barrow explains, though Shufeldt did not name her specifically, the article's subject was obvious to those who knew the couple, and it amounted to a vicious character assassination. He gave out hundreds of free copies of the article stamped with his name and Smithsonian affiliation and then published it in the

Medico-Legal Society's journal. The leadership of the Smithsonian was outraged and dismissed him as an affiliate, and he narrowly avoided being thrown out of the American Ornithologists' Union—of which he was a cofounder—as well.[150] His career ultimately did not suffer appreciably, however. He continued to publish widely on a range of scientific subjects, including ornithology and race, his two biggest (and strangely divergent) passions, and when World War I broke out, his military service was reactivated, and he returned to his post at the Army Medical Museum, from which he retired for good in 1919 at the age of sixty-nine.[151]

Given the personal scandal in which he was embroiled, Shufeldt's condemnation of the hypersexual natures and immoral influence of African Americans as they degenerated toward extinction seems rather ironic. The irony seemed wholly lost on Shufeldt, however. The march of American civilization and the fate of the white race depended on the quality of its environment and who shared it. Anticipating the rather obvious question as to why a "clean and cultured race" would be so easily corrupted, he remarked that while blacks were inferior in every way, their negative influence was disproportionately powerful, and "with respect to the superior people, their morals and ethics are bound to suffer." Black youth in particular "set a fiendish example to the coming generation of the youth of our own race in this country," who on account of their age were especially susceptible to negative influences, Shufeldt maintained. Again, Shufeldt was most concerned about sex, as "the young blacks are distinctly lewd and sensuous in their inclinations, and at early stages develop a desire to carnally possess white maidens and women, traits in reality they never outgrow, and which are a constant menace to the gentler sex of our people everywhere. This influence is extremely bad and the knowledge of it reacts disastrously upon the minds of our daughters and wives throughout the country."[152]

Shufeldt had no patience for charitable attempts to "improve" the Negro. It was idealistic "nonsense" to think the "better race" could redeem or elevate the lower; nature simply did not work that way. Consequently, "in this country, the highest civilization, the cleanest and soundest white people flourish where the negro is not, or exists in comparatively few numbers," he argued.[153] If whites were to intervene at all, it should not be to save a doomed and dangerous race but rather to help nature finish the job it had started. The white race had evolved to the point that an "intelligent man" could control his environment and that is precisely what he called on white men of conscience to do in regards to the "lower race" in their midst. As the superior race, "we have it in our power to render the negro race extinct in the United States in very short order," Shufeldt declared, insisting that they would be acting in accordance with natural law and that "mankind in the long run would be enormously benefited."[154]

Shufeldt found a kindred spirit in William Lee Howard, a Baltimore physician and author of many books aimed toward a general audience on topics related to sex, including sexual crimes, perversion, marital hygiene, and sex education for boys and girls. In his article "The Negro as a Distinct Ethnic Factor in Civilization," published in the journal *Medicine* in 1903, Howard similarly argued that "the Caucasian as a race is moral," while "the African as a distinct race is not immoral, he is unmoral and no amount of education or training is going to change an non-existing element." His oversexed and unmoral nature was exactly that—his *nature,* rooted in his biology and thus unchanging and unchangeable. In language strikingly similar to Van Evrie who facetiously remarked nearly forty years earlier that the "Yankee school marm" and Freedman's Bureau would have to give the Negro an "equal beard"—that is, change his biology—to give him an equal education and equal voting, Howard scoffed, "When education and religious teaching change the biologic basis of his color it will also be able to change the anatomical and physiological reason for his sexuality and bestiality. . . . When education will reduce the large size of the negro's penis as well as bring about the sensitiveness of the terminal fibers which exist in the Caucasian, then will it also be able to prevent the African's birthright to sexual madness and excess." But the black race *was* changing in Howard's estimation, just not for the better. Rather than approaching the civilized status of the white race, "the negro of to-day, untrammeled and free from control, is rapidly showing atavistic tendencies. He is returning to a state of savagery, and in his frequent attacks of sexual madness, his religious emotionalism, superstition and indolence, is himself again—a savage."[155] As the white race evolved and the black race degenerated, the gap between the races grew wider with each passing year. For Howard, the black men's "ancestral" sexual aggression was reawakened and compounded by them being thrust into a false position of social equality with whites. Black individuals were biologically ill suited and mentally unprepared for the kind of self-reliance they needed to survive outside of slavery without a benevolent white master to take care of their needs *and* keep their behavior in check, let alone the privilege of full citizenship. As a race, they had not evolved to the stage of civilization necessary to coexist on an equal plane with those who had, and misguided attempts to force them to do so was only exacerbating the problem. For the good of society, "the anatomical and physiological conditions of the African must be understood, his place in the anthropological *scale* realized, and his biologic basis accepted as being unchangeable by man"—and that was where medical scientists' expertise was essential, another point on which Howard and Shufeldt concurred.[156]

Also like Shufeldt, Howard did in fact have a respectable medical background and access to a range of venues for espousing his ideas about race, including popular periodicals and medical journals as well as his numerous books on sex. Howard was born in 1860 in Hartford, Connecticut, apparently to a family of some means, and educated as a child in Europe, from tutors in France and England as well as the United States, before attending college at New York's Columbia University and England's Oxford University. Sharing Frederick Hoffman's wanderlust as a young man, though not his initial poverty, Howard left college several times to travel to Europe and Africa and finally obtained his medical degree from the University of Vermont in 1890, settling in Baltimore the following year, where he practiced for most of his career and specialized in nervous diseases.[157] The coeditor of the *American Medical Biographies,* physician Howard A. Kelly, broke from his typical dry and perfunctory style when, amid other insults, he described William Lee Howard as "a writer of books on sex subjects, and a pamphleteer" who "was held in more esteem by the laity than by the profession."[158] Kelly, like many of his contemporaries, seemed to be put off by Howard's work on sex—and both Howard and Shufeldt frequently railed against the prudery of the medico-scientific professions, as the next chapter will demonstrate—but not, apparently, his claims about race. Moreover, Kelly's complaints about Howard also underscore another important point about scientific work on race—and sex—during this period. An increasing number of physicians and scientists sought a wider audience for their ideas about society and turned to newspapers, popular magazines, and books to reach it, but in doing so, they also provoked the ire of some of the more traditional and proprietary elements within the medical profession. So Howard perhaps had both the "laity" and the "profession" in mind when he insisted that the sentimental idealists who tried to reform the black race did not understand its biology or comprehend the futility of their efforts, while medical scientists such as himself had a duty to bring their expertise to bear on pressing social issues, even if those issues concerned sex. In this sense, Howard constructed the physician as socially relevant and uniquely qualified to "render justice to civilization." But in order to do so, the medical scientist had to overcome prudery and "silence regarding sexual matters" within his own ranks.[159]

Indeed, the importance of sexual differentiation in characterizing racial difference, as well as the often-raised specter of miscegenation prompted by an allegedly oversexed black race, translated to an emergent fascination with bodily sex characteristics, including genitalia, among ethnological writers. "The African is also unlike the Caucasian in secondary sexual characters, and this being so he can never be absolutely alike in the highest psychical processes," Howard wrote.[160] The Negro suffered from an "overdevelopment of sexual energy in certain portions of the brain, which the normal power of inhibition—that which the white man possesses—cannot control," Howard maintained. He insisted that the bodies and brains of black men were inextricably linked; both their cerebral cortexes (the site of "sexual instinct" in Howard's assessment) and their penises were enlarged and easily excitable and thus contributed to their innate sexual degeneracy.

Though not the subject of his diatribe, black women did not escape this particular characterization either. Howard insisted that the average penis size of black men was "six inches" while flaccid—larger than all other races in his estimation—and according to natural law, the "dimensions of the female organs" within the race must be proportionately large, the implication being that within his biology is destiny ethos, black women were similarly driven by savage lust.[161] It was an important comparison in the context of evolutionary thought. Howard simultaneously emphasized the contrast between black and white men *and* the similarity between black men and black women—both of whom were defined by sexual excess in body and behavior, which in turn indicated their race's low rung on the evolutionary ladder.

Shufeldt quoted Howard's discussion at length, but he also excerpted a far less frequently heard voice in racial science, that of a white *female* physician, Ellen Barret Ligon, of Mobile, Alabama. Ligon was born on a Mississippi plantation in 1864 and received her medical degree from the American College of Osteopathy in Missouri in 1900 followed by several years practicing in New York City. She settled in Mobile in 1905, where she continued to practice medicine and was active in a variety of women's clubs. Though she published a number of articles in professional journals, she was most known for her piece "The White Woman and the Negro," published in the popular magazine *Good Housekeeping* in 1903, the same year as Howard's similarly themed text.[162] Therein, she made many of the same arguments as Shufeldt and Howard about black men's degeneracy outside the confines of slavery and their relentless—and relentlessly violent—pursuit of white women. Ligon also shared Shufeldt's flair for the dramatic, invoking images of battered baby girls found in pools of blood and "pure" white women with "bruised breasts" and "crushed skulls," all at the hands of "savage negroes." Her call to action was likewise gendered, emphasizing white women's vulnerability and appealing to white men's chivalry, their civilized *manhood,* to come to their rescue. And indeed, "the white woman must be saved!" For Ligon, "the vital point in the race question today is the safety of the Southern white woman, and all the helplessness of womanhood appeals to the manhood of the world to protect her with every possible safeguard."[163] Clearly a sucker for helpless womanhood, Shufeldt responded in kind, answering her closing question of "What are you going to do about it, Americans?" with,

> Yes, Americans, what *are* you going to do about it? . . . Are we of the North to remain deaf to the earnest cry for our aid in this matter from the mothers and daughters of *our own race* in the South? What in the name of all that is manly and brave and honorable has come over this American nation? Are we becoming so foully adulterated by the black scum of the earth that we can no longer be moved through the spirit of chivalry, through the dictates of honor, through the highest instincts of manliness, to listen to and to heed the distress and danger of thousands of *our own women?*[164]

For Shufeldt and many other scientific—and popular—writers on race during this period, white supremacy depended on white men maintaining their dominion over white women. They likewise believed that giving black men political equality had made them hungry for "social equality"—that is, equal access to white women. Consequently, racial scientists equated the political threat—and competition—black men represented to white supremacy through their vote with sexual threats and competition over white women. As Ligon warned, "Miscegenation is the inevitable outcome of social equality."[165]

Medico-scientific writing on miscegenation and the mulatto between the antebellum period and the turn of the century demonstrated both continuity and change over time. When ethnological writers in the antebellum period characterized mixed race people as less vigorous, possibly sterile, and thus destined to die out, they dismissed the mulatto himself as inconsequential, in part to write off the circumstances of most race mixture under slavery—the sexual exploitation of female slaves by their white masters—as inconsequential by extension. Racial theorists at the turn of the century in contrast saw miscegenation as very consequential indeed and, as Ligon characterized it, "a crime against two races."[166] For one, though in reality it was black women who remained sexually vulnerable to white men throughout the country far more than vice versa, it was the boogeyman of the savage black despoiler of white virtue that framed most white scientific and popular thought on race and race mixture during the Redemption period.[167] Just as the continued rape of black women by white men served as a means of asserting white supremacy, so too did white supremacy rely on white men's exclusive sexual access to white women, as Shufeldt's work underscored.

Miscegenation also resonated with turn-of-the-century discussions of degeneracy and racial extinction more generally. The mulatto was not an improvement on either race, Ligon wrote, adding, "Nature stamps him for extermination by making him physiologically inferior to both races, with far less endurance and vitality, according to statistics more susceptible to disease, tuberculosis, and insanity."[168] Ligon's language regarding the mulatto is remarkably similar to the assessments in the Civil War anthropometric studies. Yet such characterizations took on new meaning in the context of turn-of-the-century evolutionary theory and Hoffman's extinction thesis. Hoffman himself characterized the black race as a whole as deteriorating toward inevitable extinction, as evidenced by increasing rates of disease, insanity, and early death within the race—ideas that permeated racial science during this period (see Figure 2.10). While Ligon agreed that the race as a whole was degenerating in this fashion outside of slavery, she believed the path to the black race's extinction was through miscegenation, with the mulatto exhibiting the most extreme degree of degenerate characteristics.

G. Frank Lydston, a Chicago-based surgeon and prolific writer on race, social hygiene, sex, and eugenics, shared Ligon's belief that miscegenation with a biologically superior and dominant race would bring about the extinction of the black race, as well as her concern for the danger they represented to whites in the interim. Lydston was born in California in 1858 and received his degree from the medical school at Bellevue Hospital in New York City. After graduating in 1879, he served as a resident at the New York Charity Hospital, inhabiting the city's medical world at the same time as J. Marion Sims and John Van Evrie, Josiah Nott having passed away shortly before Lydston arrived there. In 1882, he moved to Chicago to accept a position at the College of Physicians and Surgeons as a lecturer on Genitourinary Diseases and was promoted to professor several years later while also practicing as a surgeon at Cook County Hospital beginning in 1893; he was later employed as a professor of criminal anthropology at Kent College of Law as well.[169]

Over the course of his career, Lydston published a rather staggering number of articles and textbooks on a range of medico-scientific topics that included venereal disease, sexual dysfunction and impotence, surgical techniques, miscegenation, criminology, eugenics, social hygiene, racial degeneration, and separate guides for boys and girls on sexual matters—as well as several works of fiction and a memoir. But among his biggest claims to fame during his lifetime and among historians and physicians today involved an experiment in sex gland implantation he conducted in 1914. Interested in sexual "rejuvenation" for men and potential treatments for impotence, Lydston was the first physician to experiment in grafting a whole testicle from one body into another, which he achieved by implanting a testicle procured from a young accident victim from the city morgue into his own scrotum—a procedure about which he wrote several times in medical journal articles and textbooks and which, not surprisingly, garnered him a good deal of publicity.[170]

Lydston offered his own assessment of the evolutionary dangers of race mixture in a 1914 medical journal article titled "The Negro—America's Special Problem in Eugenics." As the title suggests, miscegenation had deep resonance for the increasingly popular field of eugenics as well. Evolutionary theory and eugenics overlapped in many ways, and the issue of race mixture was central in both. In fact, the very term "eugenics" was created by Charles Darwin's cousin, British scientist and geneticist Francis Galton.[171] Eugenics relied on many of the central tenets of ethnology regarding biological difference and hierarchy between the races and applied them to the manipulation of human genetics toward the improvement of society. Eugenics was a *popular* science, impacting average Americans' everyday lives in a myriad of ways throughout the early twentieth century. Positive eugenics, which encouraged the "fit" to reproduce more, manifested in a range of wide-reaching strategies from marriage guides for picking a mate of the best "stock" to "Better Baby" and "Fitter Family" contests at county fairs, while negative eugenics sought to curtail the reproduction of the "unfit," which included poor whites, criminals, racial minorities, and the disabled, often through involuntary sterilization.

By the time Lydston published his article in 1914, eugenics had become extremely popular in the United States, pervading popular thought on race as well as public policy, and America was widely recognized as a leader in the eugenics movement.[172] Lydston complained, however, that "the most important problem in eugenics in America"—which he identified as "the relation of the blacks and the whites"—was receiving woefully little attention. He believed that the black race *could* potentially progress slowly over time, given the right circumstances, but constituted a dangerous biological and social element in America in the present. Thus preventing miscegenation benefits both races, as well as civilization overall, he argued, since "white and negro bloods" were "distinctly alien" and experience had proven mixture to be detrimental to both.[173]

Lydston affirmed the usual scientific assertion that mixed race people lacked the "physical stamina and moral fiber of either parent stock" but refuted the equally common claim that they were sterile by the third or fourth generation of race crossing.[174] Any decrease in fertility was minimal, he maintained, and could be attributed simply to the more fertile black race mixing with a less fertile white race. Because mulattoes would not become sterile, as many had previously believed, racial amalgamation would continue unchecked, and he pointed to the already large number of individuals with black blood he claimed were passing undetected as white to underscore the grave danger such amalgamation represented to American civilization. Immediate intervention was necessary, he warned, pointing to how few full-blooded blacks he saw in Chicago and, more ominously, how many could almost pass as white. (In a story about a public lecture Lydston gave on the same subject in Chicago the previous year, the founder, publisher, and editor of the black newspaper *Broad Axe,* Julius F. Taylor, coyly recommended that Lydston "should at all times firmly impress this one simple fact on the minds of all White gentleman, namely, that they are themselves solely responsible for all the mixing process which has for so many years been going on in this country between White gentlemen and Colored women."[175]) Because the white race was biologically superior, racial admixture would eventually subsume the inferior black race, Lydston maintained, but permanently pollute and degenerate the nation's white bloodline in the process—and eugenics, at its core, was most concerned with protecting and improving whiteness.[176]

As Lydston's remarks suggest, the boundaries of whiteness and how to police them represented a set of larger anxieties inextricably tied to racial scientists' construction of the bestial black man and their concerns about miscegenation. But their anxieties about whiteness in general, and white manhood in particular, framed their discussions of other racial issues as well, including immigration and U.S. imperialism. As historian Matthew Frye Jacobson explains, "Whereas the salient feature of whiteness before the 1840s had been its powerful political and cultural contrast to nonwhiteness, now its internal divisions, too, took on a new and pressing significance." He attributes this shift in the second half of the nineteenth century and first decades of the twentieth to the tremendous increase in immigration on account of the labor demands associated with industrialization in the United States and concurrent instability in Europe. This increase was met with nativist fears of foreign-born workers as a political and economic threat and scientists who in turn questioned their fitness for citizenship, just as they had with the freedmen after Emancipation.[177] The immigration increase starting in the mid-nineteenth century was truly staggering, with Irish and German immigrants representing the overwhelming majority of that first wave.[178] These were the immigrants with whom Charles Brace was most familiar when he first started working among New York's poor in the 1850s and whom he described in his widely read 1872 book, *The Dangerous Classes of New York and Twenty Years' Work among Them,* as degraded but ultimately redeemable through education, labor opportunities, Christianity, and the beneficent influence of their class superiors—that is, a change of *environment.* And they were also the immigrants whom his ethnological contemporaries Josiah Nott and John Van Evrie sought to redeem in the mid-nineteenth century. Nott suggested that the various "Caucasian" *races,* each distinct in physiognomy across time, may have originated separately from one another, though like Van Evrie, he attributed any negative biological or moral characteristics the Germans and Irish exhibited in America to the difficulties of immigration itself, which could be ameliorated with improved conditions. Van Evrie, meanwhile, saw Irish immigrant laborers as vital to bolstering proslavery sentiment in the urban North.[179] As Jacobson notes, "As the first to immigrate in huge numbers at once well within the literal language but well outside the deliberate intent of the 'free white persons' clause of the [Naturalization Law of] 1790, Irish and German arrivals of the 1840s and after drew special attention in discussions of race and its implications for assimilability and citizenship."[180] But ultimately, racial theorists writing in the mid-nineteenth century, including Nott and Van Evrie, were far more concerned with slavery than immigration, and unlike Brace (who instead portrayed "unoffending negroes" as unfortunate victims of "wild and brutal crimes" committed by "ignorant masses" of immigrant workers, particularly during the 1863 draft riots), they saw free African Americans as a bigger threat to U.S. politics and society than either Irish or German arrivals.[181]

In contrast, as immigration continued to increase and ever-greater numbers came from southern and eastern Europe, Asia, and Latin America, many of whom were Catholic or Jewish—peoples whose claims to the political category of whiteness were even less clear and their admixture with white blood more problematic to racial scientists—ethnological writers assessed the issue far more negatively than Van Evrie and Nott in the 1850s and 1860s or even Brace in the 1870s.[182] Likewise, immigration restrictions, which targeted specific nations based on racial characteristics and perceptions about the group's potential benefit or danger to native-born whites, became an increasingly popular solution. Published in the *Atlantic Monthly* in 1896, Yale political economist, census superintendent, and former Union army general Francis Walker's "Restriction of Immigration" is a case in point.[183] Therein, he described the necessity of restricting who and how many could enter the United States in order to protect "the quality of American citizenship from degradation" and native-born Americans themselves from the negative impact—culturally *and* biologically—of the "aliens" in their midst. He did not share Brace's optimism that immigrants' "natures"—physical and moral—could be changed over time. Where Brace described immigrant boys and men who could assimilate into civilized American manhood through hard work, education, and a change in environment, Walker characterized them as "beaten men from [inferior] races" and, in evolutionary terms, "representing the worst failures in the struggles for existence." They have never and would never evolve, Walker insisted, and they simply did not have the capacity to be full citizens because they were not "fit men." Consequently, while Brace describes at least some immigrants as part of "our race," if not of his *class,* Walker saw only swarthy "aliens," biologically distinct from—and inferior to—native-born Anglo-American stock, unfit for the "manhood right" of citizenship and a drain on the nation's resources.[184]

At the turn of the century, some critics of immigration considered it an even greater menace to American democracy than the freed and newly enfranchised black population, but most saw them as interrelated threats to white supremacy, domestically and globally.[185] At the same time that the United States saw a tremendous increase in immigration, the nation had also entered the pursuit of empire, with imperial involvements in the Philippines, Hawaii, Samoa, and the Caribbean and growing interest in South America, China, and Africa. Consequently, politicians and scientists often employed the scientific language of natural selection or survival of the fittest "in defense of the subjugation of weaker races" across the globe, as historian Richard Hofstadter points out. But, he adds, evolutionary theory simply provided a justification rather than an impetus for imperial pursuits.[186] Moreover, U.S. racial scientists tended to discuss immigration, imperialism, and American race relations between whites as blacks as representing different aspects of the same problem—that, as suggested by the title of one widely circulated and influential book by Harvard-educated ethnological writer, eugenicist, and avowed nativist Lothrop Stoddard (1883–1950), white supremacy on a global scale was being threatened by a "rising tide of color" from highly fertile yet profoundly uncivilized "lower races" who could, should, and indeed must be dominated, even on their own lands, for the sake of civilization and the human species' survival. There was another connection as well. In his 1920 tome, *The Rising Tide of Color against White World-Supremacy,* Stoppard insisted that the "prodigious birth rate" among nonwhite peoples would inevitably result in an "outward thrust of surplus colored men from overcrowded colored homelands"; as with the problem surrounding free and enfranchised African American men, racial theorists like Stoddard posed the growing number of "colored" people globally as a sexual threat.[187]

Furthermore, even as whiteness was complicated as a political category within the United States, "fitness for citizenship" remained a profoundly gendered concept, and the denigration of a particular race or immigrant group relied on impugning manhood within the group. Stoddard generally juxtaposed genderless, indistinguishable nonwhites with his often-invoked "white man," "the indisputable master of the planet." For example, "few regions of the earth had escaped the *white man*'s imperial sway, and vast areas inhabited by uncounted myriads of *dusky folk* obeyed the *white man's* will," he wrote.[188] White men alone had evolved enough to be fit citizens as well as fit leaders for all their biological inferiors, which included white women and nonwhites of both genders.

Among the most astute critics of scientific racism and its use of evolutionary theory to undermine the citizenship rights of nonwhites was C. V. Roman, a black physician and founding editor of the *Journal of the National Medical Association,* a national organization representing physicians of African descent. Roman was born in Pennsylvania in 1864 and received his medical degree from Meharry Medical College in Nashville, Tennessee, where he later served as a professor of medical history and ethics in addition to maintaining a private practice and a busy schedule of speaking engagements.[189] In his 1916 book, *American Civilization and the Negro,* he noted that "the fundamental error of most anti-Negro thinkers on the race problem is the unwarranted assumption that the white man is the *norm* of humanity; mentally, morally, and physically."[190] He also repeatedly challenged the central argument of turn-of-the-century ethnological and eugenic thought that all people are like their ancestors as well as the web of logic that was typically spun from that premise: "The Afro-American of today is a savage. No savage is fit for citizenship in a republic, therefore 'the war amendments to the Constitution were a huge blunder.'"[191] In contrast, he aimed "to show that humanity is one in vices and virtues as well as blood; that the laws of evolution and progress apply equally to all; that there are no lethal diseases peculiar to the American Negro; that there are no debasing vices peculiar to the African; that there are no cardinal virtues peculiar to the European; that we are all sinners and have come short of the glories of civilization."[192] He was not above constructing his own divisions and racial stereotypes, however. His glossary, for example, defined heterosexuality as "the natural affiliation" and homosexuality "an abnormality"—the latter of which he attributed in the text itself not to "the heathen denizens of darkest Africa" but "the civilized inhabitants of our city slums."[193] Ironically, here he mirrored the anxieties white medical scientists like Lydston, Howard, and Shufeldt expressed between the apparent preponderance of homosexuality among whites and the threat it posed to civilization as a result, as the next chapter will take up.

In the second half of the nineteenth century, ethnological writers often employed differences of sex to bolster their claims about differences of race, first to defend slavery as natural, then to justify continued racial inequities after slavery's demise. Facing threats to white male dominance from feminist agitators and a vastly expanded free black population, white men within and outside the scientific establishment became deeply invested in defending the existing racial and gender hierarchy as biologically determined. Without slavery to structure racial hierarchy, postbellum racial scientists looked instead to sex and gender to prevent the full inclusion of black people in American social and political domains. If black bodies did not fit these scientists' standards of sex, they need not be afforded the many rights and protections associated with gender. The scientists of race constructed the gendered standard of separate spheres as a white privilege; white men alone could vote, and only white women needed to be protected from labor and exploitation.

Toward the end of Reconstruction and into the Redemption period, the issue of black male suffrage remained potent, but how scientific experts on race approached the subject changed, largely in response to the enormous advancements African Americans made following the Civil War. Black men rushed to the polls with the active involvement of their female family members and served in elected positions in unprecedented numbers. African Americans opened businesses, schools, and their own churches throughout the South, and a small but highly visible middle class emerged. In other words, ethnological thinkers were faced with overwhelming evidence against their claims about black inferiority. In the last decades of the nineteenth century, racial theorists began to refocus their argument from black men being incapable of political participation to black men being undeserving of the franchise on the grounds that they were by nature prone to sexual crime and excess. In the context of evolutionary theory's growing popularity in the United States, sexuality and sexual instincts joined sex and gender as markers of racial difference in late nineteenth-century racial science. American scientists both echoed and enflamed public discourse on race when they reframed the political and economic threat that black men had represented to white supremacy as a sexual threat to white women's virtue during Redemption, a far-reaching ideology that often served to justify racial violence as the nineteenth century drew to a volatile close. But this racial competition for power in America also meant that the proper sexual behavior of whites was of paramount importance to scientists and politicians alike, for the fate of the nation depended on the superior race "outbreeding" the inferior. Therefore, many of these scientists framed white men and women who challenged heterosexual norms in similar terms as a threat to the white race and, in turn, American civilization.

· CHAPTER 3 ·

Inverts, Perverts, and Primitives

Racial Thought and the American School of Sexology

IN MAY 1890, CHICAGO PHYSICIAN G. Frank Lydston opined that the "society of the future" should look to "the physician, and not to the moralist or law-maker" for "better correction and prevention of vice and crime."[1] The reason, he insisted, was that many crimes were "a consequence of inherent sexual perversion."[2] The perversions that received his most frequent and sustained attention were homosexuality, inversion, and "hermaphroditism."[3] In particular, he struggled to map the relationship between these categories—the extent to which they were distinct, related, or overlapping. Typical of late nineteenth-century scientists, he viewed sexual variance as pathological and thus looked to the body for clues, particularly "malformations of the sexual organs with or without associated close approximation to the general physique of the opposite sex, male or female."[4] Also among the causes of vice and crime were "defective physique and imperfectly developed intellect, hereditary or congenital." The racial connotations are difficult to miss.[5] For nearly a century, ethnological scientists had been arguing that the "lower races" were permanently and pervasively deficient in both body and mind. Innate sexual deviance was prominent among the deficiencies to which scientists pointed, with aggressive hypersexuality attached to the black race in particular—an argument Lydston himself made throughout his career.

Moreover, such deficiencies were not just of scientific interest; they threatened the safety and order of society writ large. Indeed, Lydston posited a dynamic relationship between the "human body" and the "social body" in which corruption in one entity could jeopardize the other. Undermined by racial and sexual contagions, America's "social body" was ailing. To protect it, Lydston brought to bear all the resources of the turn-of-the-century medical and scientific establishment, including social hygiene, ethnology, criminal anthropology, eugenics, and sexology. These overlapping sciences of difference were weapons in the arsenal to defend the social body by interrogating and controlling human bodies.

Lydston's professional interests and driving concerns were characteristic of turn-of-the-century American scientists. Throughout his long and prolific career, he wrote and lectured on varied but intersecting topics, including sex education for children, the dangers of masturbation, venereal disease, immigration, sexual crime, lynching, miscegenation, homosexuality, and hermaphroditism. Sexuality and race were central to his discussions of each. Scholarship by Siobhan Somerville, Jennifer Terry, and Lisa Duggan, among others, has revealed important parallels and connections between ethnology and sexology in this period. Lydston's career, however, underscores the fact that racial science and sexology were not separate fields in the United States. Not only did they share similar concerns and cultural politics; they were literally populated by the same key scientists. Sociologist Janice Irvine defines sexology loosely as the scientific study of sex, "an umbrella term denoting the activity of a multidisciplinary group of researchers, clinicians, and educators concerned with sexuality."[6] In the United States at the turn of the century, this multidisciplinary group was largely composed of medical scientists like Lydston, William Lee Howard, and R. W. Shufeldt who were also working on race and social hygiene.

As we have seen, by the turn of the century, many white scientists were bolstering their claims about racial difference and hierarchy by insisting that the "lower races" deviated from white norms of gender roles, reproductive anatomy, and sexual behavior. Predictably, when these same medical scientists turned their attention to gender or sexual variance, they often read these differences through the lens of race. Just as gender difference had provided the earliest ethnological scientists with a foundation—and convenient conceptual shorthand—for constructing race in the early nineteenth century, race in turn framed medical scientists' varied concerns about sexuality at the turn of the twentieth century. This chapter, then, will explore both ideological and institutional convergences between racial and sexual science in turn-of-the-century America.

The overlapping sciences of difference that Lydston exemplified were driven by racial, class, and gender anxieties, themselves fueled by enormous social upheaval, including economic changes, immigration, labor unrest, and black and women's rights organizing at the turn of the century. All these changes challenged the social and political power of white middle-class men, prompting them to redefine what manhood meant. Some worried that civilized society had made white men weak and effeminate, leaving them vulnerable to competition from working-class, immigrant, and black men as well as white middle-class women; in so doing, they connected male dominance with racial dominance.[7] Meanwhile, many whites believed the races were in a Darwinian struggle for survival in which reproduction was key, while evolutionary theorists also argued that distinct gender roles and distinctly sexed bodies marked higher evolutionary stages. The disruption to gender norms posed by homosexuality and "inversion" among whites thus threatened both white manhood and the white race as a whole. An overwhelmingly white and male group, American medical scientists like Lydston had a vested interest in maintaining existing gender, racial, and sexual hierarchies.

Lydston found his ideal case study in a hermaphroditic mulatto cook who engaged in sex with men and women. The "spurious hermaphrodite" had "an affinity for women, as illustrated by the fact that he contracted a gonorrhea in the normal manner," Lydston explained. But he also "had a predilection for the passive role in the act of copulation" and in that capacity transmitted the disease to "a number of young lads, ranging from ten to seventeen years of age," who had come under Lydston's care.[8] That the cook could perform the role of the male penetrator and the passive penetrated female seemed to provoke both awe and trepidation. The cook occupied a threateningly intermediate space in every possible way, confounding the binaries of male–female, homosexual–heterosexual, and black–white. Moreover, the cook's intermediate sex, gender, race, and sexuality were all linked in Lydston's account. In his other writings and those of his scientific contemporaries, the "mulatto" embodied perhaps the most socially threatening "sexual perversion" of all: miscegenation.[9] The subject in question also fit seamlessly with their racialized and class-conscious characterization of both sexual predation and venereal disease in other works. This mulatto, hermaphroditic, and seemingly bisexual (in the current sense of the word) cook performed considerable ideological work for the scientist. This one figure, who Lydston revisited in several publications, brought together salient turn-of-the-century discourses on race, gender, sexuality, miscegenation, disease, and crime.[10]

Lydston's fascination with the mulatto hermaphrodite also sheds light on the nature of turn-of-the-century American sexology and the racial context in which such work was produced. For one, it reveals the extent to which intermediacy provided a framework for scientific considerations of race, gender, sex, and sexuality. In a society where social and political institutions were structured around binaries, it was the spaces in between that fascinated and loomed most dangerous to scientists. Second, the cook's venereal infection made literal the link scientists drew between sexual variance and both individual disease and social contagion, which in turn opened new possibilities for medico-scientific authority over America's ills. Third, like ethnology, American sexology was concerned with classifying difference, but the categories themselves sparked debate. The ambivalence Lydston exhibited about the relationship between sexual object choice and nonconforming gender performance or bodily sex was typical of the field. Historians often discuss a turn-of-the-century shift in scientific discourse from the concept of sexual inversion to that of homosexuality. However, for years the two concepts, as well as the terminology, existed concurrently—and both were fundamentally racial categories from the start.[11]

Finally, medical scientists' assessments of the causes of "sexual perversion" often broke down along racial lines. They frequently linked genital abnormalities to bodies that were already marked as racially suspect: African Americans, mixed race people like Lydston's mulatto cook, and "undesirable" immigrants. Meanwhile, they often characterized "sexual perversion" in the same population as vice and as indicative of the physical and moral degeneracy of the group. In contrast, scientists usually read cases of "sexual perversion" in middle- and upper-class whites as symptomatic of individual pathology or disease, likely congenital, which many attributed to "overcivilization." While the "lower races" could threaten moral corruption by spreading their sexual vice, especially concerning if that influence crossed racial lines, white homosexuals from respectable families threatened racial degeneration and "race suicide"—itself a concept that arose in the first decade of the twentieth century, concurrent with and influenced by the boom in American sexological literature.

In other words, while medical scientists believed that homosexual behavior could occur in any race, when it occurred among black men, it was simply emblematic of their general hypersexuality—of being sexually indiscriminate and debased overall—and not necessarily a distinct category unto itself. Sexually and gender-variant whites on the other hand were acting contrary to the norms of their race as well as imperiling it, and sexologists were far more likely to categorize them with sexual terms—such as invert or homosexual—that set them apart from other whites. Lydston's cook aside (who fit neatly into existing ideas about African Americans' predatory instincts *and* confounded every possible binary), sexologists simply *cared* much more about homosexuality among whites, and gave it far more sustained attention as a result, because of its implications for civilization and white supremacy.

The Sciences of Human Difference: Racial Science and American Sexology

American sexology, of which Lydston was representative, developed in the last decades of the nineteenth century in the context of pervasive racial anxieties among scientists. For one, the nascent eugenics movement, which aimed to produce a "superior race" by promoting the reproduction of "desirable" races or groups while limiting the fertility of "less desirables," prompted support for scientific studies of sex. Similarly, concerns over prostitution and venereal disease—which moral reformers usually associated with immigrants, African Americans, and the working class—lent some degree of legitimacy and social import to sexology as well.[12]

Moreover, racial science had become increasingly concerned with issues of sex and sexuality during the era of "Redemption." Black men were cast as a sexual threat in the South, and much of the scientific writing on race and sex during this period—including work by Lydston and his Chicago colleague, physician and psychiatrist James Kiernan, as well as prolific sexologist, naturalist, and retired U.S. military surgeon R. W. Shufeldt—focused on "furor sexualis," a biological imperative that drove black men to rape white women.[13]

Meanwhile, concerns over immigration and the impact of evolutionary theory in late nineteenth-century science gave rise to new concepts of intermediacy in regards to both sex and race. Evolutionary theorists measured the stage of a race in part by its degree of sexual differentiation and pitted the races in a struggle for survival that hinged on reproduction. And whereas earlier ethnology was literally black and white in its focus, turn-of-the-century racial scientists also turned their attention to other races. Scientists typically described these races as inhabiting a racial and evolutionary middle ground between black and white but often portrayed them as sexually intermediate as well. Scientists were focused on racial intermediacy in another regard too; white scientists now painted miscegenation as a threat to American civilization. Whereas earlier ethnological writers tended to argue that mixture between whites and blacks produced "sterile hybrids" and was thus incapable of producing a "permanent stock," racial scientists at the turn of the century represented miscegenation as a polluting force that would degenerate both races physically and culturally.[14]

But homosexuality and other "sexual perversions" were linked to these racial concerns as well. On the one hand, homosexuality, onanism (masturbation), prostitution, and other so-called perversions among whites threatened "race suicide" by diverting sexual activity from its true purpose—reproduction. On the other hand, such activities, which were linked in scientific discourse to a host of ailments, particularly neurasthenia, also threatened to weaken the race in subsequent generations.[15] At the same time, scientists interpreted sexual perversion in the "lower races"—typically, African Americans as well as immigrants from Asia and southern and eastern Europe—as further evidence of their physically and morally degenerate nature and thus their inherent propensity for "vice." Finally, racial theorists' evolutionary method of examining and ranking bodies extended beyond race to color the way scientists viewed other "deviant" bodies—even when obvious physical differences could not be found. By the late nineteenth century, this framework was firmly entrenched in scientific thought and discourse, which had also begun to turn its attention toward matters of sex and social hygiene.

In order to find the key players in American sexology, then, one need look no further than the successors of the American school of ethnology. Though a number of scholars have begun to explore the ideological links between ethnology and sexology, their work has tended to underplay American scientists like Lydston who inhabited both fields in favor of focusing on European sexologists or the reception of these theorists in the United States.[16] Scholarship on sexology prior to Alfred Kinsey in the 1940s and 1950s tends to characterize American scientists as merely following Europeans such as Karl Westphal, Richard Von Krafft-Ebing, Albert Moll, Karl Ulrichs, Havelock Ellis, Magnus Hirschfeld, and Edward Carpenter, and in general, early American scientific work on homosexuality and inversion remains relatively underexamined.[17]

Although European sexologists were indeed widely read, reprinted, and cited in the United States, it would be a mistake to assume a one-way chain of influence. Late nineteenth-century American and European scientists wrote about sex concurrently and cited each other's work. Kiernan's work featured prominently in Krafft-Ebing's *Psychopathia Sexualis,* for example, while Lydston claimed that Ellis preferred his sexual classification system to Krafft-Ebing's.[18] And U.S. scientists cited each other as well as European sexologists. American physician Allan McLane Hamilton noted in 1896 that the work of "Krafft-Ebing, Moll, Chaddock, and numerous continental and American writers" had given sexuality "a definite place in modern psychological medicine."[19] Moreover, they were also familiar with and referenced each other's case studies.[20] While U.S. scientists definitely saw themselves as part of a transatlantic conversation on sex and "sexual perversion," Americans like Lydston and R. W. Shufeldt are now far more likely to appear in scholarship on race than on sex—if discussed at all. This lacuna has obscured not only early American contributions to sexology but also the extent to which the field was driven by the country's unique racial history.

Indeed, sexology in America grew into a distinct scientific field, separate from racial science, far later than in Europe. While European sexology also had roots in social hygiene and eugenics, it quickly developed into a discipline of its own, with institutes focused on sexuality and scientists self-identifying primarily as sex scholars. In contrast, U.S. research on sex and sexuality remained firmly entrenched in the racial sciences and medicine, just as individual U.S. scientists themselves continued to work in a range of disciplines premised on human difference.[21] European scientific thought did have a profound effect on American work on sexuality, but social Darwinism may well have been more directly influential in shaping this work than the writings of European sexologists.

Rather than simply categorizing sexual variance, as several European sexologists sought to do, American sexology was never fully divorced from practical concerns over what to do with the "sexual pervert" to ameliorate his or her influence on the social body. Though American scientists shared with their European counterparts a medical and psychological framework for understanding homosexuality and inversion, criminal anthropology, eugenics, and, above all, social hygiene were brought to bear on pressing questions about the relationship between "perverted" individuals and society as a whole. It is perhaps more useful, then, to think of these turn-of-the-century studies in the United States as a patchwork science of human difference, of which race and sexuality were both components.

The patchwork science that was the American school of sexology found its ideological home in Chicago in the late nineteenth century with Lydston and Kiernan as its fathers. Born in New York City in 1852, Kiernan received his medical degree from the University of the City of New York in 1874. Like Kiernan, Lydston also obtained his medical degree in New York City from Bellevue Hospital Medical School in 1879, having moved there from California to complete his studies. Kiernan left New York for Chicago in 1878, after working for four years as an apothecary and assistant physician at the Ward's Island Asylum, where Kiernan insisted he was unfairly dismissed for, in part, refusing to help cover up the beating death of an inmate.[22] Though Lydston followed Kiernan to Chicago four years later, it is unclear if the two men knew each other during their overlapping years in New York City, nor is it clear if either met J. Marion Sims, John Van Evrie, or, in Kiernan's case, Josiah Nott (Nott passed away before Lydston arrived in New York, while Kiernan was a long-term resident) during their time there during the same period; it is likely the two younger men at least knew of these elder statesman in the medical world by reputation by that point if not personally.

Once in Chicago, the ambitious Kiernan practiced and taught psychiatry, and within three years of his arrival, he became the managing editor of the *Chicago Medical Review,* which he used as a platform for biting critiques of the city's asylum system, which he viewed as corrupt and counterproductive. As medical historian Charles Rosenberg describes, Kiernan's fervent attempts at reforming the asylum system won him enemies within Chicago's psychiatric circles, just as it had in New York, but in the process, he had also quickly built such a name for himself that he was called as an expert witness in the 1881 trial of Charles Guiteau, who had assassinated President James Garfield earlier that year and would ultimately be executed for his crime. Kiernan traveled to Washington, D.C., to serve as the defense's first witness in support of Guiteau's insanity plea. By Rosenberg's account—as well as those of Kiernan's contemporaries—he was a rather awful witness, but the media circus surrounding the trial made him a recognizable name nationally as well as back home in Chicago, and he continued to be called as an expert on insanity in criminal trials, particularly those for sexual crimes, for years to come.[23] Indeed, his status as a celebrity witness of sorts was how he seemed most remembered after his death at his home in Chicago in July 1923—the same year Lydston passed away as well. For example, the headline of his obituary in the *New York Times* read, "Noted Alienist Dies; James G. Kiernan Was Expert for the Defense in Guiteau Trial," while the text went on to hail him as "one of the country's leading alienists and psychologists" and noted his successful reform efforts in the asylum system, his various teaching positions and publications, and that "he testified regarding neurotic and insane tendencies in many famous criminal cases." The obituary closed by again referencing his role in the Guiteau trial.[24] Left unsaid in a respectable mainstream publication was the degree to which he had also become a leading authority on homosexuality during his career.

The Chicago Lydston and Kiernan inhabited was home to a large social hygiene movement, an extremely active vice commission, and a thriving gay underground. Both Lydston and Kiernan resided and maintained thriving clinical practices in the city and taught at its medical schools, and, as historian Chad Heap describes, "by the 1880s, they had made the city a center for American sexological studies, not only publishing reports on the 'sexual perverts' they encountered in the course of their medical practice, but also supplying case studies of these patients to prominent European sexologists such as Havelock Ellis." Together, Lydston and Kiernan—himself the first person in America to use the term "homosexual" in print—established "an intellectual environment where the study of sexual abnormality was seen as a proper, even urgent subject of academic inquiry."[25] The University of Chicago, widely recognized as "the birthplace of urban sociology," would also be at the forefront of American sexology.[26]

In Chicago and elsewhere, sexual vice was central to political and scientific discussions of urban problems, and debates raged whether such vice demanded medical or judicial interventions. Both Kiernan and Lydston employed a medical paradigm of sexual deviance. Kiernan, however, called for strict legal punishment and denounced sympathy for homosexuals in particular as wholly unwarranted while Lydston, a scientific paternalist throughout his career, urged greater understanding and pity for "perverts" and tended to favor prevention (though as we will see in the next chapter, few of Chicago's "perverts" were likely to prefer Lydston's preventative measure). From the 1880s on, American scientists writing on sexuality were as likely to reference Lydston and Kiernan as European sexologists like Ellis.

Refocusing our view specifically on the work of American sexologists is revealing in a number of ways. It demonstrates that U.S. scientists, while informed by and in dialogue with their well-known European colleagues, were not simply responding to knowledge produced overseas but also concurrently developing a uniquely American concept of sexuality and sexual difference shaped by the country's racial context. Indeed, America's most well-known "experts" on race were also its preeminent scholars of sex. Moreover, in the late nineteenth and early twentieth centuries, scientific work on race and on sex was not only produced by the same scientists but also targeted the same audience. Scientific studies on the two issues were generally published in the same journals, often side by side, and many individual studies looked at race and sex together. For example, "The Psychology of Modesty and Clothing" (1899), an article by sociologist William I. Thomas, a colleague of Lydston and Kiernan at the University of Chicago, cited natural, racial, and sexual scientists, including Charles Darwin; a variety of European and U.S. ethnological studies of "primitive" peoples that applied Darwinian theory to human diversity; and British sexologist Havelock Ellis.[27] The article was immediately followed by a review of a new ethnology book, *The Races of Europe,* a review that compared scientific debates over racial difference to similar considerations of sex difference. "The social philosophy which attributes everything to environment is a black number: environment is only one of the conditions," the reviewer wrote. "But the anthropologist who, on the other hand, would attribute everything to race is just as greatly in error. The question has become much like that with reference to the superiority of one or other of the sexes."[28] Kiernan himself offers another example. Under the column heading "Sexology" in the *Urologic and Cutaneous Review,* Kiernan penned two separate articles, one on homosexuality, "An Increase in American Inversion," and the second titled "Birth Control among Primitive Peoples."[29]

Lydston, Kiernan, and Shufeldt, then, rose to prominence at a time when scientific discussions of ethnology and sexology shared similar concerns about the nation's fate in the context of changing gender, sexual, and racial norms. Much of their concern about human evolution and white civilization were written onto the body. They worried that women's agitation for political rights, higher education, and social freedoms were affecting their biology, resulting in declining birth rates among middle-class whites at a time when immigrants, the poor, and an emancipated black population were reproducing in higher numbers. At the same time, they worried that growing numbers of white, middle-class men suffered from "neurasthenia," an amorphous condition brought on by an overly cerebral and decadent civilized life and marked by a weakened constitution and nervousness—which in turn left them susceptible to inversion and homosexuality, the ultimate form of impugned manhood.[30]

Homosexual women did not escape American scientists' notice either. On the contrary, female inverts also threatened middle-class manhood by usurping male privileges of employment, social deportment, and sexual dominion over women. Traditionally feminine women engaged in romantic friendships were accepted, even celebrated, because their relationships complemented rather than supplanted heterosexual marriage and their displays of emotion and sentimentality accorded with gender expectation of Victorian womanhood and domesticity.[31] Masculine women who eschewed feminine pursuits and heterosexual marriage entirely, however, were labeled pathological "inverts," who, like their male counterparts, threatened America's social body. Likewise, scientists scrutinized the bodies of female inverts for biological evidence of their deviance, remarking on square jaws, broad foreheads, or coarse hair when no obvious genital anomaly could be found.[32]

Closely linked to their concerns about sexuality and civilization, scientists also worried about human evolution. Scientific discussions of racial and sexual difference shared a common language and, often, similar central questions and debates. Evolutionary theory, which provided a conceptual framework for turn-of-the-century racial science, informed discussions of homosexuality and gender deviance as well. Because "the position of the animal in the scale of evolution" depended "upon the perfection of the reproductive apparatus," Kiernan explained in 1891, the subject of sex had "important ethical aspects" for the march of civilization.[33] As Kiernan alluded, an important dimension of evolutionary theory, and one that was consistently applied to discussions of race, held that sexual differentiation was a mark of evolutionary development. That is, the more distinct males and females were in both body and social role, the more evolved the race to which they belonged. By this standard, scientists often characterized the "lower races" as less evolved. For example, they emphasized black women's labor, muscular bodies, and supposedly enlarged clitorises, while they interpreted black men's "hairless bodies" and "scanty beards" as evidence of the race's low evolutionary stage. In contrast, evolutionary theorists described Anglo-Saxons as the pinnacle of evolution. With white women at least in theory ensconced in the home, white men in public and political spaces, and their bodies polar opposites, they were the most sexually differentiated race. Kiernan applied the same logic to physical and psychical "inverts," who were, in his estimation, an evolutionary throwback. He remarked, "It seems certain that a femininely functioning brain can occupy a male body, and vice versa. Males may be born with female external genitals and vice versa. The lowest animals are bisexual and the various types of hermaphrodism are more or less complete reversions to the ancestral type."[34]

Kiernan's remarks are indicative of the evolutionary framework of early American sexology, the far-reaching significance attached to sexually normative bodies, and the social danger that sexual intermediacy represented. For decades following Kiernan's influential work, other scientists also described homosexuals and inverts as "atavistic," that their lack of sexual differentiation in body or behavior mirrored "the very remote beginnings of the biological development of the human species," as New York City psychiatrist and psychoanalyst Clarence P. Oberndorf (1882–1954) wrote in 1929.[35]

Though he was born three decades later, Oberndorf shared similar training and institutional affiliations with Lydston and Kiernan. The native New Yorker received his BA and medical degrees at Cornell University and then interned at Bellevue Hospital, where he decided to become a psychiatrist. He served for a time on the staff at Manhattan State Hospital on Ward's Island, maintained a respected private practice, and, on one of several professional trips to Europe, worked directly with Sigmund Freud, whose psychoanalytic theories and methods Oberndorf espoused throughout his career. He spent the bulk of his career at Mount Sinai Hospital, where he organized the country's first hospital outpatient clinic for psychiatry, and taught neurology at his alma mater, Cornell. He also served as the associate editor of several leading psychiatric journals and published three books and more than one hundred twenty-five articles during his celebrated career.[36]

The implications of the homosexual's apparent evolutionary regression were troubling for scientists like Oberndorf. Following Kiernan's lead, scientists worried that the rise in homosexuality and inversion they perceived among America's middle and upper classes indicated that white civilization was beginning to degenerate. Indeed, in turn-of-the-century scientific thought, "degeneracy" could be used to characterize an early or stagnated stage of evolution—which scientists often linked to the "savage" races—but it could also mean an active process of devolution among the "civilized" races. That is, just as racial groups could evolve over time from savagery toward civilized perfection, so too could the process be reversed. Moreover, the role of the scientist was not just to understand evolutionary processes but also to help them along by mediating threats to white supremacy, whether the threat was in the form of intermixture with other races or degeneration from within.

As they did with other forms of "degeneracy"—criminality, pauperism, insanity, mental retardation—scientists often described homosexuality as a "hereditary taint" passed from one generation to the next. They typically began case studies with a detailed family history that highlighted any "bad blood" within the patient's "stock," especially other examples of homosexuality or "perversion." Attempting to ascertain whether a specific patient demonstrated "a case of inherited or acquired homosexuality," respected New York gynecologist and nationally known sexologist Bernard Talmey declared that the man's "entire clan was sexually tainted."[37] Talmey was born in Germany and graduated from the University of Munich before moving to New York City, where he practiced medicine for more than three decades, including stints as a gynecologist at Yorkville Hospital and a pathologist at the Mothers' and Babies' Hospital. A member of the New York Academy of Medicine, he also published a number of articles and books on sexual hygiene and sex disorders.[38] Unlike Talmey, Kiernan looked more generally for "neurotic ancestry" among his case studies, while Baltimore physician and member of the New York–based Medico-Legal Society William Lee Howard noted that "ninety per cent of the cases of sexual perversion" he observed "show the stigmata of neuropaths and have histories of inherited unstable nervous organizations."[39]

As these discussions of "tainted heredity" suggest, while some scientists placed homosexuality and inversion among the causes of degeneracy, others saw them as symptomatic of degeneracy. In 1904, Howard argued that "with the increase of neuropathic individuals which our high pressure living is constantly producing," the courts would continue to see rising numbers of "sexual perverts" and "degenerate acts."[40] In other words, civilization had perhaps advanced too far, creating disorders like neurasthenia that weakened individuals physically and morally; sexual deviance was just one among many manifestations.[41] Shufeldt also saw sexual variance as a manifestation of degeneracy, but through an explicitly eugenic perspective. Rather than linking degeneracy to the problems of overcivilization, he blamed bad breeding. In his 1905 "Medico-Legal Consideration of Sexual Perverts," he posited, "The reason to account for the presence of such a vast number of psycho-sexual perverts and inverts among us is, that for untold generations in the past couples have given birth to children who have been anatomically, physiologically and psychologically unfitted to bear them." For Shufeldt, "such mating stands responsible for present-day crime."[42]

Meanwhile, some scientists cautioned against making any direct correlation between degeneracy, homosexuality, and inversion at all. Philadelphia-born physician and psychiatrist Isador H. Coriat, for example, wrote that "the loose term degeneration" could not "be used to explain these conditions, because many homosexuals do not show the slightest evidence of what are usually termed degenerative stigmata, and furthermore, many so-called degenerates are absolutely free from symptoms of inverted sexuality." Coriat had moved to Boston at an early age, and after obtaining his MD from Tufts Medical College in 1900, he went on to study philosophy at Harvard while building a name for himself for his pioneering use of psychoanalysis to treat stuttering. He married in 1904 but never had children and remained busy through his work with Worchester State Hospital and the Boston Psychoanalytic Society—of which he was a founding member and the organization's first president—and penning numerous books and articles on a range of topics, including the psychoanalytic method, hysteria, the use of religion in treating mental disorders and sex.[43] Coriat challenged the congenital model of homosexuality espoused by many of his American peers in favor of the psychoanalytic approach promoted by Sigmund Freud, explaining, "A homosexual, insane, or neuropathic heredity cannot be incriminated, because the condition appears just as frequently in those who are free from any hereditary taint."[44]

Still, scientists agreed on one critical point: sexology had a vitally important role to play in improving society. Indeed, "there can be no improvement so long as crass ignorance of the science of sexology prevails," Shufeldt predicted. He asserted that science, not law or religion, should have the ultimate authority over human reproduction. His recommendation was eugenic: he stated that "would-be parents" should be "examined by a competent medical expert or experts, and pronounced sound," lest America "continue to breed millions of sexual perverts and inverts—psychopathic types."[45] Similarly, Howard argued that the legal establishment needed to look to the medical to consider adequately the cases of "sexual perversion and inversion" that so frequently came before the courts. Jurists "must be able to distinguish the congenital from the acquired form as a disease, from the vulgar vice of the male and female prostitute."[46] Although Howard insisted here on carefully distinguishing between congenital homosexuality, vice, and prostitution, other scientists saw considerable overlap among these categories.

By describing homosexuality as degeneracy, disease, or hereditary taint, scientists often created a catch-22 for homosexuals, who they also accused of race suicide on account of their propensity for nonprocreative sex. Homosexuals and inverts could potentially pass their pathology on to future generations, thereby weakening the race physically and morally. But scientists and social critics alike also worried that the number of "inverts and perverts" among middle-class whites was increasing at the same moment that birth rates in that population were falling. In fact, Brooklyn-based neurologist and psychiatrist John F. W. Meagher explained in 1929 that the harsh treatment of homosexuals derived in part from society's understandable resentment that they had turned away from their "biological duty in regard to race preservation."[47] Born and raised in Brooklyn, Meagher received his medical degree from Columbia University in 1901 and began his career working at Manhattan State, Ogdensburg State, and Bellevue Hospitals and specializing in neuropsychiatry. He went on to open a thriving private practice in Brooklyn while continuing to work as a consulting neurologist or psychiatrist to several local hospitals and serving as a medical officer in World War I. Like Kiernan, the married father of two was frequently called as an expert medico-legal witness in criminal trials, in Meagher's case by the Brooklyn district attorney's office.[48]

According to Meagher, society's concerns about homosexuality and race suicide were not unfounded, noting, "One authority said that only three per cent of active homosexual men had the natural impulse to preserve the species." Women were under greater pressure to marry, and few had the economic resources to live independent of men and foster the kind of elaborate subcultures found among homosexual men in urban enclaves. But even if they married and had children, Meagher implicated lesbians as well, for, he claims, "we often see homosexual women who love neither their husbands nor their children, but crave only power or social success, even at the expense of a happy family life."[49]

American sexologists connected homosexuality to race suicide in a less direct way as well by linking both to masturbation. Their case studies of homosexual men and women almost always made note of the patients' masturbatory habits.[50] Others argued that homosexuality was a form of masturbation and vice versa—loving oneself and one's own sex—and later, with the influence of Freud, characterized homosexuality and masturbation as sexual "narcissism."[51] For instance, "with the mother image, autoerotic practices and narcissism shutting off the gate to womanhood, we find our patient well on the road toward homosexuality," Benjamin Karpman (1886–1962), the Russian-born founder of the Washington, D.C., Psychoanalytic Institute, remarked in 1923.[52] After receiving his medical degree from the University of Minnesota in 1919, Karpman moved to the District of Columbia to intern at St. Elizabeth's, a psychiatric hospital where he served as chief psychotherapist for many years, and then to teach at Howard University, where he also directed the Department of Psychiatry. Known among his colleagues for a penchant for working with even the most difficult patients, as well as his delight in courting controversy, the prolific psychoanalyst penned books and articles, many on sexuality and crime, and contributed his expertise to several popular periodicals.[53]

Kiernan meanwhile alleged that chronic masturbation could cause homosexuality, particularly among the "neurotic" and "insane." He wrote, "The female masturbator of this type usually becomes excessively prudish, despises and hates the opposite sex, and frequently forms a furious attachment for another woman, to whom she unselfishly devotes herself. The same phenomenon may be observed in male masturbators."[54] Thus as Oberndorf pointed out, "as a disease, homosexuality, quite as much as autoeroticism, with which it is so often associated, warrants the attention of physicians dealing with abnormalities of conduct."[55] Amid declining marriage rates, the subject of homosexuality, like masturbation, could not be ignored, for "the actual conditions of modern civilization give it a great social significance."[56] In short, homosexuals imperiled the race if they reproduced and if they did not.

While scientists worried about homosexuality's significance for the march of civilization, they often had more immediate fears as well. In addition to connecting sexual variance with venereal disease, as Lydston did with the "mulatto cook," scientists linked "perversion" with violence, crime, and vice.[57] Howard asserted in 1896 that it was a well-known fact that the "insane jealousy of homosexuals" and "young women" engaged in "lesbian love" and mutual masturbation drove many crimes, particularly murder.[58] Similarly, Kiernan connected sexual perversion to primitive "blood thirst" and "tribal fetishism."[59] Unlike Lydston and Howard, however, he believed that although most "perverts" suffered from congenital defects or "imperative conceptions," they were nonetheless legally responsible and should not be immune to prosecution. "While the victim of congenital sexual inversion cannot be regarded as a lunatic, nor as criminally nor civilly irresponsible, still there exists a peculiar psychical state closely akin to that of the hysteric or sexual neurasthenic," he remarked in an 1894 article, focusing there on the sexual perils of overcivilization rather than primitivism.[60] Still, Kiernan maintained, the criminal justice system critically needed a better scientific understanding of sexual perversion.

Accordingly, across the board, American sexologists asserted the broad social relevance of their medical knowledge, and many urged that their expertise on sexual matters be brought to bear on the law. Beginning in the 1890s, Howard urged that the "unfortunate class of sexual perverts be dealt with in a more humane manner than at present."[61] "The law should understand crime arising from psycho-sexual causes," he reiterated in 1904, and "this knowledge must be given to the law by the medical profession"; indeed, the social and legal application of sexology was "absolutely necessary if we wish to render justice to our fellow-man."[62] It was a sentiment he shared with a number of other American sexologists who similarly maintained that most sexual perverts, particularly inverts, suffered from a congenital disease or biological defect beyond their control and thus deserved pity rather than condemnation.[63] Others concurred with Kiernan that perverts and inverts were criminally responsible regardless of the cause of their "anomaly." Oberndorf, for instance, saw homosexuality simultaneously as a disease and a "social menace . . . to the psychic health of the community," warranting both "legal criminal restrictions" and "the attention of physicians."[64]

Indeed, many American sexologists worked at the nexus of medicine and law. Nearly all held medical degrees, many belonged to the Medico-Legal Society based in New York, and several also specialized in neurology or psychiatry, often working as consulting physicians at local mental hospitals or institutions for the criminally insane. Kiernan is a case in point. The frequent expert witness in criminal trials worked as a physician and clinical psychiatrist in Chicago, taught at several medical schools, was a fellow of the Chicago Academy of Medicine, and served for some time as a professor of forensic psychiatry at Kent College of Law, where G. Frank Lydson also held a professorship in criminology and sociology (and like Kiernan, continued to teach at medical schools as well).[65] Also like Kiernan, Allan McLane Hamilton (1848–1919), a New York neurologist and grandson of Founding Father Alexander Hamilton, frequently testified in court about the nature of sexual inversion and its criminal culpability as well as insanity more generally.[66] In fact, the native New Yorker, consulting physician to the Manhattan State Hospital and professor of mental diseases at Cornell University, also served as an expert witness in the Guiteau trial, though in his case for the prosecution to rebut Guiteau's insanity plea. He also consulted in the pretrial examination of another presidential assassin, Leon Czolgosz, who killed President William McKinley in 1901, for which he was put to death—though Hamilton believed that unlike Guiteau, Czolgosz was indeed insane. Hamilton was called as an expert witness in so many high-profile trials that he published a well-received memoir about his experiences bringing his medical expertise to bear on the law.[67] And Bernard Talmey titled two of his books *Woman: A Treatise on the Normal and Pathological Emotions of Feminine Love, for Physicians and Students of Medicine and Jurisprudence* (1904) and *Love: A Treatise on the Science of Sex-Attraction, for the Use of Physicians and Students of Medical Jurisprudence* (1915), the latter of which was published by the New York–based Eugenics Publishing Company and went through four editions. Despite sexologists' medical backgrounds, as Irvine suggests, American sexology was truly an interdisciplinary endeavor and one with considerable resonance for the law.

Moreover, since racial and sexual science shared much the same scientists and audience in turn-of-the-century America, it is hardly surprising the overlapping fields also shared a common language and set of concerns.

Indeed, for many scientists, questions of sex and "sexual perversion" resonated for the ethnologist as much as for the sexologist. Under the heading "Medical-Legal Contributions," the *American Journal of Neurology and Psychiatry* included brief discussions of recent scientific work. In the 1882 issue, the section first highlighted Kiernan and then went on to praise German sexologist Krafft-Ebing for his work on "the disgusting, but from an anthropological, ethnological, and medico-legal point of view important topic of sexual perversion." Part of what made Krafft-Ebing's work so noteworthy and of such ethnological interest to the editors was that his assessment of the "morbid character" of sexual perversion "call[ed] attention to the coexisting somatic evidence of degeneration."[68] As previously demonstrated, degeneration was a key concept in evolutionary theory's application to racial science. "Sexual perversion," then, was of considerable import to the ethnological scientist because it threatened the race (implicitly the white race), racial progress, and racial hierarchy. Likewise, in his 1905 article, "Medico-Legal Consideration of Perverts and Inverts," R. W. Shufeldt railed against suppression or censorship of sexological work in the United States as prudish and dangerous to "any one having the interests of his race continuously in hand."[69]

Sexology was not only legitimate science, Shufeldt argued; it was vitally important to the interests of the white race. "How can we ever hope to improve the tissue of which our race is composed unless we are familiar with every cell now included in its composition?" Shufeldt asked. "This demands the very closest study of every type in existence and particularly the abnormal and objectionable types, and an enormous number of these are to be met with among the psycho-sexual perverts." He lauded the recently published *Autobiography of an Androgyne* as a fascinating insider account of the life and mind of a self-professed sexual invert. At a time when anthropology was largely devoted to ethnographic studies of race, he declared, "It is an object lesson to every doctor of medicine and lawyer in this or any other civilized nation. It should be read by every sociologist, anthropologist, and professional person in this country."[70] In his antiblack missives, *The Negro: A Menace to American Civilization* (1907) and *America's Greatest Problem: The Negro* (1915), he would construct African Americans as a threat to civilization and white supremacy—the two were irrevocably linked in his work and that of many of his scientific contemporaries—but here, he focused on the threat from within. Shufeldt called for better scientific understanding of the problem of perversion not just for the sake of knowledge but for its application in improving the race. Clearly, U.S. scientists did not view race and sexuality as presenting separate problems but rather as interlinked threats to the status quo.

Congenital or Acquired Pervert? Homosexuality, Inversion, and Sexual Taxonomy

Taxonomy of human difference was one of the most visible scientific traditions that sexology borrowed from ethnology, as was the basic premise that such differences were grounded in the body. Ethnology had long devoted much of its attention to classifying the races, and each scientist promoted his own set of racial categories and subcategories. Moreover, these categories had deep social and political resonance that denoted not just difference but also hierarchy. Similarly, much of sexology was devoted to classifying and defining sexual variances. Sexology texts such as those by Kiernan, Lydston, Howard, and Coriat typically began with carefully qualified definitions of various "perversions."[71] Like racial classification, sexual categories were defined differently and often debated among scientists, who were particularly divided on how to classify inversion and homosexuality. Many sexology texts began by differentiating or connecting the two phenomena, and numerous articles in American scientific journals were devoted to exploring the relationship between sexual object choice, gender deviance, and bodily sex. Thus the sexual taxonomy each scientist devised was generally indicative of how he viewed this relationship.

While sexologists sometimes differed on terminology and often insisted they were proposing an entirely new schematic for understanding sexuality, a similar classification scheme ran through these late nineteenth- and early twentieth-century scientific texts. For one, sexologists usually divided sexual perversion into two categories: those with abnormal "sexual object choice" (which primarily referred to homosexuality, though some scientists also included bestiality or pedophilia under this heading) and those with abnormal "sexual aim" (which usually referred to a variety of fetishes or sexual crimes).[72] Although many American sexology texts included a titillating array of fetish case studies, the first category received scientists' most sustained attention. After all, in disrupting gender norms, frequently rejecting the sociopolitical institution of marriage, and impeding reproduction, scientists viewed homosexuality as endangering white civilization. Furthermore, just as adherents of ethnology had debated the origin of the races, sexologists debated the causes of homosexuality and inversion. Sexual perversion, particularly homosexuality, could be either (1) congenital or (2) acquired. Homosexual object choice could either (1) correspond to deviance in bodily sex or gender or (2) be entirely independent of such deviance. And finally, a man or woman could either be (1) the active agent in pursuing a homosexual relationship or (2) the passive recipient of another's homosexual advances. The passive–active dichotomy had implications for roles in the sex act itself as well. Scientists characterized homosexuality as a phenomenon that disrupted Victorian gender binaries, but, ironically, they created new binaries to describe homosexuality.

Among historians, homosexuality is now widely viewed as a relatively recent—and Western—invention, one that emerged when heterosexuality and homosexuality became constitutive of exclusive and opposite sexual identities rather than acts.[73] As Michel Foucault famously stated in *The History of Sexuality,* "The sodomite had been a temporary aberration; the homosexual was now a species." Moreover, sexual behaviors that were once thought of as criminal acts were increasingly described instead as diseases or pathologies—bodily deficiencies rather than simply misdeeds. Consequently, scientists attempted to locate clues about sexuality and character on the corporeal body. Also central to historians' narrative of the invention of homosexuality is the notion that scientists initially subsumed same-sex object choice under the broader category of gender deviance or "inversion."[74]

However, for scientists in the late nineteenth- and early twentieth-century United States, the relationship between gender performance, bodily sex, and sexual affinity was far from settled. The classification systems they proposed attempted to sort out this relationship as well as the myriad causes of sexual perversion. For example, Lydston used an elaborate schematic (Figure 3.1) in at least three publications, and it was reprinted by several other sexologists, including Kiernan and British sexologist Havelock Ellis.[75]

Notably, the primary division in Lydston's schematic was between "congenital" and "acquired sexual perversion," which he and other sexologists saw as an important distinction and one that resonated in the field well into the twentieth century. In an 1889 lecture, Lydston argued that

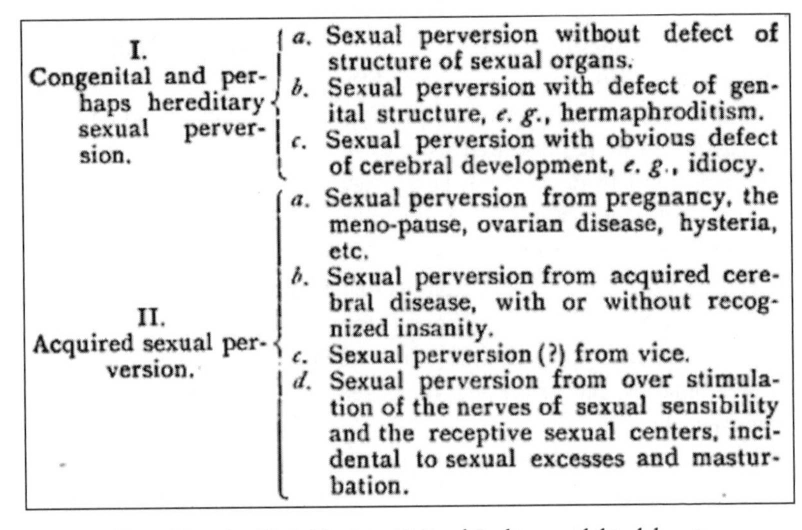

I. Congenital and perhaps hereditary sexual perversion.	a. Sexual perversion without defect of structure of sexual organs. b. Sexual perversion with defect of genital structure, *e. g.*, hermaphroditism. c. Sexual perversion with obvious defect of cerebral development, *e. g.*, idiocy.
II. Acquired sexual perversion.	a. Sexual perversion from pregnancy, the meno-pause, ovarian disease, hysteria, etc. b. Sexual perversion from acquired cerebral disease, with or without recognized insanity. c. Sexual perversion (?) from vice. d. Sexual perversion from over stimulation of the nerves of sexual sensibility and the receptive sexual centers, incidental to sexual excesses and masturbation.

Figure 3.1. Perversion classified. *Physician G. Frank Lydston used this elaborate classification schematic of the forms and causes of "sexual perversion" in at least three publications, and it was reprinted in texts by several other sexologists, including an article by his colleague in Chicago, famed psychiatrist James Kiernan. For many sexologists at the turn of the century, the relationship between bodily sex, gender performance, and sexual object choice was an open question, and this schematic, which divided perversion into "congenital" and "acquired" categories, represents Lydston's attempt to make sense of it. Some of his scientific contemporaries, Kiernan included, found Lydston's schematic overly complicated, but most subscribed to his basic distinction between congenital and acquired forms of sexual variance. There was considerable debate among them, however, over which form was more common; what, if anything, could be done to treat the problem; and whether "perversion" should be legally punished (Lydston, "A Lecture on Sexual Perversion, Satyriasis, and Nymphomania").*

sexual perversion with "congenital and perhaps hereditary" causes was more palatable.[76] "Even to the moralist there should be much satisfaction in the thought that a large class of sexual perverts are physically abnormal rather than morally leprous," he remarked. In other words, the congenital pervert should be pitied rather than persecuted and, ideally, treated medically. However, Lydston cautioned that it was often difficult to distinguish "between physical and moral perversion," and they could in fact be mutually dependent in some instances. Nonetheless, he concluded, "this does not affect the cogency of the argument that the sexual pervert is generally a physical aberration—a lusus naturae."[77] Sexologists largely followed Lydston's distinction between acquired and congenital perversion, but some argued that the acquired form was far more common. His colleague Kiernan, for example, maintained that "sodomy" was "often but simple vice," though he acknowledged "sometimes it results from an imperative conception," or from "vice . . . conjoined with congenital defect."[78]

Lydston's distinction between acquired and congenital perversion also had implications for his understanding of race. He and other scientists frequently associated "sexual perversion from vice" with the lower classes and, often, the "lower races." Moreover, sexual "over-stimulation" was considered by many scientists, including Lydston, as an innate feature of all black men; white scientists argued that this general hypersexuality could drive them to rape white women, or it could lead them to unnatural relations with other men. Elsewhere, Lydston argued that black men's supposed sexual excess was a feature of their biology that was beyond their control, but here he associated sexual "over-excitement" with "acquired" rather than "congenital" perversion.

Whether sexual perversions were congenital or acquired was not the only murky territory for Lydston. Individuals with "a predilection for their own sex" presented additional challenges. "The Precise Causes of sexual perversion are obscure," he wrote, but "just as we may have variations of physical form, and of mental attributes, in general, so we may have variations and perversions of that intangible entity: sexual affinity."[79] Throughout his publications on the subject, Lydston struggled to unpack the relationship between homosexuality and physical and "psychical" hermaphroditism, or what was often called "sexual inversion." Indeed, he attributed many "sexual perversions" to atypical genitalia and saw the problem of same-sex affinity as a form of physical or "psychical" hermaphroditism. Still, he could not avoid observing that in many cases, men and women who sought out members of their own sex demonstrated no corresponding deviance of gender or bodily sex.

Some sexologists claimed that only homosexuals of the most pronounced type displayed visible bodily differences, but most sexologists nonetheless spent considerable time examining the bodies of homosexuals and "inverts" with the expectation of physical indicators. However, the fact that such difference was often invisible to the naked eye did not make it nonexistent for these scientists. For many sexologists, the deviant behavior or gender expression of homosexuals and/or inverts was evidence that a congenital abnormality must exist. For example, in an 1896 article published in the *American Journal of Insanity,* Allan McLane Hamilton conceded, "While it is true that some of these subjects of the contrary sexual instinct present a physical departure from the ordinary standard, approaching that of the other sex, there is no arbitrary rule to guide us." But, he added, they nonetheless had "a dominant mental defect" that controlled their relationships with others.[80] Other scientists followed a similar model; cases of homosexuality or inversion in which no bodily differences were ascertained could still be considered congenital but indicated an inherited mental defect rather than a physical one. "The genital organs of the pervert are almost without an exception normal in appearance and function," Howard observed in 1897, and "the condition of these individuals is a decided psychical morbid entity." Still, he insisted, "I do not believe that true inversion is ever an acquired condition; it is congenital."[81]

Over time, with the growing influence of Freud and psychoanalysis, an increasing number of American scientists viewed homosexuality and inversion as a mental disorder rather than a strictly physical one. Far from a universal trend, however, it also reflected a disciplinary divide, with physicians and biologists continuing to search for bodily clues and psychiatric professionals looking instead (or in addition) to the mind. Writing in 1929, John Meagher approached the subject of homosexuality "from the standpoint of the psychiatrist." He remarked, "The urologist and gynecologist would most likely stress the physical rather than the psychic factors. But nearly all of these patients owe their status to psychic rather than to physical mal-development."[82] Even then, however, many scientists continued to link homosexuality with physical abnormalities, particularly in regards to primary or secondary sex characteristics. Writing in the same issue of the *Urologic and Cutaneous Review,* Oberndorf commented, "Medical literature and anatomical museums abound in curious examples where this bi-sexual predisposition of either sex manifests itself in the various forms of glaring, physical pelvic anomalies contrasting with the predominant physical characteristics of the opposite sex." He in contrast would be discussing "only those forms of homosexuality which appear in persons who deviate physically in no appreciable way or only very slightly from the average person."[83]

Indeed, the belief that character and morality were inscribed upon the body was so central to nineteenth- and early twentieth-century American cultural and scientific discourse that any feature of a body presupposed to be deviant could be read as evidence of its abnormality. In his 1896 article, Hamilton noted, "Of the sexual female examples that have come under my notice the offender was usually of a masculine type, or if she presented none of the 'characteristics' of the male, was a subject of pelvic disorders, with scanty menstruation, and was more or less hysterical or insane." This teleological approach to the body marked both racial and sexual science and blended physical and psychological features in constructions of difference. Likewise, Hamilton's discussion of female inverts focused on deviation from gender norms as well as masculine appearance, menstrual disorders, and insanity: "The views of such a person were erratic, 'advanced,' and extreme, and she nearly always lacked the ordinary modesty and retirement of her sex."[84]

Like racial taxonomy, then, the physical body clearly played an important—though complicated and often contested—role in classifying sexual variance. The distinction between congenital or acquired forms of sexual perversion or among homosexuality, inversion, and "true hermaphrodism" often hinged on physical features. As they had long done to distinguish the races, scientists examining homosexuals and inverts remarked on the shape, size, or symmetry of their crania, for example.[85] In this regard, the bodies of the "lower races," inverts, and homosexuals were described in remarkably similar ways.

The similar ways in which scientists described the "lower races" and "perverts" also focused on perceived sexual difference. Scientists often described black women as having masculine features, enlarged clitorises, and a pathologically voracious sexual drive. This characterization was echoed in many scientific considerations of lesbians, regardless of race. A summary of medical literature in the *Journal of the American Medical Association,* for example, pointed to a link between "hypergenitalism" and "unnatural sexual acts," which was "not without importance with regard to the responsibility of these perverts."[86] Describing a female patient who left her husband, renamed herself "Joe," and took up residence in the woods with a woman she called her wife, Kiernan wrote, "She had an enlarged clitoris, covered with a large relaxed prepuce. She had periodical attacks of sexual furor."[87] Scientists at the turn of the century typically characterized black men and women alike as suffering from "sexual furor," though the former's sexuality was cast as especially dangerous. Indeed, on the same page that Kiernan discussed the white lesbian, "Joe," he also turned his attention to the "blood thirst" and sexual furor that drove the "criminal assaults made by negroes in the South" upon white women.[88]

Scientists thus used similar language and logic in characterizing homosexuals and nonwhites. They described homosexual women of all races and African Americans of both sexes as overly masculine and dangerously hypersexual.[89] At the same time, there were also parallels in scientists' assessments of the bodies of nonwhite men and male inverts and homosexuals of all races, which they similarly painted as not masculine enough. Ethnological scientists had long distinguished black, Asian, and Native American men from their white counterparts by their "scanty beards"; this secondary sex trait simultaneously marked their racial difference and their status as somehow less than male. Likewise, "scanty beard" was often used to mark the sexual difference of homosexual and "inverted" men as well as the biological nature of their deviance.[90] Jennifer Terry writes, "In this respect, race functioned as both an analogous and synonymous rubric for conceptualizing sexuality in its deviant homosexual form."[91]

Scientists also categorized racial and sexual "intermediate types" in strikingly similar ways. Lydston's "spurious" mulatto cook demonstrated the threat of "passing" embodied in racial and sexual intermediacy. The fertility of both biracial people and physical and psychical "inverts" drew scientific attention as well. Scientists had argued since the early nineteenth century that biracial people were incapable of producing offspring, often as a means of proving the black and white races were separate species. Similarly, Lydston argued that "fortunately for society," hermaphrodites were sterile and "so far as procreation is concerned" could not function "as either male or female." And Douglas McMurtrie, who penned a regular column on sexology for the *American Journal of Urology,* recounted numerous inverted women he encountered who, pressured to marry, suffered miscarriages or stillbirths when capable of conceiving at all. For McMurtrie, the apparent infertility of "inverted women" demonstrated physical abnormality, not their likely resistance to marital relations.[92]

Sexual and racial taxonomies overlapped in other ways as well, with certain "perversions" occasionally attributed to a specific racial or ethnic group. In one article, Kiernan remarked that sexual perversions may frequently "reappear in certain races under the influence of tribal customs."[93] Meanwhile, in a lengthy paper presented before the Chicago Academy of Medicine, where he was a fellow, Kiernan recounted more than thirty case studies drawn from his own practice and the published reports of other sexologists. These case studies described a range of sexual "abnormities" or affinities outside the prescriptive norm of married, heterosexual, and procreative sex, beginning with hermaphroditism and sexual inversion and building to increasingly exotic fetishes.[94] Of these thirty studies, two were cases of physical intersexuality; race was specifically noted in both cases and neither was white. One was "a twenty-eight year old San Salvador domestic [who] was arrested for prostitution" and the other was Lydston's mulatto cook. The San Salvador domestic had apparently alternated between male and female identities, and "sexual gratification was equally distributed between the two sets of organs." The domestic, whom Connecticut physician C. W. Fitch (Kiernan excerpts Fitch's study in his paper) referred to as "it," was living as a man when authorities discovered her pregnancy and forced her to don female attire. The case studies of the domestic and the cook shared several common elements: their engagement in sexual activity with men and women alike, their representation as deceitful and predatory, their links to crime, and their status as racial "others." In contrast, Kiernan's case studies of "psychical" inverts, whose race or ethnicity was not noted, painted these individuals as pitiable, insane rather than deliberately deceitful, and self-destructive rather than predatory.[95]

While the role of racial difference was implicit in Kiernan's analysis of the two domestic laborers, it was explicit in the reprint of Kiernan's talk in the *Alienist and Neurologist.* There, his paper also included a summary of the discussion that occurred afterward in which respondents considered possible links between sexual perversion and racial "primitivism." Lydston was in attendance at Kiernan's talk and found his own work on sex and race frequently referenced in Kiernan's paper. In addition to the case study of the mulatto cook, Kiernan also featured Lydston's widely reproduced schematic that attempted to categorize the various causes of "sexual perversions." Also in attendance at the discussion, a Dr. Shobal Vail Clevenger pointed out that atypical "sexual appetite" was linked to "primitive desire" and reversion to lower evolutionary stages, while Kiernan expressed his disagreement with another scientist's characterization of "sexual perversion" as always a "moral insanity" by countering that it was essential to take into account "racial and religious customs and the influence of vice."[96]

Of Vicious Classes and Good Births: The Race and Class of Homosexuality

Ironically, while many American scientists worried that homosexuality and inversion threatened civilization, some suggested that these sexual problems were also caused by civilization. Kiernan, for example, wondered if "over-civilization" could be at fault for the "rise in American inversion," while Howard posited that "sexual perversion both in its congenital and acquired form seems to be correlated with the cycles of civilization." The problem was not limited to the lower classes, where the uninformed scientist might expect to find it. Quite the opposite was the case, Howard argued: "The last decade has demonstrated that sexual perversion is much more frequent in America than the general practitioner realizes, and that its victims do not belong to the vicious classes but to those of good birth—socially and mentally—and to those who have had educational advantages; many belonging to the professions." The scientist had to be careful to distinguish between the causes, however, and recognize that sexual perverts were "divided into *classes*—the congenital and acquired," he remarked.[97]

For Howard, the line between acquired and congenital perversion was often one of class and, implicitly, of race. The congenital pervert was Howard's concern, and that category did not refer "to the vicious, to the morally depraved, to the male prostitute, to that class which comes under police notice, or to the degenerate whose actions are decidedly anti-social." While acquired perversion was a vice of circumstance among people who cared little for morality and respectability—the "banal and vicious classes"—for Howard, the congenital "class" was found among the most respectable circles of civilized society. The "unfortunate psychical pervert" could especially be found "among the aesthetic class . . . among the painters, musicians, poets, and the writers of erotic fulmination." Indeed, Howard argued, "Ninety percent of these abnormal individuals are engaged in artistic pursuits."[98] These "unfortunates" should be pitied for their sexual abnormality, yet they seemingly held some redeeming social value through their artistic contributions.

In Howard's account, it was the trappings of advanced civilization and modern life—not primitiveness—that could be most closely tied to homosexuality and inversion. Some scientists exposed drag balls in urban black neighborhoods, which they usually associated with the "acquired class" of sexual perversion—the concern of vice squads and police rather than physicians and threatening mainly when their influence crossed the color line.[99] However, like Howard, most scientists were far more concerned with the sexual proclivities of middle- and upper-class whites and the import of sexual deviance for the future of the race. Indeed, American scientists often argued that modern civilization made men effeminate, and the advancement of women's rights led women out of the home and into each other's arms. But whether they turned a critical eye toward America's underclass or, like Howard, maintained that homosexuality and inversion were most readily found among the upper echelons of white society, race and class profoundly shaped scientific explanations for homosexuality's cause as well as recommendations for its control. White homosexuals, who made up the vast majority of case studies described by American sexologists, were often described as suffering from a congenital disease or abnormality, likely a product of civilization or overcivilization. When scientists described homosexuality and inversion among "primitives" or nonwhites, they represented the phenomena as an unsurprising indicator of the generally degraded morality and sexually indistinct character of that race—as vice rather than disease.

Given scientists' generally bleak assessment of black sexuality, it is perhaps not surprising that some of the most sensationalistic accounts of homosexuality in American sexology revolved around its incidence in African American communities. Most famously, in an 1893 article often quoted by queer historians, St. Louis physician Charles Hughes reported, "I am credibly informed that there is, in the city of Washington, D.C., an annual convocation of negro men called the drag dance, which is an orgie [*sic*] of lascivious debauchery beyond pen power of description." Police in New York City had recently raided a "similar organization," he added. The group drew its members from across the social strata of the black community, Hughes carefully noted, thus linking this "lascivious debauchery" to racial rather than class status. His description of the scene blends disruption of sex and gender with markers of primitive ritual and savageness: "In this sable performance of sexual perversion all of these men are lasciviously dressed in womanly attire, short sleeves, low-necked dresses and the usual ball-room decorations and ornaments of women, feathered and ribboned head-dresses, garters, frills, ruffles, etc., and deport themselves as women. Standing or seated on a pedestal, but accessible to all the rest, is the naked queen (a male), whose phallic member, decorated with a ribbon, is subject to the gaze and osculations in turn, of all the members of this lecherous gang of sexual perverts and phallic fornicators."[100] Understandably, queer historians, particularly those interested in race, have made much of this brief article—all of three paragraphs in length.[101] But Hughes's text may be even more notable for its exceptionality. Texts like this represent an extremely small proportion of sexology compared to work that focused implicitly or explicitly on homosexuality among middle- and upper-class whites. Moreover, those texts generally drew on case studies from the sexologists' own practice or that of their peers, discussing pathology or disease among specific individuals in contrast to Hughes's description of an anonymous black "gang of sexual perverts." And tellingly, Hughes's text is not even an article per se but rather a "Postscript to Paper on 'Erotopathia,'" a piece included in the same issue that examined perversion among whites. The phenomenon among African Americans, while clearly fascinating to Hughes, remained a "postscript" to his larger analysis of homosexuality and inversion.

Scientists paid far more attention to homosexuality and inversion among whites of the "respectable classes" for two primary reasons.[102] One, for these white scientists, this was the group on which the future of American civilization depended. And to be sure, in the late nineteenth- and early twentieth-century United States, sexology was almost entirely a white, male profession.[103] Second, in addition to their own racial affiliations and politics, numerous scientists concurred with Howard that homosexuality and inversion actually occurred more frequently among the "cultured classes."[104] Meagher echoed Howard's class-based characterization of homosexuality two decades later, remarking, "Many homosexuals are intellectual and cultured, though sexually infantile." The scientific methodology behind his conclusions is entirely unclear, but, he continued, "Indulgent male inverts like pleasant, artistic things, and nearly all of them are fond of music. They also like praise and admiration. They are poor whistlers. Their favorite color is green (red, complementary) whereas most individuals prefer blue or red."[105] One can only speculate what

Meagher's favorite color was. Ultimately, though, like many of his bourgeois peers, Meagher worried less about the middle class than the extremes of American's class strata. He described the country's wealthy elite—the "cultured classes"—as overly indulgent and hedonistic and seemed to imply the lower classes were simply too ignorant and degraded to think about sexual propriety. "Riches and poverty are more apt to favor homosexuality," he wrote, while "the development of the average individual is more apt to follow the normal course."[106]

Others suggested that general sexual dysfunction was an unfortunate product of civilized cultural mores. Shufeldt, for example, argued that "distressing prudery" had led to a broad range of sexual problems among "too large a proportion of the descendants of the original stock of the United States—that is, that proportion of the present population derived from Anglo-Saxon ancestry." Such dysfunction was not limited to homosexuality, and as an example, Shufeldt highlighted men who instead made unwanted sexual overtures to women in public. But there, too, "those guilty of such behavior are frequently men of high social position, and represent such professions as artist, singers, actors or even the clergy, rather than . . . the humbler walks of life, or still less the ignorant class." Shufeldt seemed to have sympathy for such men who were dragged into court and public notoriety whenever the "annoyed lady" complained, which stands in sharp contrast to his discussions elsewhere of sexually aggressive, "savage" black men and "outraged" white women.

Another frequent theme in American sexology was that while individuals with homosexual instincts were plentiful among refined whites, they sought outlets for their abnormal desires among the underclass, where opportunities for sexual debauchery of varying stripes were readily available. In his 1922 article, "Homosexuality," Oberndorf described one patient, a seventy-four-year-old Civil War veteran, as a "refined gentleman interested in art and literature" and much respected by his peers. For sexual partners, though, the man in question had long "preferred rough, coarse men, like longshoreman, husky and full of vitality," and found many willing participants among that group. In the same article, Oberndorf described a second patient, a physician, whose many sexual trysts were with "casual strangers of low social caste." The physician had a large circle of friends who shared his homosexual proclivities as well as his class status, but no sexual activity occurred within this group, Oberndorf noted. Instead, they seemingly all acted out their sexual inclinations among the working class. In this account, the longshoremen and men of low social caste were not characterized as homosexual so much as sexually indiscriminate, in contrast to the gentleman and doctors who sought sex among them.

Of course, not only did homosexual men and women cross class lines, but they sometimes crossed racial boundaries as well, which did not escape scientists' attention. Irving Rosse, a Washington, D.C., physician whom historian Jonathan Katz suggests was Charles Hughes's informant on the city's "colored erotopaths," was particularly disturbed by a series of arrests "under the very shadow of the White House." There, black and white "moral hermaphrodites" were found "in flagrante delicto."[107] He concluded that what happened in Washington was likely representative of other cities. Indeed, Rosse's friend Charles Hughes noted a similar phenomenon in St. Louis in his article "Homo Sexual Complexion Perverts in St. Louis—Note on a Feature of Sexual Psychopathy," published in the *Alienist and Neurologist* in 1907. He reported, "Male negroes masquerading in women's garb and carousing and dancing with white men is the latest St. Louis record of neurotic and psychopathic sexual perversion." The black men were "gowned as women at the miscegenation dance and the negroes called each other feminine names," reinforcing scientists' frequent characterization of the black race as both sexually debased and indistinct in sex and gender.[108]

Along with a 1913 article in the *Journal of Abnormal Psychology* by psychologist Margaret Otis that described interracial sexual relationships among female boarding school students, Rosse's and Hughes's texts have received considerable attention from queer scholars. But like Hughes's brief earlier discussion of "colored erotopaths," these writings make up a very small proportion of American scientific work on homosexuality.[109] Moreover, both Hughes and Otis seem less surprised by or concerned with black people engaging in homosexual sex than the fact that they were doing so with whites. Homosexual activity across the color line received considerable attention from vice commissions as well as the press during this period, but medical scientists were far more concerned with white homosexuals specifically or interracial sex—particularly interracial sex with procreative potential.[110] As Hughes himself ultimately concluded, "Homosexuality may be found among blacks, though this phase has not been so recorded, as between white males or white females," a sentiment common among scientists.[111] Most sexologists did not find instances of homosexual sex among black people especially noteworthy, nor did they seem to describe such activity in terms of a distinct category of behavior or pathology, since medical scientists widely viewed African American sexuality in general as inherently debased and indiscriminate. The same sexologists also rarely used the term *homosexual* or *invert* to describe them, instead referencing them by names—"complexion pervert," for example—that emphasized their *racial* difference and proclivity for interracial sex, the true source or their classification as "perverts." That is, while members of either race could engage in homosexual sex, the *homosexual* was implicitly white.

While scientists in the late nineteenth and early twentieth centuries frequently argued that nature made men and women less sexually differentiated at the low end of the evolutionary scale, some worried that bourgeois culture might also result in the breakdown of gender roles. Writing in 1929, Meagher seemed to hold particular disdain for the woman "who yearns only for higher education and neglects love."[112] Indeed, "so-called emancipated women are usually frigid, and usually have little unselfish maternal feelings."[113] American civilization had gone too far in Meagher's account, allowing women to pursue masculine paths and making men effete. In both men and women, "indulgence" in homosexuality could lead to a host of other physical and psychological problems, particularly impotence and frigidity, respectively.[114] "Needless to say, there are far too many frigid women and impotent men in our cultured classes," he remarked, raising implicit concerns about the future of the race. Race progress depended on the strict policing of gender and sex. "The best biological and social asset to society are the complete she-women, and the complete he-men and not their substituted opposites," Meagher declared.[115]

Implicit in some scientists' association of (primarily male) homosexuality and inversion with the arts and intelligencia lurked the possibility of homosexuals' social value. William Robinson—a prominent New York doctor; chief of the Department of Genito-Urinary Diseases and Dermatology at Bronx Hospital Dispensary; and the author of numerous books on sexual disorders, venereal disease, eugenics, and social hygiene—commented to fellow physician E. S. Shepherd that "perverts" should be given cyanide. Shepherd responded that "such wholesale elimination" might very well remove "a number of valuable citizens." Robinson in turn suggested he "sketch for the Journal a few intermediates who were of social value," a call Shepherd heeded in "Contribution to the Study of

Intermediacy," published in 1918 in the *American Journal of Urology and Sexology,* of which Robinson was the editor. First of all, Shepherd noted, "intermediacy and perversity [are not] synonymous," despite their correlation in popular and scientific thought. Partly to blame, "physicians, for the most part, see only the lower class of intermediates—those neurotics who are both perverse and mentally inferior or the spiritually feebleminded," from whom they generalize about all inverts. He conceded "that our streets and beaches are overrun by male prostitutes (fairies)," but "there is no more warrant for judging the intermediates by their lowest manifestations than for judging our womanhood by the lowest class of prostitutes."[116] And thus Shepherd set the stage for a distinction—between inverts who were socially valuable and those who were socially threatening—that rested explicitly on class status and implicitly on race as well. He wrote, "Intermediates are not only very common but also frequently men and women of great social value. Many of our national leaders have been not only intermediates but even sexually perverse, yet gave valuable service to their fellow men."[117] He even suggested that sexual intermediates were particularly inclined toward humanitarian and reform work because of their psychical "combination of feminine sympathy with the masculine initiative."[118] But, he added, only reformers "of the better sort" fit this definition. Indeed, for Shepherd, the social value of homosexuals and inverts was directly tied to their social status. It was precisely this association upon which one self-declared intermediate would base his own claims when he reached out to the scientific establishment for sympathy and support.

On Sex, Science, and Sympathy: An "Androgyne" Speaks

Between 1918 and 1922, Alfred W. Herzog, a New York physician and the editor of the *Medico-Legal Journal,* published three books by self-professed "androgyne" Ralph Werther, who also went by the additional pseudonyms of Earl Lind and Jennie June.[119] Under Herzog's editorship, the journal had published a number of articles on homosexuality.[120] He also penned lengthy introductions for each of Werther's books—*Autobiography of an Androgyne, The Female-Impersonators,* and *Riddle of the Underworld*—in which he explained that he was "persuaded that androgynism was not sufficiently understood and that therefore androgynes were unjustly made to suffer."[121]

Born in Vienna in 1866, Herzog moved to the United States as a child, though he received his secondary education back in Austria before returning permanently to America to obtain his medical degree from New York University. He worked at a variety of hospitals in New York City and northern New Jersey, specializing in diseases of the ear, eye, and throat, on which he also taught as a professor at New York's Eclectic Medical College. A renaissance man of sorts, the physician, surgeon, and author also dabbled in hypnotism and therapeutic electricity and went on to practice law in both New York City and New Jersey, alongside his continued work in medicine, remaining very busy indeed until his death in 1933 at the age of sixty-seven.[122]

Werther recalled that he completed *Autobiography of an Androgyne* in 1899 but was unable to procure a publisher for his sympathetic insider account of New York's sexual underground—that is, until he encountered Herzog and convinced him of the book's scientific import and "missionary" value.[123] The book, finally published in 1918, was "not intended as a defense of all those who indulge in homosexual practices," Herzog clarified; it was "published in an endeavor to obtain justice and humane treatment for the Androgynes, that class of homosexualists in whom homosexuality is not an acquired vice but in whom it is congenital."[124]

Toward that end, Werther consistently emphasized the biological roots of his "inversion." His publications interspersed textual descriptions of his anatomical difference from "normal" men with nude photographs of himself that underscored his fleshy, feminine form juxtaposed with classical sculptures of muscular masculinity (Figure 3.2).[125] But his self-representation as "sexually abnormal by birth"—and thus worthy of sympathy from scientists and society as a whole—relied on another important juxtaposition as well.[126] Largely embracing scientific understandings of sexuality and race, in all three books, Werther was careful to distinguish himself from those for whom homosexuality was an acquired vice. He emphasized his education and good birth to a respectable white family, which he contrasted with the lower-class—and, more implicitly, racially suspect—"toughs" and "degenerates" he encountered in the city's bars and meeting places. He also sought to refute "the chief charge against androgynes" that "they are guilty of 'the awful crime of race suicide.'" It was "the fault of nature alone" that "androgynes" did not perpetuate the race, he insisted.[127]

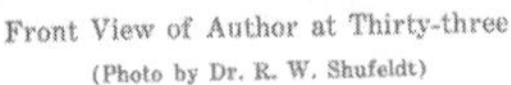

Rear View of Author at Thirty-three
(Photo by Dr. R. W. Shufeldt)

Figure 3.2a–b. Androgyny in black and white. *Included in a memoir explicitly written to elicit sympathy from scientists and society, these photographs of Ralph Werther, a self-professed "androgyne" or "invert," reinforced his physical divergence from classical models of masculinity while also signaling his whiteness. (Werther,* The Female-Impersonators*).*

One of eleven children, Werther was born in 1874 and raised in "the most refined section" of a small town in Connecticut, where his parents "were eminently respectable people."[128] Likewise, the childhood photograph he included was a formal portrait that visually underscored his bourgeois status (Figure 3.3). According to Werther, he exhibited feminine characteristics from a very young age, for which he was often teased, but he also recounted a large number of sexual experiences with other willing children. "My addiction [to fellatio] was common knowledge among the boys, and others sought it," he wrote in his autobiography, adding that he "told these boy playmates to call [him] Jennie."[129] He emphasized the time he spent playing with girls, who accepted him as one of their own, and described himself at length in terms that nineteenth-century Americans would have associated with women—emotional, weak, mercurial.

Figure 3.3. An invert by (respectable) birth. *With this childhood photograph, Werther underscores his respectable class background, a frequent theme in the memoir itself. His clothing, as well as his family's ability to afford an individual formal portrait, marked his middle-class status. Moreover, Werther referred to himself as "Fairie Boy" in this photo's caption to underscore that his inversion was not a moral failing he incurred as an adult but rather an innate characteristic that was evident even in childhood (Werther,* The Female-Impersonators*).*

He further recalled that as an early teen, he frequently fantasized about castrating himself "in order to bring [his] physical form more in accordance" with his mind.[130] Deeply religious and an excellent student, Werther left Connecticut when he was sixteen to attend a university in New York City. There, the intense crushes on male classmates he had experienced since childhood continued, and he reached out to various physicians for a cure. There began his long and complicated relationship with the medical establishment.[131]

From the beginning, Werther attempted to shape the terms of his medical care. Between the ages of sixteen and nineteen, he submitted to a variety of treatments, including electroshock therapy, among others, to no avail.[132] He wrote, "After several months treatment, I was rendered almost a physical and nervous wreck by the powerful drugs administered, but my amorous desires showed no change. I now repeatedly appealed for castration." His physicians refused, insisting that Werther might later regret the decision. Their refusal to perform the procedure is particularly interesting in light of the fact that contemporary scientists were recommending—and carrying out—sterilization for a variety of people who were not of the same "good stock" to which Werther laid claim.[133] One wonders whether the idea of castrating an intelligent white man, regardless of his sexual "abnormality," was simply anathema to the doctors at a time when scientists and politicians alike were increasingly concerned with falling birth rates among middle-class whites. Werther was persistent, however, and set out to conduct his own research on his condition as well as possible solutions. "I had recently read in a medical journal of a man similarly but not identically afflicted who was placed in possession of the normal procreative instinct through castration," he explained, and he recalled that he poured over articles from "American and foreign journals" at the New York Academy of Medicine to find more information.[134]

Werther also found a sympathetic psychiatrist who encouraged him to accept what "Nature" made him rather than attempt to change his biological fate. Whereas he initially sought the help of science to emancipate him from his "abnormality," this psychiatrist offered emancipation of a different sort from the guilt and shame Werther experienced about his sexual desires. As he recounted in his third book, *The Riddle of the Underworld*, edited by Herzog and serialized in the journal *Medical Life* in 1921, it was not until he obtained a third medical opinion, from a psychiatrist named Robert S. Newton, that he was finally told the "true nature of [his] malady." Newton's "frankness put an end to my chronic melancholia," Werther wrote, and "opened my eyes." Medicine had finally offered Werther the comfort he sought: "He taught me that the androgyne's proclivities are not the depth of depravity that every one previously had given me to understand. I now accepted thankfully from Providence the career of an androgyne."[135] This pivotal moment in his life likely prompted him to reach out to the scientific establishment again several years later with his literary efforts to end the persecution of sexual variants.

With his newfound acceptance of his "nature," at the age of nineteen, Werther began his double life as "Jennie June," a "fairy" street hustler, on the weekends and a legal secretary during the week.[136] "In 1893, at the age of nineteen, I count my adult life to have begun," he wrote in *Riddle of the Underworld.* "In 1893 I finally concluded that medical science was helpless in the matter of rescuing me from the hands of Destiny." And with the support of Dr. Newton, he then "ceased to struggle against Destiny."[137] He first solicited sex partners in a working-class Italian neighborhood in lower Manhattan, followed by several years working the Rialto for wealthier clientele. There, he developed a close network of fellow "androgynes," which he called the "Cercle Hermaphroditos," with the expressed intent of protecting each other against persecution.[138] As Henry Minton writes in *Departing from Deviance,* "What stands out in Lind's writings is his emancipatory spirit. Reflecting his self-perception, he equated homosexuality with congenitally based physical and emotional characteristics of the opposite sex. 'Androgynes' were thus men with male genitals, but with the souls of women."[139] When he was twenty-eight, Werther was finally castrated, a wish he had repeatedly expressed to doctors. In earlier years, he had claimed to seek castration in the hopes of obtaining a "normal" sex instinct; at twenty-eight, the procedure was ostensibly performed to correct spermatorrhea. However, Minton suggests that "his desire to be castrated may have actually satisfied a transsexual urge to become physically transformed into a woman," a point that is well supported by Werther's recollections of his early childhood.

In his introduction to *Autobiography of an Androgyne,* Herzog too expressed doubt that spermatorrhea was the true cause for the operation, remarking, "My belief is, that, feeling as a woman, desiring to be a woman and wishing to seem as much as possible like a woman to his male paramours, he hated above all the testicles, those insignia of manhood, and had them removed to be more alike to that which he wished to be."[140] Furthermore, by his late twenties, he had been adopting his Jennie June persona for years—to much local acclaim, Werther often boasted—and seemed to have little desire to change his sexual instinct.[141] Werther's castration, then, likely reflects his understanding and strategic manipulation of the medical establishment toward his own sexual emancipation and that of other congenital sexual variants.

Werther was keenly aware of scientists' power to shape popular discourse and thus their potential to save individuals like himself from both social ostracism and arrest. And so, during the height of his part-time life as Jennie June in the 1890s, he began to work on his autobiography, which he hoped would promote sympathy for inverts among the medical community. This was his target audience—and the only audience obscenity laws would allow him to reach. Indeed, sale of his books was restricted to members of the medical and legal professions and, perhaps to ensure that the most titillating portions would be understood only by scientists even if the texts fell into the wrong hands, he frequently described sexual acts in Latin.[142] He finished the first book in 1899, but it was "fated to wait eighteen years for publication primarily because American medical publishers—on the basis of the attitude of the profession—have had an antipathy against books dealing with abnormal sexual phenomena."[143] He eventually found a reticent but ultimately willing editor in Herzog and, despite his long road to publication, maintained his faith in the book's potential to recruit scientific allies to the androgyne's cause. "The author trusts that every medical man, every lawyer, and every other friend of science who reads this autobiography will thereby be moved to say a kind word for any of the despised and oppressed step-children of Nature—the sexually abnormal by birth—who may happen to be within his field of activity," Werther wrote in 1918.[144]

Although *Autobiography of an Androgyne* did not sell as well as Herzog hoped, he agreed to publish Werther's second book, *The Female-Impersonators* (1922), which was written "in a popular style for the general reader," followed shortly after by *Riddle of the Underworld,* an exposé of the characters he encountered in several urban "underworlds" "suitable for the scientific investigators to whom *Autobiography* was addressed." Still,

Werther insisted, "The scientific reader who is interested in my psychology and life experience should read all three." The trilogy, he explained, "together set forth all phases of the life experience of a bisexual university 'man.'"[145]

In these publications, Werther demonstrated that he was quite well versed in the scientific literature on homosexuality. He discussed the work of American and European scientists alike, finding much he liked in their biological models of sexual variance, while taking other scientists' theories—and, often, their moral condescension—to task. Shufeldt, who Werther hailed as "one of America's Foremost Medical Writers," was quoted at length in *The Female-Impersonators.*[146] Having received much practice in photographing anatomy in his *Studies of the Human Form: For Artists, Sculptors, & Scientists* (1905), Shufeldt also contributed the nude photographs of Werther that peppered the second book of the trilogy. Moreover, even when Werther was not directly citing scientists' work, he frequently used their language to describe himself or his compatriots. He used the familiar terms "invert" and "intermediate" as well as his preferred "androgyne," and like many scientists, he sometimes distinguished between "inversion" and "homosexuality" while using the terms interchangeably at other times. On several occasions, he also referred to himself as atavistic or an "atavic."[147] But most often by far, he used the term "congenital" to characterize his sexual variance, which he insisted was a product of biology and "Providence," not "moral depravity."

Werther grounded his claim to congenital "androgynism" on two factors: his bodily difference from "normal" men and his racial and class status. Throughout his writings, he emphasized his feminine physical features as well as his affinity for activities gendered female. He insisted that since boyhood, he was frequently told that he "markedly resemble[d] a female physically, besides having instinctive gestures, poses, and habits that are characteristically feminine." He provided an extensive list of the ways he differed from other men, including a high-pitched voice, skin "as soft as a woman's," silky hair, small features, "sacral dimples," a broad pelvis, sparse body hair, and enlarged breasts. The popular influence of anthropometrics was still evident in the early twentieth century, and he noted that he possessed "the feminine slope of shoulders and the feminine angle of arm." In contrast, his penis was small "but entirely normal" and his "testicles were pronounced of normal appearance" by the surgeon who performed his castration. Overall, he concluded, he was physically attractive with a healthy and robust appearance. Werther's presentation of his own physiognomy did not just mark his androgyny, though; it also underscored his whiteness (see Figure 3.2). From his creamy, much-admired complexion to his silky hair and "deep red" lips, his body disrupted sex binaries while simultaneously painting a picture of racial purity.[148]

Further refuting any possible argument that his inversion was a product of degeneracy, he traced his family's history in detail to demonstrate he was of "good stock"—a term loaded with racial and class implications in early twentieth-century America. He reported that the stock from which he came was both physically and morally sound and, over many generations, produced no criminals, only a few "half-wits" and one "mildly insane" cousin.[149] Insisting that "androgynism is not moral depravity or degeneracy," he held up his own bloodline as an example: "I myself—an extreme type of androgyne—spring from the most puritan stock. I was brought up to consider that on Sunday, reading anything but Christian doctrine or walking a hundred feet for mere pleasure were heinous sins. In addition to springing from the most puritan stock, both my paternal and maternal stock are of unusually strong build. . . . My stock and early environment are indeed the last that any one would pick out as likely to bring into the world a homosexual or androgyne as a result of moral degradation."[150] He pointed to examples of congenital sexual deviance among respectable whites outside his own family as well. According to Werther, one of his closest platonic friends in the "Underworld" was a voyeur who, like himself, could not help his innate sexual proclivities. Also like Werther, his voyeur friend hailed from an "ultra-Puritan" background, held a good job, and was well educated and respectable in every other way. Indeed, "apart from his voyeurism and related idiosyncrasies, his entire life had probably been *perfect* ethically," and he had never committed "the *least* offence against his fellow man."[151] Distinguishing between sexual desire and morality, he maintained that inverts were like normal people except in one small way that took up relatively little of their time, and thus "it is not a whit more immoral or irreligious to live in the manner ordained for 'him-her' by Destiny than for the normally sexed to live as ordained for them."[152] An admirably emancipatory message to be sure, but in divorcing ethical and religious levels from sexual proclivities, he instead tied them to race and class. In the racial logic of the early twentieth century, the very fact that Werther and his friends were white and middle class proved their sexual variance could not derive from degeneracy or general immorality and must instead be a congenital anomaly.

In contrast, Werther juxtaposed himself and his friends with the "vicious classes" he encountered in the "Underworld"—namely, immigrants, African Americans, and the poor. Whereas he, the other androgynes in his "Cercle Hermaphroditos," and his voyeur friend were all congenital sexual variants, part of—rather than apart from—Western Civilization (as underscored by his frequent allusions to classical Greece), and "perfect ethically," sexual deviance among other groups was simply indicative of the general moral degeneracy of that race and/or class (Figure 3.4). Of course, the fact that linking "degeneracy" to an entire race would seem to denote an inherited trait of some kind—a congenital phenomenon—was an irony lost on Werther. In another ironic note, he negatively portrayed his frequent and much desired partners, Italian immigrant "toughs," as morally lax and sexually indiscriminate precisely because they had had sex with *him*. Werther engaged in sex with men because his biology dictated it; his "virile" partners did so because they cared little how or with whom they found sexual release. Also in keeping with contemporary racial stereotypes, he recounted that his voyeur friend would often happen on "a picnic of the uncultured . . . particularly of the emotional African race,

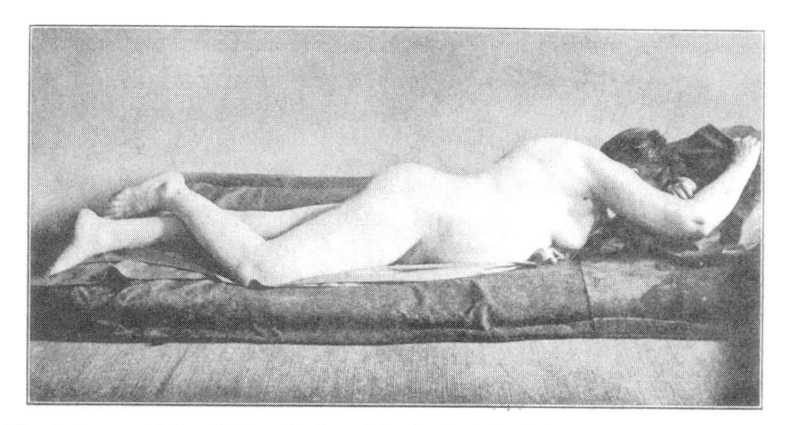

The Author—a Modern Living Replica of the Ancient Greek Statue, "Hermaphroditos"
(Photo by Dr. A. W. Herzog)

Figure 3.4. A classical form. *By again invoking the ancient Greek world, Werther represents the "androgyne" as a long-standing, even honored part of civilization rather than a threat to it (Werther,* The Female-Impersonators*).*

who were numerous in the large city in which we then resided" in the park he frequented. There, he would hide in the bushes with his binoculars "to make a study of the young bucks and wenches" who were "rendered delirious on the dance-platform." The same park grove was a popular sexual trysting spot for "love-sick couples of the lower class of Caucasians," Werther added. In this account, it was the black "bucks and wenches" and lower-class whites who were morally suspect, even if his friend was admittedly spying on them from the bushes.[153]

Having firmly established his whiteness, Werther sought to disprove "the claim that intermediates constitute a horrible menace to the perpetuation of the race."[154] Homosexuals were often charged with "race suicide" by scientists and, he noted, sometimes even by the law. In *The Female-Impersonators,* he described an incident in which a friend, brought up on sodomy charges, reported, "The judge said: 'It is as heinous as murder, because it strikes at the very existence of the race! No one but a criminal of the deepest dye could descend to it! Frank White, you have been convicted of the awful felony of race suicide!'" Notably, Werther generally created pseudonyms for his acquaintances as well as himself; it is unlikely an accident that he gave the man accused of "race suicide" the surname White.[155] "There is no ground for the charge that the passive invert's practices are aimed at the very existence of the race," Werther asserted. "In the first place, Nature made him psychically impotent from birth. In the second place, his practices could not be spread by example. They are regarded by all normally constituted males with such disgust and aversion that practically no one would stoop to them except those born with peculiar cravings."[156] He also suggested that it was not "high-minded homosexualists" who imperiled the race; it was the physicians who all too frequently encouraged them to marry the opposite sex as a cure for their "homosexual tendencies." "Chronic and extreme homosexuality is congenital and incurable," Werther maintained, and "it is monstrous to advise even a mild androgyne to marry, and thus contribute to propagating a line of unhappy and unwelcome bisexuals down through the centuries."[157]

Seemingly contradicting this bleak assessment of "unhappy and unwelcome bisexuals," elsewhere Werther asserted that androgynes reflected the pinnacle of civilization. They comprised civilization's most accomplished artists, musicians, and writers and were thus an asset to the race, not a threat or detriment. His "own impression" was that "sexual intermediates occur far more often among the brainy than among the brawny." Likewise, he "once read the declaration of a sexologist that 'sexual inversion is particularly common among authors'"—a rather ironic claim considering how many of the turn-of-the-century scientists discussed here tried their hands at popular writing.[158] Shufeldt, for example, wrote birding guides and contributed work to the *Century,* among other periodicals. A contributor to the *Saturday Evening Post* and the *Ladies Home Journal,* Howard also pitched an eleventh-century Icelandic romance to the *Century* as a serialized novel and in 1901 published a novel drawn from his scientific work on sex, titled *The Perverts.*[159] F. E. Daniel, discussed at length in the next chapter, wrote a folksy memoir of his Civil War experiences as well as a science-fiction novel, while Lydston wrote a travel memoir, a novel, and a play in support of eugenics.[160]

Werther's own writings illustrate the complex and dynamic relationship he had with the medical establishment. Facing public ridicule and the constant threat of arrest, Werther reached out to scientists as potential emancipators. The "vast majority of even the medical profession" remained ignorant of or resistant to the androgyne's plight, so he appealed to them as modern, civilized men and rational scientists: "In former centuries, the 'high-brows' burnt the unfortunates at the stake or buried them alive, just as today the benighted savages of Africa, animated by horror and loathing, bury *alive,* as soon as born, an albino infant. Only since the latter part of the nineteenth century, a handful of sexologists have been bold enough to proclaim that intermediates are often victims of birth, and irresponsible for their idiosyncrasy."[161] In no doubt deliberately provocative terms, Werther asked scientists to cast their lot with the most progressive among their peers instead of mimicking the superstitious derision of difference exhibited by "savages"! Indeed, "only bigoted pseudo-scientists have pronounced androgynes degenerates. Only medieval medicine, not modern medicine."[162]

To be sure, the scientific paternalism of sexologists like Herzog and Lydston hardly offered a positive depiction of sexual variation and presented its own set of problems that continue to resonate today. Moreover, Werther's texts, as well as the work of the American sexologists upon whom he drew for legitimacy and support, reveal that scientific sympathy for some depended on the demonization of others, often along racial and class lines. Still, it is not difficult to understand why medical paradigms of homosexuality as a pathology, defect, or disease were for some a welcome alternative to the police harassment, legal prosecution, religious condemnation, and physical violence that Werther describes in his work. And as the final chapter will show, it was also not the last time that scientists and their human subjects would form strategic alliances.

Racial scientists in the late nineteenth and early twentieth centuries were deeply concerned about sex. They were particularly worried about homosexuality among whites, which threatened the race from within, and, as the next chapter will discuss, black male sexual aggression, which posed a threat from outside. Racialized language abounded in scientific studies of sex, and for American scientists, homosexuality had clear implications for the fate of the races and their relationship to each other. Indeed, inversion and homosexuality—which numerous scientists interpreted as a form of evolutionary regression or degeneracy—imperiled the forward march of civilization and the stability of the middle class. And these scientists often accused white homosexuals of "race suicide," a damning charge at a time when the fate of mankind was largely understood as a racial competition. In this context, then, America's prominent scientists of race became the country's first sexologists.

Speaking to intersecting concerns over race and sex, these scientists sought to cement their authority in American society. "The subject of sexual perversion," in particular, "although a disagreeable one for discussion, demands the attention of the scientific physician, and is of great importance in its social, medical, and legal relations," Lydston wrote in 1917. With the very future of civilization at stake, scientists insisted that the nascent field of sexology must be taken seriously in the United States. "Inverts" and "perverts" were frequently "victims of a physical, and incidentally of a psychic, defect" and thus required study by the "scientific physician," not the "moralist" to whom authority over such matters was far too often given.[163]

During the same period, scientists—in fact, many of the *same* individual scientists—applied a similar model to another form of "sexual perversion" they saw as threatening American culture, that of black male sexual aggression. As with the issues of inversion and homosexuality, these medical scientists attempted to wrest authority over such matters from the law, as well as extralegal lynch mobs, by describing black male sexuality in terms of biological pathology for which they alone held the cure. But like their concerns about homosexual sex and gender nonconformity, even their anxiety about the allegedly hypersexual black man, the subject of the next chapter, was fundamentally about *whiteness*—that sexual and gender deviance among whites would contribute to "race suicide" and degeneration and black men would taint the purity of the white race by raping white women. Lydston's work in particular illustrates that sexual and racial threats were linked, if not rhetorically interchangeable, in American scientists' construction of a vulnerable social body.

· CHAPTER 4 ·

Unsexing the Race

Lynching, Castration, and Racial Science

THE YEAR 1893 PROVED A watershed moment in the history of lynching. It saw the first modern spectacle lynching, as historians have come to characterize the brutal, ritualized torture and murders, most frequently of African Americans, in front of large, enraptured and often cheering crowds.[1] The lynching of Henry Smith, accused of raping a four-year-old white girl, took place on February 1, 1893, in Paris, Texas, and was attended by approximately ten thousand people, thousands of whom had come by train from neighboring counties and states to witness Smith burned alive. The lynching was announced in advance, businesses in Paris were closed, and schools dismissed by the mayor so that everyone might attend.

However, 1893 was also an important year in black mobilization against lynching. The year marked intrepid journalist Ida B. Wells's first tour of England and Scotland to raise awareness about American lynch law. There, she turned civilization discourse on its head, calling into question whether white men were themselves more savage than civilized for their brutal displays of violence.[2] And born that same year into a middle-class black family in Atlanta, Georgia, Walter White would later grow up to inherit Wells's legacy as an antilynching crusader. His work with the National Association for the Advancement of Colored People (NAACP) in the 1920s and 1930s brought unprecedented attention to lynching, which he linked in part to America's tradition of scientific racism.

By the 1890s, white American scientists had spent nearly a century attempting to prove that racial hierarchy was biological and permanent. They invariably described people of African descent as irrevocably inferior in mind and body and thus placed them at the bottom of this natural order. Moreover, by the turn of the century, they characterized black bodies not just as inferior but also as threatening. Black men in particular were described as overly libidinous and aggressive, prone to a racially specific sexual perversion that drove them to rape white women. No idle chatter, this scientific discourse on race had all too tangible effects on living human beings. American medical scientists played a key role in constructing the popular image of black men as a sexual menace and grounding it in the physical body; in turn, this negative assessment of black men was often used outside the scientific establishment to justify lynching. Largely condemning lynch violence, however, scientists and physicians offered their own suggestions for how these threatening bodies should be treated or controlled.

Representing a wide array of fields, including medicine, biology, criminal anthropology, eugenics, psychiatry, sexology, and social hygiene, American physicians and scientists frequently commented on lynching. And beginning in 1893, numerous medical scientists began to suggest that castration would be more effective, humane, and progressive than lynching for dealing with the "Negro problem." With activists like Wells embarrassing America on a world stage for its complicity in extralegal violence, U.S. medical scientists came down on the side of law and order. But their frequent recommendation of castration as an alternative to lynching was as much about their concern for the authority of science over American race relations as it was about the authority of the law. The "castration remedy" represented a culmination of racial scientists' long-standing use of gender, sex, and sexuality to bolster their claims about the innate and permanent inferiority of black people. Thus while mostly condemning lynch law, these scientists nonetheless saw the rape of white women by black men as a serious—and growing—threat that needed to be addressed by those with proven expertise in the intricacies of race. They often asserted that lynching was not effective as a preventative and that the rape instinct was a pathology that needed medical intervention. Moreover, because scientists saw black men's alleged propensity toward rape as rooted in the body, specifically the sexed body, they recommended a bodily—and distinctly genital—solution.

America's foremost experts on race and sex contributed to the decades-long conversation among scientists about castration and lynching, which also extended into the popular press. G. Frank Lydston and F. E. Daniel in particular published on the issue for nearly twenty years. Most of these scientists offered at least a mild rebuke of the lynchers themselves but rarely challenged the stereotypes of black male sexuality that drove public discourse on lynching; quite the opposite, they helped create and lend scientific authority to such stereotypes. Indeed, there was no clear divide between scientific and public discourse on lynching or, as Walter White argued in the 1920s, even between scientific theories and public action. Castration in particular represented a point of intersection between science and the street. Even as scientists like Lydston and Daniel tried to present the two as separate domains, racial science and popular discourse on lynching shared language, imagery, and a highly charged set of assumptions about race, gender, sexuality, and power in America. Meanwhile, these scientists themselves broadened the scope of their work, increasingly publishing their recommendations for dealing with America's "race problem" in popular sources such as newspapers and pamphlets to reach a wider audience. Their efforts seem to have been successful; writings by an array of individuals outside the medico-scientific establishment, some of which I examine here, reflected a familiarity with this scientific discourse on race and racial violence. And more viscerally, the "surgical procedure" that medical scientists advocated as an alternative to death at the hands of a lynch mob instead became a frequent component of the lynching ritual.

It is important to recognize that this was not a fringe conversation. Not only did the decades-long "debate" reach a broad audience; it drew on the popular appeal of eugenics, which primed the American public to be receptive to sterilization as a cure for a variety of social ills. Similarly, it also overlapped with a broader—and ultimately successful—campaign to transform circumcision from a primarily religious practice among people of Jewish descent to a dominant *medical* procedure performed on most male infants by the mid-twentieth century for reasons of social and sexual hygiene. Because numerous influential physicians saw the foreskin as sexually stimulating, they believed men's moral character could be improved through modification to the genitals, and if this relatively minor medical intervention could control the sexual impulses of otherwise civilized men, they reasoned, by extension the more drastic surgical castration could curtail sexual aggression among the most savage. In short, from the late nineteenth century through the 1920s, emasculation was central to scientific and popular discussions of lynching, the brutal practice itself, and racial scientists' proposed medico-legal alternatives to mob violence.

"Castration Instead of Lynching": Science Confronts Violence

On March 11, 1893, one month after Henry Smith was burned alive in Texas, Hunter McGuire, a respected Richmond physician, professor, and president of the American Medical Association, wrote an open letter to G. Frank Lydston, then professor of genitourinary surgery at the Chicago College of Physicians and Surgeons, seeking "some scientific explanation of the sexual perversion in the negro of the present day."[3] The Smith lynching demonstrated that the punishment for such perversion was swift and severe, McGuire noted matter-of-factly, and, he added, entirely inevitable. In response, Lydston proposed castration as both a solution to sexual crimes and an alternative to lynching, which he deemed ineffective as a rape deterrent. McGuire's letter and Lydston's lengthy response five days later were published in the *Maryland Medical Journal* and later reprinted in book form, under the title *Sexual Crimes among the Southern Negroes, Scientifically Considered.*

McGuire and Lydston's correspondence launched a conversation among scientists that lasted three decades and spanned medical—and some popular—publications across the country. Lydston himself returned to the question repeatedly in numerous publications. Writing explicitly in dialogue with one another, scientists considering the "castration versus lynching" question demonstrated near consensus on several key points. For one, these scientists represented a uniform front in perpetuating the rhetoric of black hypersexuality common to popular discourse on lynching. Second, most scientists in what I loosely call the "castration or lynching" debate explained black men's supposed propensity toward rape as rooted in their biology. Third, as criticism of American lynch law mounted amid the antilynching publicity campaigns launched by Ida Wells and later the NAACP, scientists offered at least a perfunctory rebuke of extralegal violence. Few did so out of purely humanitarian concerns or belief in racial equality, however. These scientists questioned not so much the morality of lynching but its effectiveness as a punishment or preventative. And most notably, nearly all the scientists agreed that castration was preferable to lynching for dealing with what they saw as an ever-growing threat—black sexual predators. Castration, then, was a twofold solution to interrelated problems; it addressed the troubling cause, black rape, and its disturbing effect, lynching.

Like Lydston, other scientists and physicians who advocated castration as an alternative to lynching viewed themselves as progressive reformers, saving society from black men and black men from themselves. By pathologizing the "black beast rapist," turning a popular racial trope into a disease, scientists created a cultural need for their medical expertise. Even more than lynching's ineffectiveness or inhumanity, scientists' objections to the practice revealed their concern for scientific authority, particularly over racial matters. Just as ethnologists half a century earlier had presented themselves as experts on the pressing racial issue of the day—slavery—so too did turn-of-the-century racial scientists weigh in on lynching. Scientists' proposition that surgical castration might offer an alternative to lynching laid bare similar issues of authority—the authority of the scientist on social and political problems versus the power of the public—by suggesting that medicine rather than the law or vigilantism offered the solution to America's "race problem."

Scientists in the "castration or lynching" debate diverged, however, on precisely what sexual modification should be performed on black rapists and toward what ends. Some scientists used the term "castration" broadly to refer a range of possible genital surgeries. Others hailed the benefits of one specific procedure over others, including removal of the testicles, vasectomy, or even circumcision. The exact surgical procedure a scientist advocated correlated to how much emphasis he placed on individual punishment or cure versus large-scale prevention. Over time, scientists who advocated surgical remedies for black rape adopted increasingly expansive definitions of "prevention" and often increasingly eugenic aims. That is, some scientists believed castration could not only offer *rape* control in the short term but also contribute to a eugenic agenda of *race* control in the long term. Still, scientists in this "debate" shared one thing in common: they saw black rape as a social problem demanding a medical solution.

Racial science could make strange bedfellows, and McGuire's correspondence with Lydston was no exception. The two men had both served as military surgeons, Lydston, a native of California (who later lived in New York, Maine, and Chicago), during the Spanish-American War and McGuire for the Confederacy in the Civil War, during which he gained prominence as the physician who operated on Stonewall Jackson and tended to him on his deathbed. McGuire remained fiercely loyal to the South after the war.[4] Perceiving anti-Southern bias throughout the world of medicine, he devoted much of his career to enhancing the scientific authority of Southern physicians. As the president of the Southern Surgical and Gynecological Association in the late 1880s, he rallied Southern physicians to the cause of elevating the reputation of Southern medicine by appealing to their manhood and sense of honor. "As Southern men, let us show to the world that, under changed conditions, we have still the stamina of our forefathers," he implored. "As members of our beloved profession, let us strive to be first in scientific attainment, first in integrity, first in high purpose for the good of mankind."[5]

Given McGuire's veneration of Southern medicine, his decision to seek Lydston's expertise on the South's "race problem" may seem surprising. But the two men traveled in the same scientific circles and belonged to many of the same medical associations. Moreover, like McGuire, Lydston worried about his professional authority; throughout his career, Lydston expressed frustration that his studies of sex were not recognized as legitimate science. Most importantly, as their correspondence indicates, given Lydston's well-known expertise in criminal anthropology and social hygiene, McGuire felt confident that he would be sympathetic to the South on issues of race, even if he was not a born Southerner himself. Scientific concern about the "Negro rapist" transcended political affinities or regional cultures.

Both scientific paternalists, the two men of very different upbringings publicly joined forces in 1893 to save white civilization from the Negro and the Negro from his own destruction. The flip side to antebellum paternalists who claimed African Americans were thriving under slavery, McGuire—and many of his contemporaries—maintained that the race was dying out in the wake of emancipation. McGuire's emphasis on the "frightful survival of the fittest" in his discussion of "Negro rape" suggested that black men's alleged hypersexuality rested at least in part from a biological drive to fend off their own demise, all at the expense of white chastity.

McGuire had read Lydston's essay, "Sexual Perversion," with much interest. With Lydston's expertise on matters of sexual deviance, McGuire hoped he would have some insight on a phenomenon that he considered a form of sexual perversion: "rape by a negro of a white woman." In his letter to Lydston, McGuire insisted the phenomenon was "almost unknown . . . before the late war between the States." Since then, he maintained, "the newspapers tell us how common it is," taking the validity of such reports at face value. Swift and certain death was the punishment. He explained, "This is the unwritten law of every community in the South; from it there is no appeal. It is immutable, and is sustained by every living white in the community in which the crime occurs. I am not engaged here in defending this law, although it is easy to do it." Instead, he was simply providing Lydston with "some facts" to inform their "purely scientific discussion."

The South had its own way of dealing with the problem of "sexual crimes among the Southern negroes"—lynching—but what McGuire sought was a scientific explanation for the crimes. "It is not the legal, social, moral or political aspect of this perverted sexuality in the negro upon which I ask your opinion," McGuire stated. These perspectives had shed little light on the problem, so McGuire asked Lydston to "investigate it as a scientific physician—one who has devoted much to this and kindred matters." Quick to throw around the label of "perversion," McGuire accused northern journalists who reported on lynching and saw "only the fearful spectacle of a hung, burnt, or shot negro" and not the "innocent, mutilated, and ruined female victim" as themselves suffering from a "perversion of mind and heart." McGuire trusted that Lydston, as a fellow man of science, would know better and focus on the real issue at hand: the black man's "sexual perversion" rather than the nature of Southern punishment.[6]

Lydston, for his part, did indeed offer a scientific explanation—"furor sexualis," characterized by "abnormal passions" compounded by a lack of self-control—for the problem of "sexual crimes among Southern negroes." Lydston insisted that "furor sexualis," a sexual disorder first described in 1885 by Lydston's Chicago colleague, sexologist James Kiernan, was innate.[7] He suggested that black men were especially prone to rape by the specific nature of their biology compounded by current environmental conditions. As we have seen, many racial scientists in the late nineteenth century—including Lydston—characterized black men as simultaneously too much man and not man enough. That is, they embodied brute masculinity without the manly restraint and rationality of civilization. Summarizing this body of scientific thought on black male sexuality under the rubric of "furor sexualis," he listed the following environmental and biological factors, each of which he later elaborated on at length:

1. Hereditary influences descending from the uncivilized ancestors of our negroes.
2. A disproportionate development of the animal propensities incidental to a relatively low degree of differentiation of type.
3. A relatively defective development of what may be termed the centers of psychological inhibition.
4. Physical degeneracy involving chiefly the higher and more recently acquired attributes, with a distinct tendency to reversion of type, which reversion is especially manifest in the direction of sexual proclivities.
5. The removal of certain inhibitions placed upon the negro by the conditions which slavery imposed upon him; these were removed by his liberation.
6. An inherent inadaptability to his environment both from a moral and legal standpoint, the result of his inadaptability being an imperfect or perverted conception of his relations to his environment—i.e. to the body social.
7. An incapacity of appreciation of the dire results to himself of sexual crimes.[8]

Scientists referenced Lydston's summary of black sexual degeneracy for decades while generally ignoring his frequent call for color-blind justice. As he had so frequently said on behalf of congenital "inverts," Lydston argued that black men needed treatment, not punishment, for their own benefit and the protection of society as a whole.

Consequently, Lydston saw the "'Southern method' of dealing with the criminality of the negro" as deeply problematic. He assured McGuire that he recognized "the liberality of [his] Southern friends" and sympathized with the problem they faced.[9] Lydston however was "opposed to capital punishment, legal or illegal," but not for "sentimental" reasons. "From a utilitarian standpoint," it simply did not achieve its stated goal of decreasing crime. "On the contrary," he maintained, "capital punishment seemed to have a direct effect in increasing the savagery of, and lessening the respect for human life entertained by, the body social." Referencing the well-publicized lynching of Henry Smith the previous month, Lydston continued, "I am not inclined to captiously criticize, mind you, the typical Southern method of dealing with negro ravishers—for I would probably be as quick to act similarly under like circumstances—but do you think that any reasoning whatever could justify the recent roasting of a negro ravisher in one of the Southern States?"[10]

Lydston then offered a scientific solution to "negro ravishers" and its attendant problem, lynching. Castration was the "only one logical method of dealing with capital crimes and criminals of the habitual class." He proclaimed, "Executed, they would be forgotten; castrated and free, they would be a constant warning and ever-present admonition to others of their race." Though it is hard to imagine that the black community of Paris, Texas, would quickly forget Henry Smith's horrific death, Lydston insisted that castration would better serve as a "wholesome warning to criminals of like propensities." It was a common belief in social hygiene, criminal anthropology, and eugenics—all scientific endeavors with which Lydston was associated—that criminal behavior was passed from one generation to the next. Thus castration of "negro ravishers" had an additional benefit: "It prevents the criminal from perpetuating his kind." Furthermore, he added, "A few emasculated negroes scattered around through the thickly-settled negro communities of the South would really prove the conservation of energy, as far as the repression of sexual crimes is concerned." For Lydston, the most efficient way to control an entire race was to inhibit the sexual capacities of a few individual men.

By castrating "negro ravishers," Lydston avowed, "the murderer is likely to lose much of his savageness; the violator loses not only the desire but the capacity for a repetition of his crime, if the operation be supplemented by penile mutilation according to the Oriental method."[11] Here, Lydston hinted at the complex genealogy of castrating human beings. Castration has meant different things in different times and places. In many historical contexts, a eunuch's inability to reproduce did not convey powerlessness or peripheral social status. In fact, they occupied a vital position in royal courts.[12] Even then, however, the eunuch was often associated in Western culture with a racial "other" (usually "oriental") and with servitude.[13]

Castration in America was a part of this genealogy, but it was also shaped by the country's unique racial circumstances. And in this U.S. context, castration was never a source of power. It was mandated as a punishment for slaves in several colonial statutes, and antebellum planters occasionally castrated their slaves to control unruly behavior.[14] Planters also castrated weak, small, or unhealthy slaves to prevent them from producing similarly impaired children, although this practice was relatively rare.[15] Furthermore, as Lydston himself noted, Gideon Lincecum, a Texas physician, proposed that castration was more humane and appropriate than (legal) execution for a host of crimes.[16] Lincecum argued, "It is the animal and not the intellectual portion of our organic structure that commits crime and does violence."[17] He shared his views in an 1854 essay he disseminated to more than six hundred legislators, newspapers, and doctors after which he was widely ridiculed. However, just ten years later, a black man convicted of rape received a sentence of castration in the same state.[18]

Castration, then, had already enjoyed a long history in America by the 1893 lynching of Henry Smith and the publication of Lydston's often cited article. The growing popularity of eugenics in the 1890s ushered in a more receptive environment for castration proposals. As scientists discussed castration's possible application to the problems of black rape and lynching, involuntary sterilization was being debated and enacted in other American contexts. At the height of eugenics' popularity in the late nineteenth and early twentieth centuries, state institutions oversaw the sterilization—often involuntary—of more than sixty thousand people suffering from mental illness or retardation.[19] Between 1907 and 1921, fifteen states—none of them Southern—passed legislation allowing for the sterilization of "defective persons," and 2,558 procedures were performed during that period in Lydston's home state of California alone.[20] A number of the same scientists, most notably Texas physician and eugenicist F. E. Daniel, who advocated castration as an alternative for lynching, had also lobbied publically on behalf of state sterilization legislation. Some scientists, including Lydston and Daniel, also recommended castration for masturbation, which many saw as either a form of insanity or a practice that exacerbated mental illness. In fact, Daniel advocated castration "as a penalty for all sexual crimes or misdemeanors, including masturbation."[21]

And castration was not just talk for medical scientists. Before any legislation had been passed to permit compulsory sterilization, Dr. Hoyt Pilcher, superintendent of the Winfield (Kansas) Asylum for Idiots and Feebleminded Youths, admitted in 1894 that he had "castrated" fourteen girls and forty-four boys under his care.[22] The ensuing controversy cost Pilcher his job, but a number of vocal physicians also spoke out on his behalf; among them, again, was Daniel.[23] And numerous scholars have shown that the United States exported eugenic ideology to its imperial interests, a product of the common belief that poor and racially inferior peoples were reproducing at a much higher rate than middle-class American whites.[24] A range of reproductive or genital surgeries—again, many involuntary—were recommended and performed in a variety of contexts in America and its territories in the late nineteenth and early twentieth centuries. It is little wonder, then, that Lydston's recommendation of castration as an alternative to lynching prompted so much discussion among scientists.

Indeed, the impact of Lydston's open letter to McGuire was immediate and long lasting. Soon after the McGuire–Lydston correspondence was published, Daniel, a vocal critic of both racial and moral degeneracy, enthusiastically endorsed Lydston's surgical solution in "Should Insane Criminals or Sexual Perverts Be Permitted to Procreate?," a paper he read before the Joint Session of the World's Columbian Auxiliary Congress Section on Medical Jurisprudence, the International Medico-Legal Congress, and the American Medico-Legal Society, New York, in 1893.[25] Approvingly, he recalled that Lydston "would castrate the rapist, thus rendering him incapable of a repetition of the offense, and of propagating his kind, and turn him loose—on the principle of the singed rat—to be a warning to others."[26] In addition to being a mainstay on the scientific lecture circuit in 1893, Daniel also published the paper, under the title "Castration of Sexual Perverts," that same year in the *Medico-Legal Journal* and the *Psychological Bulletin,* as well as the *Texas Medical Journal,* of which he was the editor, publisher, and proprietor. As with the McGuire–Lydston correspondence, Daniel's work had appeal for and reached a national audience; his publications and lectures were not limited to the South.

Concerned with justice rather than vengeance, Daniel argued that civilized society demanded the calm rationality of the informed medical scientist working in conjunction with the legal establishment. "The aim of jurisprudence should be, in addition to the repression of crime, a removal of the causes that lead to it, and reform, rather than the extermination of the vicious," he observed. Daniel, like many of his scientific contemporaries, believed that criminality was inborn and hereditary. Thus "the offender should be rendered incapable of a repetition of the offense, and the propagation of his kind should be inhibited in the interest of civilization and the well-being of future generations." Also referencing Henry Smith's recent lynching, he added, "These ends are not fulfilled by hanging, electrocution, or burning at the stake."[27]

For Daniel, castration would be "prophylactic and protective, both to society and to posterity." It would protect (white) women from the immediate threat of rape, and the (white) race from the threat posed by criminal sexual behavior being passed on to future generations or degenerating the white race through miscegenation. That is, castration had the added eugenic benefit of "race improvement."[28] Daniel argued that not nearly enough was being done to address sexual crimes and ruefully pondered, "Is it not a remarkable civilization that will break a criminal's neck, but will respect his testicles?"[29] However, when it came to black men, lynch mobs were all too willing to break their necks and emasculate them. Clearly considering his article a classic, Daniel reprinted it again in the *Texas Medical Journal* in 1912.[30] He also returned to the issues of rape, lynching, and castration in at least two additional articles, "The Cause and Prevention of Rape" (1904) and "Elements of Decay in American Civilization" (1909), published in the *Texas Medical Journal.*[31]

While none was as prolific as Lydston and Daniel, other scientists—most of them physicians—also weighed in on castration's potential as a remedy for both black rape and lynching.[32] The medical scientists who followed them all took as a given that the rape of white women by black men was an ever-growing threat that needed to be addressed immediately. In so doing, they employed the same racist language as Lydston and similarly attributed the black men's sexual aggression to his biology. P. C. Remondino (1846–1926), a San Diego physician, proponent of circumcision, and publisher of the *National Popular Review,* is a case in point.[33] Remondino reinforced the popular stereotype that black men had overly large penises. Moreover, from childhood on, black males were "more subject to nervous disorders than girls"; like women, they were emotional, passionate, and irrational. This unfortunate combination—hypermasculine bodies with the weak minds of women—made them particularly susceptible to rape, Remondino maintained.[34] He was convinced that the foreskin of the penis and its secretions could often prove "irritating" or "stimulating," resulting in an "over-exuberant and impatient virility," a problem that was exacerbated in the overly endowed black man. He characterized African Americans as "a race proverbial for the leathery consistency, inordinate redundancy, generous sebaceousness and general mental suggestiveness and hypnotizing influence of an unnecessary and rape, murder and lynching breeding prepuce."[35] In Remondino's web of logic, the blame for lynching could be placed not on the mob but on the black man's foreskin. He did not condone lynching, however. On the contrary, the mob response to America's race problem was "not at all creditable to us" or in keeping with "civilized nations."[36]

"A Far Greater Authority on Racial Matters": Science and Civilization

That Remondino and other turn-of-the-century scientists characterized lynching as contrary to "civilized nations" was due in large part to the efforts of antilynching activist Ida B. Wells in the early 1890s. According to historian Gail Bederman, Wells "brilliantly and subversively manipulated dominant middle-class ideas about race, manhood, and civilization in order to force white Americans to address lynching. . . . Wells, in short, convinced nervous white Northerners that they needed to take lynch law seriously because it imperiled both American civilization and American manhood."[37] Her impact extended outside the North, however. Between 1892 and 1894, Wells lectured across the United States and Britain, making it increasingly unfashionable for public figures on either side of the Mason-Dixon Line to condone lynching. American discourse on "'civilization' positioned African American men as the antithesis of both the white man and civilization itself."[38] But in her editorials, pamphlets, and lectures, Wells masterfully inverted civilization discourse by arguing that it was the white members of the lynch mob who were the savages. In her speaking tours of the British Isles in 1893 and 1894, she put the "white man's civilization on trial," indicting the United States as a whole for allowing such barbarism as lynching.[39]

In this context, even the most virulently antiblack scientists were faced with increased pressure to condemn lynching, at least publicly. Scientific authority hinged on the scientists' status as rational and civilized. But even as growing numbers of Americans were denouncing lynching between the 1890s and the first decades of the twentieth century, the "black beast rapist" trope maintained widespread acceptance among white Americans. This gave scientists the space to offer a medical solution to both social ills: black rape and its frequent result, lynching. Remondino, for example, stated that the surgeon's knife would be more "humane" than the "many burnings, hangings, shooting and stonings that have of late taken place."

The actions of the lynch mob in contrast were "rather more in keeping with thought and actions that prevailed before the revival of Medicine," he added. Lynch mobs were unnecessary when a far greater authority on racial matters existed: the physician and the scientist.[40]

Scientists' own words often revealed that their primary objections to lynching were not moral. Like Lydston and Daniel, Remondino questioned lynching's effectiveness as a deterrent. Indeed, medical intervention instead was needed to control black male sexuality, but Remondino insisted that circumcision offered a better solution than castration. Castration only addressed individual criminals after they had already destroyed the lives of their victims, Remondino pointed out. But circumcision was a true preventative measure because it would prevent rape rather than recidivism. Moreover, because Remondino and other scientists believed that all black males were potential rapists by the nature of their biology, mandated circumcision would reach black men as a whole. In a very backhanded compliment, he pointed to the low incidence of rape among Jewish men to reinforce his claim that circumcision could serve as a preventative. "Although the male Jews are much given to unholy and unedifying carnal pursuits," he noted, "we never hear of a Jewish rapist."[41] Thus Remondino felt "fully warranted in suggesting the wholesale circumcision of the negro as an efficient remedy in preventing the predisposition to discriminate raping so inherent in that race."[42]

Remondino's recommendation of circumcision for all black men received a warm reception in several medical journals at a time when growing numbers of physicians advocated the procedure more generally for similar reasons of sexual purity and control.[43] The *Boston Medical and Surgical Journal* favorably reported in February 1894 that circumcision would be an "efficient and gentle means of stopping the sexual crimes and improving the moral system of the negro race."[44] The same month, an editorial in the *Maryland Medical Journal* concurred that unlike universal black circumcision, "legal castration," which so many had endorsed as a remedy for black rape, could not reach a large enough number of men and could not be "enforced with any degree of justice or humanity that would prove practical or efficient."[45]

Meanwhile, one supporter of Lydston's castration proposal came from a most unlikely and unusual source: a black man who had served with the U.S. Colored Troops during the Civil War. But then, there was nothing usual about William Hannibal Thomas. Of mixed-raced heritage, by the turn of the century the preacher, journalist, and legislator had transformed himself from an optimistic black leader into one of the most critical anti-black voices in America. And in 1895, he attempted to put the "castration remedy" discussed by white scientists into action. Thomas drafted a congressional bill, "An Act to Prevent and Punish Criminal Assaults on Female Chastity, and Other Felonious Acts," which aimed to stem the rising tide of lynching by addressing what Thomas saw as its cause—the rape of white women by black men.[46] Simultaneously condemning lynching and blaming Southern freedmen for their own deaths, he argued that the growing incidence of sexual assaults by black men in the South threatened "the character of our civilization" because they inspired extralegal violence by whites in response. Therefore, he proposed that men convicted of rape or attempted rape be castrated by "Executive Surgical Bailiffs" "to deter others from committing like crimes, and to remove all incitement to lawlessness consequent thereon."[47]

Thomas's biographer John David Smith characterizes his "proposal to punish both rapists and lynch mobs" as "one of the most unusual approaches of his day for solving the South's race problem."[48] But as we have seen, by 1895 the idea that castration might ameliorate mob violence was not unusual at all in scientific circles. While Thomas was unable to find a congressional sponsor for the bill, one wonders whether the congressmen he approached objected to his proposal regarding castration or his insistence on holding lynch mobs accountable for their extralegal—but tacitly condoned—response. Despite the bill's failure, Thomas did not abandon his belief that castration could ameliorate the "negro problem."

Thomas returned to the idea in his 1901 publication, *The American Negro: What He Was, What He Is, and What He May Become,* proposing that any male over the age of fifteen who was convicted "by due process of law" of sexual assault (or attempted sexual assault) be castrated.[49] Like the various white medical and scientific experts before him, Thomas argued that castration had "deterrent features, as well as punitive functions," while also protecting the rights of the accused and preventing "lawless usurpation of authority." Without referencing these scientists by name, Thomas used almost identical language as Lydston and Daniel in hailing castration's preventative benefits. "Our knowledge of negro nature convinces us that one living example of judicial emasculation would be worth, as a deterrent object-lesson to the race, a thousand summary executions of appalling barbarity," Thomas wrote.[50]

However, in other regards, the often conflicting and conflicted Thomas echoed black critics of lynching. Strongly condemning lynch violence, he insisted that the government had a duty to protect all its citizens and to uphold law and order within the United States.[51] Although Thomas did not cast doubt on the guilt of black men lynched for alleged sexual assaults, he nonetheless followed Ida B. Wells's lead in pointing out that not all lynchings of African Americans involved even an accusation of rape. Still, sentiments such as these were overshadowed by Thomas's scathing indictment of black character. Reinforcing the claims of white supremacist politicians and scientists, he stated, "The negro is of a preeminently sensual race, and one whose male members have an inordinate craving for carnal knowledge of white women."[52] He attributed the problem to both environment and heredity but also pointed the finger at Southern whites for setting such a poor example of civilized behavior. Black people were, after all, "imitators pure and simple, and inevitably adopt the manners of those around them." But ultimately, he decided, "lynching will stop when [blacks] cease to commit heinous crimes, and when the freed men and women . . . set a higher estimate on morality and chastity."[53]

One vocal white supremacist, R. W. Shufeldt, found Thomas's assessment of the "negro problem" very compelling, though he disagreed that castration offered the solution. Shufeldt quoted at length from *The American Negro* in his own published attacks on black America. Devoting a full chapter to lynch law in both *The Negro: A Menace to American Civilization* (1907) and *America's Greatest Problem: The Negro* (1915), he argued that only Anglo-Saxons could properly be termed "man."[54] Citing Thomas as an authority, Shufeldt characterized black people as "purely animal" and without morals or restraint. They sought sex for pleasure, not posterity, he maintained, and civilization was meaningless to them.[55]

Although Shufeldt praised Thomas as an otherwise "keen and thoughtful observer," he rejected Thomas's castration proposal as "too puerile to be worth of sober consideration" and "simply idiotic and impossible." It was only good in (eugenic) theory. "It would doubtless be a righteous thing," he posited, "if it could be done, to emasculate the entire negro race in this country and effectually stop the breed right now, thus prevent any further danger from them and their crossing continually with the Anglo-Saxon stock."[56] But as a punishment for rape, castration simply did not go far enough. "No, when a respectable white woman, in any plane of society, has been brutally assaulted and outraged by a negro, it will be of little satisfaction to her, or to any of her relatives, to know that the brutal raper had simply been carefully and kindly operated upon by a surgeon," he mused. For the potential white rape victim and presumably for Shufeldt himself, the most galling part of Thomas's castration "scheme" was "the fact that the bestial creature lives after he has accomplished the crime."[57] He offered a brief and truly perfunctory condemnation of lynching, to which he was "of course, morally opposed," but Shufeldt's indignation that the "brutal raper" might live after his "kind" operation made his objection to lynching seem hollow indeed.[58]

Shufeldt's "moral opposition" to lynching was further called into question by an anecdote he described in *The Negro* in 1907. Therein, he recounted his own participation in a near lynching in Washington, D.C., after a "great burly black negro" allegedly accosted the twelve-year-old daughter of a military officer with whom Shufeldt was acquainted. The man was discovered by several witnesses who were alerted by the child's screams, but he escaped amid the ensuing commotion. Shufeldt recalled, "With the father, I hunted for that negro until long after dark, both of us being armed with revolvers. The father undoubtedly would have shot him on sight, as he frequently so expressed himself, while, for my part, I undoubtedly would have halted him and allowed the people to take him." Presumably, "the people" referred to the lynch mob, but the "would-be raper" was never found.[59] Shufeldt's story underwent a rather noteworthy revision in his 1915 tome, *America's Greatest Problem.* Describing the same incident, Shufeldt underplayed the extent of his own involvement. There, he simply stated, "The father and I hunted for that negro until long after dark, and he would, undoubtedly, have been shot on sight. The would-be raper was never discovered in this case."[60] Between the publication of the two books, it was becoming increasingly unpopular to defend lynch law outright, due in large part to the onslaught of publicity generated by antilynching activists.

Inhabiting a blurry middle ground between science and lynch mob, Shufeldt is also noteworthy for demonstrating how the same racist logic could lead to different conclusions. A dissenting voice among white medical scientists who addressed lynching, Shufeldt argued that it was precisely because black men's impulse to rape white women was rooted in their biology, innate and unchangeable, that castration would not work. For Shufeldt, black men's embodied threat was as self-evident as it was pervasive. Shufeldt's first text, *The Negro,* contained a series of photos of Henry Smith's lynching in Paris, Texas. In the chapter on lynch violence in *America's Greatest Problem* eight years later, however, Shufeldt replaced the pictures of Smith's lynching with two photos of an unnamed, nude black man (Figure 4.1). Neither of the two new photos is referenced in the body of the text, and the caption of the second simply reads, "Photo from life by the author. Note the prognathous jaws, the length of the arms, and the comparatively slight development of the gluteal region."[61] Nonsexual body parts like jaws and arms indicated the black man's threatening nature as well. Notably, there is nothing to indicate that the unidentified subject had ever been accused of any crime. For Shufeldt, it did not seem to matter. Any black male body could demonstrate an ever-present potential as a sexual threat. He need only point out a few important features, the language and implications of which would have been familiar to a scientific audience or the lay reader who had made it through Shufeldt's earlier chapter on "The Ethnological Status of the Negro." "Prognathous jaws" and long arms had been invoked by ethnologists for decades as indicative of the black race's proximity to animals, an important allusion for Shufeldt, who was arguing that black men were hypersexual beasts.

As a consequence of their deviant bodies and brains, black men were driven by deviant biological imperatives. Shufeldt asserted that these impulses were so strong, so instinctive that nothing could make black men overcome them—not even the immediate threat of castration. In a rather overblown claim even by Shufeldt's standards, he stated, "With the surgeon's knife actually pressing upon his scrotum; with the blazing fagots so near him that he could actually feel the heat of their flames, he would nevertheless seize his victim and outrage her if it lay within his power to do so."[62] Using examples from the animal kingdom, Shufeldt added that castration, even when properly performed, rendered a male sterile but not necessarily impotent. Moreover, sexual aggression did not necessarily require genital function and the Negro's "brutish passion" and amoral character lay at the very core of his being. No surgeon could "change the leopard's spots," Shufeldt insisted.[63]

In contrast, Lydston remained adamant throughout his career that castration would make black men "docile, quiet, and inoffensive," as

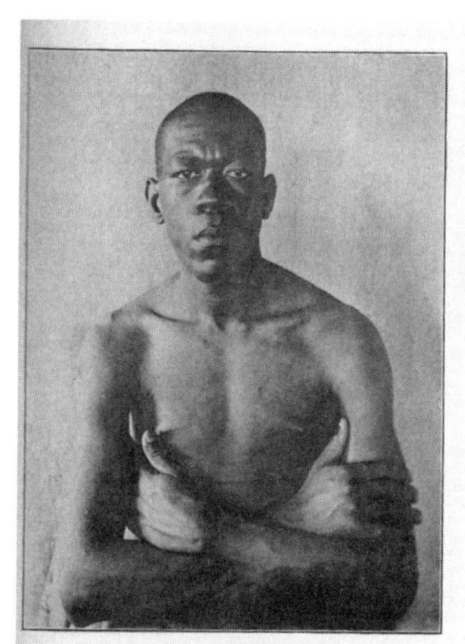

Fig. 39. Pure African Negro. (New York City.)

Male. Born in Demarara, British Guiana. Parents unmixed. Africans. Age, 24; height, 5 feet 11 inches; weight, 165 lbs. Came to the United States when eighteen years of age. Victim of syphilis. (From life, by the author.)

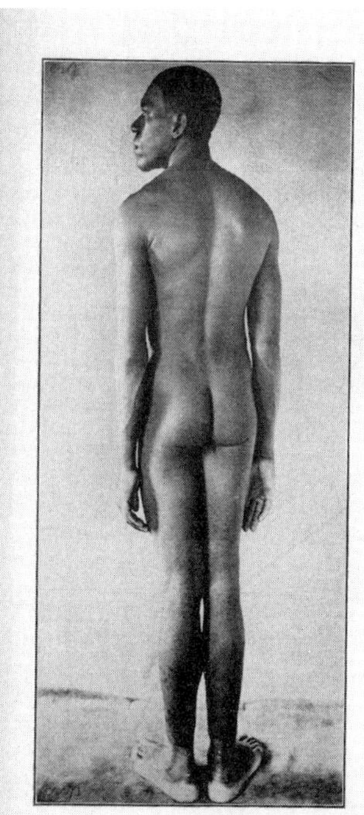

Fig. 41.—Same Subject as Shown in Figs. 39 and 40.

Photo from life by the author. Note the prognathous jaws, the length of the arms, and the comparatively slight development of the gluteal region.

Figure 4.1. Visualizing the "black beast." *In contrast to ethnological illustrations of the nineteenth century, which allowed scientists to shape the reader's interpretation of the image by exaggerating or highlighting specific characteristics, photography offered new opportunities—and challenges—for representing bodily deviance. Even seemingly "benign" pictures such as these depicting black physiognomy informed medico-scientific theories of racial difference and character. Retired military surgeon R. W. Shufeldt included these two photos of an anonymous black man amid a discussion of lynching in his 1915 invective,* America's Greatest Problem: The Negro, *with the caption, "Note the prognathous jaws, the length of the arms, and the comparatively slight development of the gluteal region." Shufeldt defended "lynch law" as an unfortunate response to black sexual crimes. He did not indicate that the man in the photographs was even accused of rape, but Shufeldt nonetheless read his body as bestial and hypersexual. Indeed, for Shufeldt and many white Americans at the turn of the century, sexual threat was embodied in* all *black men (Shufeldt,* America's Greatest Problem*).*

he stated in "Castration Instead of Lynching," published in the *Atlanta Journal-Record of Medicine* in 1906.[64] Moreover, Lydston often noted that white men were hardly immune to rape themselves. If anything, he repeatedly remarked, white rapists were even more contemptuous because, as a race, whites were intelligent enough to know better. Whereas white sexual crimes were generally a product of poor character, black offenders were driven by defective biology and environmental conditions beyond their control, Lydston pointed out.[65]

The scientists who endorsed Lydston's surgical remedy for rape had largely adopted his model of black biological inferiority while generally ignoring his more reasoned calls for color blindness in responding to sexual criminals—much to Lydston's frustration. "It is my opinion there is but one logical method of dealing with the rapist, and that is the total ablation of the sexual organs," Lydston reiterated in his 1904 opus, *Diseases of Society: The Vice and Crime Problem.* But, he reminded his readers, "to be effectual, asexualization should be enforced against rapists of whatever color. Unjust discrimination against the blacks merely serves to defeat the purpose of the method."[66] The point still apparently lost on other scientists, Lydston was even more emphatic in 1912 in *Sex Mutilations as a Remedy for Social Ills:* "THERE SHOULD NOT BE ONE LAW FOR THE BLACK AND ANOTHER FOR THE WHITE."[67]

By far the most progressive among the scientists in the "castration or lynching" debate, Lydston consistently emphasized rational and humane treatment of criminals and "degenerates." He thus challenged historians, particularly historians of science and medicine, to think of "progressive" as an entirely relative term. In addition to his continued calls for castration, for example, Lydston recommended "a hermetically sealed apartment with a secret pipe for the admission of deadly gas" for repeat murderers and death by chloroform for the "driveling imbecile." "The social cancer should be removed by the social surgeon—in effect, the executioner—quietly, humanely, and unexpectedly," he remarked.[68]

Lydston passed away of pneumonia at home in California, where he had returned to retire, on March 14, 1923, at the age of sixty-five and thus did not live to see some of the kinds of eugenic recommendations he made come to full, horrifying fruition in Nazi Germany in the 1930s and 1940s. It is difficult to say whether he would have approved or what if any limits he might have imagined for the role of the "social surgeon" in improving society. What we do know is that when he died, he was not remembered by his contemporaries as a quack or a fringe scientist. The *New York Times,* for example, memorialized Lydston as a "famous surgeon, physician, and author . . . of novels, medical works, and travel books" and praised his teaching, medical service, and vision, particularly in regards to his experiments in gland transplantation.[69]

By today's standards, most of Lydston's proposed treatments for society's ills seem anything but progressive or humane. But compared to the torture meted out by lynch mobs during the same period—or perhaps even Shufeldt's assertion that the solution to the "negro problem" was to colonize all African Americans somewhere, anywhere, outside U.S. borders where they would inevitably die out—surgery under anesthesia could indeed be seen as a more "humane" option. More important, Lydston *saw himself* as progressive and was optimistic about science's potential to reform society. For Lydston and many of his peers, "the medical doctor" was "the best social doctor."[70]

"Dealt with in a Peculiar Manner": Lynching and Castration on the Ground

Scientists were not speaking in isolation. In the late nineteenth and early twentieth centuries, they were proposing solutions to social issues while race relations on the ground were influenced by and reflected scientific discourse. As scientists debated the effectiveness of lynching versus castration, countless newspapers and eyewitnesses reported on the character of racial violence in the streets. Rather than describing a "surgical operation" performed on hypothetical subjects, these stories recounted in often grisly detail the suffering endured by living human beings, including many men who lost their lives as well as their sexual organs. A 1917 article titled "Boy Unsexes Negro before Mob Lynches Him," published in the black newspaper the *Chicago Defender,* offers a revealing and tragic example. The incident took place in Houston, Texas, on October 12 of that year at an oil field where the victim, Bert Smith, was employed as a cook. Prior to his vicious murder, Smith had complained to the head of his camp about the sexual remarks and insulting behavior his mother and sister faced at the hands of several white workers when they came to visit him. Smith went so far as to identify the men responsible. A week after Smith made his complaint, three white men ambushed his sister as she walked home and bound, gagged, and raped her. They left her propped against a tree, where she was later found by several small boys. According to the article, one of the men approached Smith the following day, asking him, "Hey, nigger, did you see that ugly black wench they picked up in the woods yistidy?" and then bragged about the details of the attack. In response, "Smith dealt this white a vicious blow that felled him." Unfortunately for Smith, the incident was witnessed by several other oil drillers. Without hesitation, they descended upon him: "[The oil drillers] placed a rope around his neck, hammered his mouth in with a sledge and pierced his body with sharp instruments, and then forced a 10-year-old white lad who carried water around the camp to take a large butcher knife and unsex him. Smith, who was still alive, begged that all his feelings be taken from him. He was dragged down the main thoroughfare near the camp houses and viewed by citizens, including women."[71] The young boy's gruesome role in Smith's lynching and the presence of women at the public display of the mutilated body represented a powerful and evocative indictment of white supremacy for the black paper, but neither was particularly unusual. Spectacle lynchings often attracted a true cross section of the community—cutting across lines of class, gender, and age—among the audience and active participants. White women were not simply passive victims whose image was invoked in lynching rhetoric. Rather, as historian Grace Elizabeth Hale notes, "White women often directed the very rituals by which white men recaptured their own masculinity through the castration of the black male. After all, the black man's supersexual image was often the result of their testimony." Some women were even given the "honor" of delivering the final, fatal blow.[72] And children of both sexes frequently collected twigs for the funeral pyre in which the terrified black victim would be burned alive. Male children were initiated into the world of white supremacy through their observation of or participation in lynchings. And reluctant young participants were sometimes goaded by older male relatives into physically maiming or taunting the lynching victim. In the Bert Smith case, the young white boy became a man by literally taking the manhood of an African American.

As was the case in many lynchings, the Bert Smith case ran counter to the standard narrative of lynching in contemporary racial science. Smith had not been accused—let alone convicted—of raping a white woman.

Instead, he had protested the rape of his own sister at the hands of white men. And castration and lynching were not mutually exclusive in that oil field in Houston. However, the Bert Smith lynching also demonstrates that scientific discourse on lynching and its enactment in the streets, as well as lynching's representation in the popular media, intersected in important ways. For one, scientific and popular discussions of lynching employed the same gendered language and sexually charged racial tropes. Whether white scientists and concerned citizens condemned or condoned lynching, most invoked the standard image of black men as sexual threats, products of inferior biology, insufficient morals, or both. Second, the "castration or lynching" debate that raged in medical journals from the 1890s through the 1920s spilled over into newspapers and the mainstream press. Third, despite racial scientists' discussions of castration and lynching as an "either/or" proposition, sexual mutilation became an increasingly central component in the practice of lynching. Moreover, lynching and castration served the same ideological and practical functions: intimidation, containment, and social control.

With these ends in mind, citizens and some members of the legal establishment weighed in on the "castration versus lynching" question in the popular press. However, while scientists like Lydston and Daniel saw castration as a progressive, therapeutic solution to the "negro problem," many newspaper letters and editorials focused on castration's punitive benefits. For example, an 1895 letter to the *Los Angeles Times*, signed "Grandmother," declared punishment "altogether inadequate" for "those horrible crimes of outrage . . . that seem to be sweeping over our country." She explained, "I am an old woman, and I have always thought imprisonment alone too light a punishment for such a crime." But lynching was no solution either; "torturing the villains" was "a disgrace to the whole United States." "Grandmother" argued that lynching was no extralegal substitute for capital punishment, and its attendant torture was uncivilized and un-Christian. Her concern was not with the potential lynch victim, however, for "these wretches are not men, and it is a disgrace to animals to call them beasts." Her solution—castration—was the same as scientists writing during the decade.[73]

While "Grandmother" advocated castration as a suitably brutal punishment befitting "wretches [who] are not men," others maintained that castration was more humane than lynching. Simeon Baldwin, a

Connecticut Supreme Court judge and former governor, suggested in 1899 that castration would save face *and* lives. The surgical procedure would satisfy the desire of lynch mobs for retribution and America could avoid being characterized as "uncivilized," an accusation increasingly lobbed by critics within and beyond the United States following Wells's anti-lynching campaign.[74] Writing in the *Atlantic Monthly* in 1904, Clarence Poe of North Carolina also argued that the aim of castration was protection, but he focused on protecting society from sexual criminals rather than protecting black men from the lynch mob. Poe pointed to a recent Wilmington, Delaware, lynching of a black man who had already served time in prison for attempted assault before being accused of a second crime that resulted in his lynching. Had the "surgeon's remedy" been employed, his next crime—and subsequent lynching—would have been prevented. "Set free with the same lustful mania, a wolf in human form, he brought death to himself and to a pure-hearted victim, and shame to a great state," Poe wrote. He acknowledged that some may label "the proposed legal remedy" barbaric, but Poe insisted that the Negro's "peculiar crime" must "be dealt with in a peculiar manner."[75]

Not all newspaper letters and editorials supported castration as an alternative to lynching, however. Ex-congressman William H. Fleming presented himself as a voice of reason in an "open letter" addressed "To the White People of Georgia." Published on October 12, 1906, as "Race Problem Reviewed, Its Evils and Its Curses: A Diagnosis of the Rape Spirit and the Mob Spirit" in the *Macon Weekly Telegraph* and reprinted the following week in a South Carolina newspaper, the lengthy essay passionately condemned lynching. Like the numerous scientists who denounced lynching, Fleming did not reject the popular depiction of black men as a growing threat, "sexual degenerates whose will power is too weak to resist the force of their passions." But Fleming also characterized lynching as a social disease and employed medical metaphors to describe incidents of mob violence. "Only by making a correct diagnosis of those maladies [lynchings] can we succeed in prescribing effective remedies," he implored.

For Fleming, castration was no remedy. He summarized and critiqued a series of letters that appeared in the *Atlanta Georgian* advocating the "procedure" for the problem of "negro rapists." These letters shared "Grandmother's" desire for swift punishment, but they also hinted at the eugenic aims that inspired scientists to advocate castration as well as their emphasis on its preventative effects. The discussion began with an editorial by John Temple Graves, who reiterated the support of castration he first voiced in a speech in the North several years earlier. Graves's editorial sparked a prolonged succession of responses from the *Georgian*'s readers, the first of which "commend[ed] the castration suggestion but insist[ed] it did not go far enough; that all male descendants of the rapist should be diligently sought out and castrated also, since they might inherit the evil tendency from their father." As Fleming recounted, the next letter concurred but added that "what we needed most was not punishment but prevention"; thus the letter urged "as an additional precautionary measure, that all male negro children should be castrated before they were eight days old." Another letter endorsed the previous ones but insisted that hypersexuality and aggression were not limited to African American men. "Negro girls had passions, and tempted white men and bore mulatto children," Fleming reported, "and this writer then urged that the castration of negro boy babies should be supplemented by the spaying of negro girl babies within eight days after their birth." For some, the only real solution to the "negro problem" was to unsex the entire race.

Fleming saw nothing progressive in such propositions. He lamented, "It is scarcely conceivable that such brutal sentiments could be entertained among civilized people by any one outside of an insane asylum. It is utterly incomprehensible to some of us how a great newspaper like the *Georgian,* edited by a brilliant Southerner and owned by a wealthy Northerner, could give publicity to such vile rot without a word of condemnation." Rather than presenting a viable alternative to lynching, the castration proposals that Fleming described fueled the very mentality that underlay lynch violence. "The point of the matter is this," Fleming began. "If one of our leading papers publishes such letters treating negroes on a level with hogs, whose new-born babes are to be castrated and spayed and follows that up with noble authority for slaughtering all the tribe of the rapist, how can any one be surprised at the members of the mob for murdering a few of those being classed with hogs?"[76] While Fleming failed to challenge the trope of the black rapist, he nonetheless offered a rare moment of moral clarity among whites considering the question of castration by pointing out that human beings should not be treated like livestock.

The topic of castration was still making the pages of the *Macon Weekly Telegraph* a year later, and as one letter therein described, a similar discussion was occurring in Pennsylvania as well. Rev. Dr. McCook of Philadelphia had advised "surgical amendment of the body" for the crime of sexual assault, the letter reported. Declaring McCook's proposition as "worthy of the most serious consideration," the *Philadelphia Record* noted that castration—for black men specifically—as a punishment for sexual assault was rooted in early criminal statutes and Pennsylvania's Quaker tradition of judicial reform. Referencing a paper delivered by lawyer Benjamin Nead before the Bar Association, the *Philadelphia Record* noted that a black man convicted of raping "any woman or maid" would be executed "under the administration of the benign Quaker founder of Pennsylvania" and "for an attempted rape upon a white woman or maid the punishment was castration." The *Macon Weekly Telegraph* reported, "The *Record* adds that 'the very common occurrence of the offense in this corner of Pennsylvania, where there are large numbers of negroes, goes far to justify the wisdom of the colonial statues." The *Telegraph* editors added, "So far as we have observed, the Philadelphia newspapers of both political parties in recent times have been more persistent than any others in urging this form of punishment, and now that they know that under the influence of the 'benign' William Penn a law providing such punishment was actually put in force in the colony of Pennsylvania, they will doubtless urge it more strongly than ever." However, the editors of the *Macon Daily Telegraph* maintained that while "certainly such a punishment would 'fit the crime,'" death or life imprisonment with hard labor would be "more decent and civilized" than Pennsylvania's medico-legal advocacy of castration.[77]

The debate in popular and scientific publications was not without consequence on the ground. On at least one occasion, the question of "castration or lynching" turned from words to deed. On October 6, 1899, the *Fort Worth Morning Register* reported that "a party of white men today castrated a negro named Jenkins at Anderson, S.C., for making indecent proposals and exposing his person in an indecent manner to a white girl." Titled "Cured One Negro," the article added that "the negro was given the choice of being lynched or submitting to a surgical operation. He decided on the latter." The article does not indicate whether any doctors were present for this "surgical operation" or if Jenkins received anything to dull the pain. Instead, the brief article concludes matter-of-factly, "After the operation he was carried to his home and told that as soon as he got well he must leave the place."[78] It is difficult to know how exceptional the incident was. As with lynching in general, racial violence was all too commonplace and not always reported. However, while white mobs often heeded the popular and scientific call for castration, for the most part, the mobs did not treat "unsexing" as a substitute for death or limit the mutilation to black men accused of sexual crimes.

Indeed, as scientists, judges, and grandmothers debated the merits of castration as a substitute for lynching, "unsexing" became an increasingly frequent part of the lynching ritual. The 1890s marked the peak of lynchings in the United States, with 118 African Americans killed in 1893 alone.[79] The number of lynchings per year remained high between 1890 and 1918, dipping below 50 only once and reaching triple digits several times. While generally on the decline, the number of black lives lost at the hands of "Judge Lynch" remained in the double digits through much of the 1920s and 1930s, the height of Walter White's antilynching work.[80] Moreover, as historian Grace Hale notes, "Although after the peak decades of the 1890s the number of lynchings decreased even in the South, the cultural impact of the practice became more powerful. More people participated in, read about, saw pictures of, and collected souvenirs from lynchings even as fewer mob murders occurred."[81]

More and more, castration took center stage in the spectacle lynchings that large numbers of people witnessed, participated in, or read about, and the victims' genitals were often among the souvenirs they collected. Newspaper accounts of America's first spectacle lynching in February 1893 did not report that Henry Smith had been castrated, so we cannot know for sure whether or not he was.[82] However, in July of that same year, four months after Lydston suggested that "negro ravishers" be castrated instead of lynched, a black Memphis man, Lee Walker, was subjected to both for approaching two white women in their buggy, possibly to rob them or to beg for money.[83] Physical emasculation featured prominently in accounts of other well-publicized lynchings as well, including Sam Hose in Newman, Georgia, in 1899 and Jesse Washington in Waco, Texas, in 1916. In both cases, members of the mob saved the victims' genitals as postmortem souvenirs.[84] In 1922, three black men were castrated and lynched in Kirvin, Texas, for the murder of a seventeen-year-old white girl, even though the girl's own father pointed the finger at a white neighbor, with whom he had long been feuding, and implored law enforcement to prevent the lynching of local African Americans for the crime.[85]

One of the most well-known—and certainly one of the most horrific—spectacle lynchings took place in Marianna, Florida, in 1934, the same year R. W. Shufeldt passed away in a sanitarium after a long illness.[86] Claude Neal stood accused of murdering a white woman, Lola Cannidy, who had been his neighbor since childhood and with whom he had been engaged in a romantic relationship for months, possibly years. The NAACP sent "a young southern white man" Howard Kester, "who is 'right' on the race question," to Marianna to investigate, Walter White explained in an appeal for funds to have the lengthy report printed in its entirety and disseminated as part of the organization's antilynching campaign.[87] In his report, Kester recounted a conversation he had with a member of the lynch mob who "described the lynching in all of its ghastliness, down to the minutest detail." Kester was later able to corroborate the details through other interviews. "After taking the nigger to the woods about four miles from Greenwood, they cut off his penis. He was made to eat it. Then they cut off his testicles and made him eat them and say he liked it," the unidentified man reported to a sickened Kester.[88] According to Kester's report, Neal was tortured for "ten to twelve hours" before being killed, and when his body was dragged behind a car to the Cannidy home, a "mob estimated to number somewhere between 3000 and 7000 from eleven southern states were excitedly waiting his arrival." There, his dead body was repeatedly run over, stabbed by women, and pierced by small children waiting with sharpened sticks.[89] A month after Neal's death, the NAACP sent copies of the investigative report to President Roosevelt and Attorney General Homer S. Cummings.[90]

Widely characterized by scholars as America's last spectacle lynching, Claude Neal's death in 1934 and the enormous publicity surrounding it marked a turning point for the practice of lynching. To be sure, extralegal executions of African Americans did not end, but the carnival-like atmosphere and enormous crowds that had accompanied many such deaths since the 1890s largely fell out of a favor, with occasional exceptions. Several factors can be attributed to this shift from public spectacle to more private, if still brutal, affairs. The tremendous publicity generated by the NAACP in response to Neal's horrific death shocked large numbers of white Americans out of complacency—or complicity—on the issue of lynching. The publicity surrounding the Neal case as well as antilynching legislation before Congress generated more sympathy from the White House than ever before.[91] The Neal lynching culminated nearly fifty years of community-sanctioned racial violence and represented the pinnacle of human brutality, but it was also that very excess that finally turned the tide of majority public opinion against lynching, at least in its most public form.[92]

Even after the character of lynch violence had changed and the number of victims decreased, castration still remained central to both the practice of lynching and its place in the American imaginary. In 1943, for example, two fourteen-year-old boys, Charlie Lang and Ernest Green, were lynched in Mississippi after being taken from the jail where they were being held for allegedly attempting to rape a thirteen-year-old white girl. Though the sheriff said the two boys had confessed, an investigation by the NAACP later revealed that the three children often played together.[93] The NAACP investigative report further noted that on the day of the alleged attempted rape, the boys "were running and jumping when the girl ran out from under the bridge and the boys behind her. A passing motorist saw them and the result you know." The two boys were later strung up from the very bridge where they had played with their white friend. They were not simply hanged, however. Lynch violence may have lost its cheering crowds by 1943, but it did not abandon the sexual mutilation of its victims, along with other ritualized tortures. According to the NAACP report, "The boys were mutilated . . . [and] their reproductive organs were cut off."[94]

It is difficult if not impossible to know exactly how often castration was practiced as part of the lynching ritual, though the two were clearly linked in America's cultural imaginary in the early twentieth century. Not all lynchings were reported, and even when they were, reporters were not always close enough to the front of the crowd (or interviewed witnesses who were) to report the particulars of the victim's torture, while other newspaper accounts likely omitted sexual details or, as was often the case, cloaked them in euphemisms. However, as historian W. Fitzhugh Brundage points out, "The most relevant measure of the importance of [sexual] mutilation during lynchings for sex crimes was never the percentage of black victims who were mutilated, but rather the lasting impression that each incident left upon the observers."[95] While castration was also a part of lynchings in which the victims were not accused of sexual infractions, Brundage's point is well taken. A shocking detail in the most lurid of lynching accounts, one can only imagine the trauma experienced by African Americans who witnessed or read about such events.

The specter of castration, as a painful and humiliating prelude to death, loomed large in the African American imaginary and was a powerful part of the culture of fear lynching produced.[96] But black contemporaries also often represented lynching itself as a symbolic emasculation. Acclaimed black writer Richard Wright, for example, described the sense of powerlessness black men felt at the hands of their white counterparts, a figurative unsexing that could at any moment be made material. "The white death hung over every black male in the South," Wright wrote, adding, "I had already grown to feel that there existed men against whom I was powerless, men who could violate my life at will."[97]

Furthermore, lynching and castration—whether prescribed as a medico-legal solution by "progressive" scientists or performed by a lynch mob—served the same function as a form of social control. Just as Lydston suggested that a "few castrated negroes" would serve as a powerful example to other potential rapists of his race, so too did lynching function not just to punish individual victims but to keep an entire race from stepping out of line. Writing at the turn of the century, Ida Wells often asserted that the threat black men represented to white society was not sexual, as so often portrayed, but political. Black men's enfranchisement challenged white political, social, and economic dominance, hence the real reason for violence against black men. "'The Negroes are getting too independent,' they say, 'we must teach them a lesson.' What lesson? The lesson of subordination," she wrote in her 1892 pamphlet, "Southern Horrors: Lynch Law in All Its Phases."[98] African Americans learned this lesson early and viscerally. As Richard Wright wrote in 1941, "Fear is with us always, in those areas where we black men equal or outnumber the whites fear is at its highest."[99] Consequently, he linked lynching to both emasculation and attempts to thwart black progress. As he poignantly described, displays of pride or self-assertion by black men were often met with violence, and death was the price for a voice raised in protest.[100]

As with lynching, the threat of castration was not simply about emasculation but about silencing all African Americans, made clear by a rather ironic response to Ida Wells's critique of racial violence. On May 24, 1892, Wells, writing under a pseudonym, published a scathing editorial in her Memphis newspaper, *The Free Speech,* in which she challenged the "threadbare lie" that black men were driven to rape white women and suggested that the charge of rape so often associated with lynching might actually cloak consensual interracial sex.[101] Immediately, a flurry of angry responses flooded Memphis newspapers. One response, published in the *Evening*

Scimitar the next day, threatened the editorialist, whom most readers assumed to be male (a reasonable assumption at a time when journalism was overwhelmingly dominated by men), with castration if "he" did not recant. "If the negroes themselves do not apply the remedy without delay," the letter warned, "it will be the duty of those whom he has attacked to tie the wretch who utters these calumnies to a stake at the intersection of Main and Madison Sts., brand him in the forehead with a hot iron and perform upon him surgical operation with a pair of tailor's shears."[102] Here, castration also meant censorship.

In the aftermath of her editorial, the office of the *Free Speech* was burned to the ground, and Wells was forced to leave Memphis, but she was not silenced. Instead, she continued to publish and speak on the issue of lynching. At the turn of the century, Wells was the most prominent and persistent antilynching voice in America, and her publicity tactics set the stage for future black activists.[103] Later in life, she herself worked—and often clashed—with Walter White on the NAACP campaign against lynching and was active in black protest politics until her death in 1931. Despite the ever-present threat of violence and "castration," both literal and figurative as a form of censorship, black activists like Wells and White continued to speak out, even after narrowly evading lynching themselves.

Scholars have attributed castration's central place in the lynch ritual to numerous factors, including sexual jealousy amid frequent claims that black men had larger genitalia and greater potency; anxieties over the changing meanings of white manhood; and reassertion of white patriarchal power in the face of black political, economic, and social advancement during Reconstruction. Others have pointed out that spectacle lynching, so often justified with rhetoric about the "black beast rapist," was itself a sexual assault in which the lynch victim was publicly exposed, humiliated, and brutalized. And more practically, in a still largely agrarian South, neutering livestock was an everyday procedure with which many men had experience.[104]

To be sure, all these factors likely played a role in castration's prominence in the lynch ritual. But generally overlooked is that the scientific and popular press alike frequently featured suggestions that "negro ravishers" be either "treated" or "punished" with castration. While these editorials and scientific articles usually advocated castration *instead* of lynching, the volatile racist rhetoric they employed to justify the "surgical procedure" fueled the lynch law mentality and perhaps offered some mob members a tempting addition to their ritual of horrors. Lynch mobs read past the articles' pleas to spare lives to their affirmation that black men were sexual threats in need of emasculation. Just as lynch mobs acted outside the legal establishment in executing their victims, so too did they perform physicians' often-suggested "surgical procedure" outside the medical and scientific establishment.

In late nineteenth- and early twentieth-century America, lynching represented a nexus among scientific discourse on race and sex, racism enacted on living bodies in the streets, and black mobilization in response. Racial scientists reinforced and legitimized the popular notion that black men were an ever-growing sexual threat to white women by describing the problem in terms of pathology and defective biology. Scientists like Lydston and Daniel saw themselves as part of the solution to the "negro problem" of rape and the extralegal violence that was its result. The issue of authority was an important factor driving scientists' advocacy of castration or other surgical procedures instead of lynching. After all, lynching would certainly achieve the same ends as castration in preventing rapists from repeating their crimes. But lynching was not a medical solution to black rape. And due in large part to the antilynching activism by Wells at the turn of the century and the NAACP beginning in 1909, lynching was an increasingly unpopular practice to endorse publicly even if it was still tacitly accepted and allowed.

Lynching, then, embodied the complex and mutually informing relationship between science and society around issues of race. Throughout the nineteenth and early twentieth centuries, scientists played a crucial role in constructing race as a biological category of difference and offered advice on the problems that race presented in American society. In so doing, these scientists frequently used gender and sex difference to bolster their claims about black inferiority. Furthermore, racial science shaped and was shaped by American race relations and political rhetoric. However, scientific claims about racial difference and hierarchy were always contested. While scientists like Lydston and Daniel had presented themselves as offering a solution to lynching, Walter White insisted racial science incited racial violence. With the clear overlap between lynching and castration in science and on the street, it is difficult to disagree with him. Scientists' engagement in the question of "castration or lynching" brought their work in the public eye more than ever before and garnered much support in the short term, but it also marked the beginning of scientific racism's demise in America. Indeed, by the time of White's antilynching work in the 1920s and 1930s, the subject of the next chapter, science offered as much potential to dismantle racial hierarchy as to maintain it.

· CHAPTER 5 ·

Walter White, Scientific Racism, and the NAACP Antilynching Campaign

IN A CHAPTER OF HIS autobiography titled "I Learn What I Am," National Association for the Advancement of Colored People (NAACP) leader Walter White (Figure 5.1) described a formative event early in his life that awakened him to the harsh realities of racial violence and precipitated his career as an antilynching activist. Therein, White depicted his harrowing experience of the 1906 race riots in Atlanta when he was thirteen years old. That September, while accompanying his father George on his mail delivery route, young Walter watched in horror from their buggy as a crippled black man tried to outrun an angry white mob before being beaten to death and left in a pool of blood in the street. As they drove home through the streets of Atlanta, their white appearance afforded them a modicum of safety from the mobs attacking black citizens at random. That night, however, they huddled in their house with the lights off as the mob headed toward their neighborhood, the threat of more violence imminent. Tipped off that the house belonged to a "nigger mail carrier," the mob approached the White home. White recalled, "In the flickering light the mob swayed, paused, and began to flow toward us. In that instant there opened up within me a great awareness; I knew then who I was. I was a Negro, a human being with an invisible pigmentation which marked me a person to be hunted, hanged, abused, discriminated against, kept in poverty and ignorance, in order that those whose skin was white would have readily at hand a proof of their superiority."[1]

Though the mob was driven off by the Whites' neighbors, the close call left Walter shaken and keenly aware of his racial identity and the social position it afforded him. In that terrifying and visceral experience of racism, he realized that the bodies of all black men were "marked" as a threat, even an elderly, crippled, and unarmed man accused of no crime. By then, prominent physicians like G. Frank Lydston had been advocating surgical

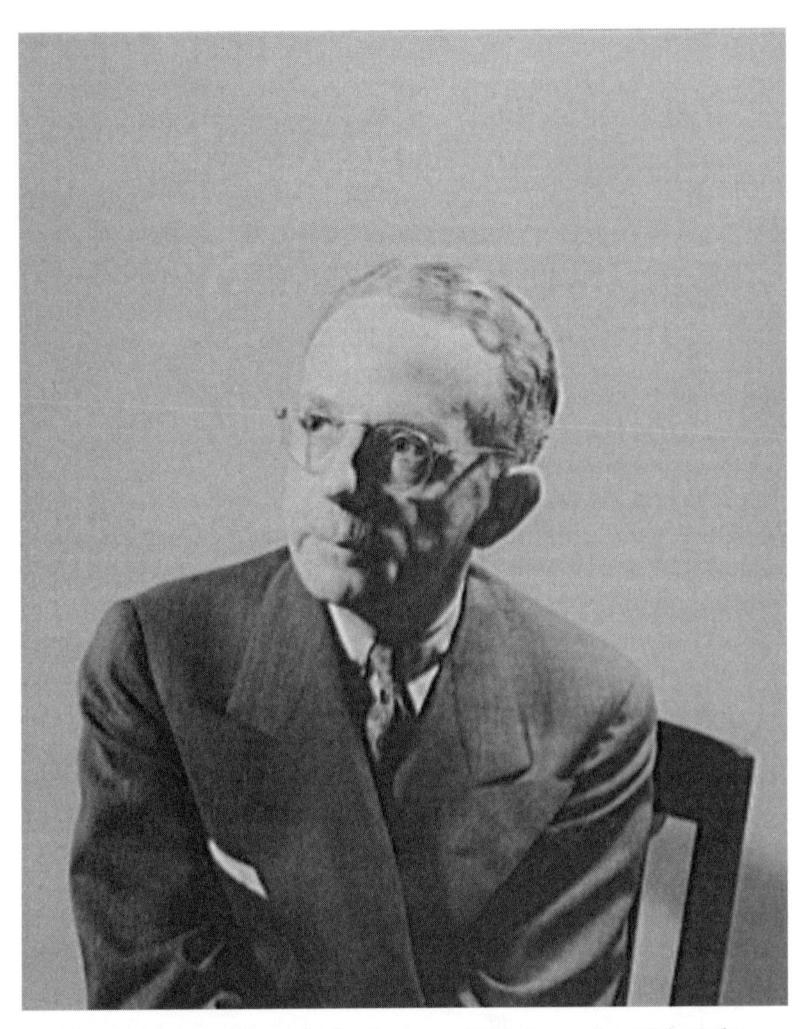

Figure 5.1. Combating violence, challenging race. *Antilynching activist Walter White, seen here in 1942, was not only a critic of "the sciences of race"; he also used his own physiognomy—and ironic surname—to challenge racial categories. While White's light skin occasionally earned him suspicion from other black activists, it also allowed him to safely pass as white when investigating lynchings in the Deep South and gain access to powerful players in American politics and science (Farm Security Administration—Office of War Information Photograph Collection, Library of Congress Prints and Photographs Division [LC-USF34–013343-C]).*

castration instead of lynching for curtailing that perceived threat for more than a decade—and there is nothing to suggest such proposals had any appreciable impact on the level of mob violence against blacks in America. Instead, the scientific credence they lent to popular fears about the menacing black man fueled the kind of violence White experienced that night as a boy. White would think of the incident often in his work with the antilynching campaign launched by the NAACP, an organization he led for more than twenty years.

An important but overlooked part of that work was challenging scientific racism and its incendiary impact on race relations in the United States. While racial scientists had presented themselves as offering a solution to lynching, Walter White saw them as part of the problem. In his 1929 exposé, *Rope and Faggot: A Biography of Judge Lynch,* written the same year he took the helm of the NAACP, White described the "various ingredients of lynching" as "economic forces, race prejudice, religion, sex, politics, journalism, and theories of racial superiority and inferiority based upon faulty or insufficient scientific evidence."[2] He further noted that ethnological writing that championed white supremacy, which he called "Nordicism," incited "prejudice and fear among those who would not consider joining such a movement as the Ku Klux Klan." "Nordicism" was simply the respectable face of the same worldview that drove the Klan. "Thus the Klan recruited the more rowdy element," he explained, "and the Nordic movement those of slightly higher mental caliber—and between them profoundly influenced the already tense racial situation in the United States and added to the antagonisms from which such a phenomenon as lynchings arises."

In the wake of Franz Boas's influential work in cultural anthropology earlier in the century, racial science was increasingly challenged at the time of White's writing in 1929. Still, biological models of racial difference—which by then were completely imbricated with presumptions about sexual and gender deviance—lingered on in some scientific circles in the United States, and their continued influence outside the scientific establishment was far-reaching. For one, innate sexual degeneracy had become fundamental to many racial scientists'—and the white public's—understanding of black inferiority. The 1920s and 1930s also saw the widespread popularity of eugenics, an applied form of racial science that was predicated on improving America's racial stock by controlling sex. At the same time, cultural models of race proved easily adaptable to existing racist stereotypes and beliefs. Racist writers, scientific and otherwise, found that blaming African Americans' cultural practices for high rates of disease, mortality, vice, or crime was just as satisfying as blaming their biology and remained just as disinclined to take factors such as poverty or the effects of discrimination into newer frameworks of black inferiority and threat. Nor were cultural and biological models of racial difference completely distinct or incompatible. As White noted, "The lyncher, the Klansman, the Nordicist, the disenfranchiser, [and] the opponent of advancement of the Negro or other dark-skinned race" reinforced for each other "the fundamental soundness of their prejudices." Moreover, "'scientific jargon' which they did not know was jargon assured them that the Negro is inferior and that it is for the general good to 'keep him in his place.'"[3]

As we have seen, White was hardly the first black thinker to challenge the logic of racial science, but he did so at a time when scientific authority over race was beginning to crack. He also had the benefit of something most of his predecessors—black men like David Walker rallying his "brethren" against slavery in the antebellum period or physician C. V. Roman writing about race at the turn of the century—did not: a position of broad influence, a national audience that crossed color lines, and, as we shall see, powerful allies in the scientific establishment.

Scientific theories about racial difference and black inferiority played an important role in Walter White's public and private writings on lynching throughout his career with the NAACP. Most notably, he maintained a professional and friendly relationship with white biologist and former eugenicist Raymond Pearl for years, soliciting his advice on scientific matters and securing for Pearl a position on an NAACP-convened committee charged with investigating discrimination at a Harlem hospital, a committee that also included W. E. B. DuBois and Adam Clayton Powell. The two men and their spouses also occasionally socialized, though their lengthy correspondence seems to indicate more missed encounters and canceled visits than successful gatherings.[4] White's belief that the widespread acceptance of scientific theories of racial hierarchy enabled, even promoted, violence against black people figured prominently in *Rope and Faggot,* an entire chapter of which was devoted to racial science.

Like Ida Wells before him, White frequently attacked the "thread-bare lie" that lynching was a response to the rape of white women by black men, addressing head-on the issues of sex that were pervasive in popular discourse on race.[5] But by and large, when White discussed racial science in his public work, he focused on scientists' claims about black inferiority that were not explicitly about sex. That is, he often critiqued studies of brain weight, head size, and skull shape, for example, from which ethnologists extrapolated about intelligence rather than medical writing on "furor sexualis" or the supposedly enlarged genitalia of black men and women, claims he never lent the authority of science by attributing them to physicians. In *Rope and Faggot,* for example, White noted that "sex and alleged sex crimes have served as the great bulwark of the lyncher" and devoted a chapter to "Sex and Lynching."[6] Listing both "sex" and "scientific theories of racial superiority and inferiority" among the "ingredients" of lynch violence, he approached the two as separate discourses; "questions of sex" do not appear to have colored scientific considerations of race in White's analysis.[7] His personal correspondence with Pearl indicates he was very much aware that popular ideas about black sexuality permeated scientific work on race, but for a number of reasons, he distanced them from racial science and instead associated them with a misguided public.

Although Walter White insisted that scientific racism had precipitated lynch violence, he also saw potential for racial science to be used against itself. Thus he enlisted the help of prominent white scientists to *disprove* black inferiority with the explicit goal of combating lynch law. An incisive critic of the United States' racial "fiction" yet always pragmatic, White placed much importance on exposing and debunking ethnology's central tenets during his involvement in the NAACP's antilynching campaign. Racial science may have reinforced widespread beliefs about black inferiority and danger that made lynching possible, but it also provided a lightning rod for black challenges to racism and mobilization against racial violence.

An Activist Awakened

White was born in 1893 in Atlanta to a "happy middle-class family." Though his father's salary as a postman was modest and the family's resources stretched thin, White's thrifty parents were able to provide a comfortable home for their seven children. George White's work with the postal service also placed the family firmly within Atlanta's thriving black middle class. He and his wife emphasized education and religion as integral parts of the family's respectability and vital to the success of the black race as a whole.[8] Like his father, White attended Atlanta University, the prestigious black school founded after the Civil War by Northern whites; an average student, he graduated in 1916.[9] He went to work for the Standard Life Insurance Company, the largest black-owned business of the time, in the summer before his senior year of college and continued with the company after graduation. The racism he witnessed as he combed the countryside for customers that first summer later prompted him to write to the national headquarters of the NAACP about starting a local branch in Atlanta. He received a response from James Weldon Johnson, who supported the idea, and White was soon elected secretary of the nascent Atlanta branch. Against Johnson's urging, however, White built a branch that saw only men in its executive and working committees, the first of many times White would run into conflicts with or regarding women in leadership positions in the organized civil rights movement.[10]

Gender was not the only source of conflict for White; so, too, were race and class. Some black activists viewed White as elitist and believed he manipulated his ability to pass as white for his own advantage. White noted that his parents "were both so light-skinned that either could have passed for white." White's own light skin would prove an asset to his involvement in the NAACP's antilynching campaign, even if it did occasionally earn him the suspicion of his African American peers.[11] White's appearance saved his life in 1906 and later enabled him to infiltrate the white South during his lynching investigations and gain an unprecedented insight into the mind-set of the lynch mob; that is, White's racial passing was strategic and situational rather than permanent. As biographer Kenneth Robert Janken notes, "His appearance gave him the option—which he did not exercise—to pass for white, but it nonetheless stamped the way he looked at the world and the way the world looked at him."[12] Moreover, "he used his anomalous condition to advantage," Janken explains. "His popular writings on his complexion and 'passing' exploded racial stereotypes and challenged the idea of race as an immutable category. At the same time, he exploited his position as a voluntary Negro; his exotic status paradoxically afforded him a social standing and a wealth of contacts that would have been denied him had he been white." White's ability to pass not only allowed him to investigate lynchings with "insider" access; it also proved a valuable tool in his lifelong fight to challenge racist assumptions and racial categories themselves.[13]

However, Walter White was always careful to assert that he did not wish to be white and that he identified with the struggles of his black brethren. Furthermore, in his sharp and innovative analyses of race, he also drew on a long tradition of black protest thought. Describing later in life his childhood racial awakening following the 1906 race riots, White repeatedly echoed the concerns and language of nineteenth-century black ethnologists. Like these predecessors, White both employed religious metaphors and bemoaned the fact that religion had been used to justify racial inequality. "It made no difference how intelligent or talented my millions of brothers and I were, or how virtuously we lived. A curse like that of Judas was upon us, a mark of degradation fashioned with heavenly authority," he lamented. "There were white men who said Negroes had no souls, and who proved it by the Bible. Some of these now were approaching us, intent upon burning our house."[14] Likely shaped by his adult experiences and knowledge, his recollection of the traumatic boyhood incident also reflected a familiarity with ethnological discourse, which itself drew heavily on biblical lore and language dating back to the early nineteenth century.

White also echoed black ethnological writers in his gendered assessments of the black and white races, juxtaposing the trope of the "angry Saxon" with that of the "redeemer race."[15] Though he lamented the discrimination and violence blacks faced at the hands of whites, it made him all the more inclined to identify with the former, to fight alongside the oppressed rather than align himself with oppressor. "I was sick with loathing for the hatred which had flared before that night and come so close to making me a killer; but I was glad I was not one of those who hated; I was glad I was not one of those made sick and murderous by pride," he intoned. The white race was hypermasculine, aggressive, and power hungry; he did not wish to be "one of those whose story is in the history of the world, a record of bloodshed, rapine, and pillage." He cast his lot instead with "the races that had not fully awakened," who had the potential to "write a record of virtue" rather than a record of conquest.[16]

Toward that end, he accepted an invitation in 1918 to join the staff of the NAACP as assistant secretary under James Weldon Johnson, just as the organization was launching its antilynching campaign. With his ability to pass for white and his passionate critiques of racial violence, White was immediately put to work in the field investigating lynchings. With the encouragement of writer and social critic H. L. Mencken, whom he met in 1922, White wrote two novels, *The Fire in the Flint* (1924) and *Flight* (1926), which dealt with lynching and racial passing, respectively. Awarded a Guggenheim Fellowship, he moved to France in 1926 to work on a third novel, but haunted by his work with the NAACP, he instead began writing *Rope and Faggot,* an analysis of the causes and character of lynching in the United States. It was during this period that White reached out to a prominent white biologist, Raymond Pearl, enlisting his help with a chapter on scientific theories about race.

Walter White confronted racial science at a transitional moment in its history. Biological understandings of race dominated nineteenth-century science, but the early twentieth century saw the introduction and gradual ascendency of cultural models of human variation. Franz Boas, a German immigrant and professor at Columbia University, dealt the most serious blow to scientific racism over the course of his long and influential career in the nascent field of anthropology—a field that became professionalized largely as a result of his efforts. As a young scholar, he had received training in anthropometrics, but in contrast with the military scientists who measured soldiers during the Civil War, Boas ultimately did not conclude that biology was destiny. Instead, he argued that culture was the primary source of differences among groups of people and that even human biology was malleable by environmental factors.[17]

Still, biological paradigms of racial difference—and hierarchy—maintained a strong foothold in American science until after World War II, prompting frequent counterchallenges by black intellectuals like Walter White as well as sympathetic scientists. That "Mr. NAACP" could enlist a white scientist to assist his attack on racial science in the 1920s and 1930s exemplifies this period of flux, in which scientists could be both friend and foe to African Americans. As anthropologist Lee Baker explains, Boas defined cultures as "particular to geographic areas, local histories, and traditions." More revolutionary still, Boas insisted that "one could not project a value of higher or lower on these cultures." Boas's cultural relativism eventually supplanted evolutionary frameworks of race among scientists, though stalwarts of biological determinism remained vocal in America. By the time White reached out to Pearl in the 1920s, then, the belief that racial differences were natural and immutable, which had been used to legitimize racial hierarchy for over a century, was under attack. As

Baker points out, "Boas's contributions were singularly significant, but he did not work alone. Without the wider social and political efforts of [W. E. B.] DuBois, the NAACP, and scholars at Howard University, Boas's contributions to the changing signification of race would have been limited to the academy."[18] Indeed, Walter White saw his own challenges to scientific racism as relevant far beyond the Ivory Tower.

White's writings were also part of a long genealogy of black protest thought critiquing the basic logic of ethnology since its development in the nineteenth century. For example, in his 1854 speech, "The Claims of the Negro, Ethnologically Considered," Frederick Douglass maintained that ethnology was fundamentally flawed, biased, and rife with racial double standards. "This is, you know, an age of science, and science is favorable to division," he told the audience. He referenced the most influential white racial scientists of his era, including Josiah Nott and Samuel Morton, and critiqued their common strategy of disassociating the black race "from every intelligent nation and tribe in Africa," which he saw as "proof that they have staked out the ground beforehand, and that they have aimed to construct a theory in support of a foregone conclusion."[19] For Douglass, ethnology developed as a means to legitimize racial discrimination. "Pride and selfishness, combined with mental power, never want for a theory to justify them—and when men oppress their fellow-men, the oppressor ever finds, in the character of the oppressed, a full justification for his oppression," he pointed out.[20]

Writing in 1916, black physician C. V. Roman was similarly critical of ethnology as an irredeemably biased venture by white men to justify their domination over others. Ethnological arguments about racial difference were "contrary alike to science, common sense, and daily experience" but nonetheless "foisted upon the public as the pure gold of truth in the currency of argument on the race question," Roman insisted.[21] At the same time, as a medical scientist himself, Roman shared White's faith that science could be used to combat the legacy of scientific racism, for "science is the enemy of prejudice. Knowledge dispels superstition." In contrast to Douglass's assessment that "science was favorable to division," Roman pointed to several instances in which science had challenged previously held beliefs about biological difference among humans in regards to sex and class while also acknowledging that it could just as easily be used to reinforce them:

> Anatomy demolished the numerical superiority of woman's ribs, and physiology undermined the social prestige of "blue" blood. History shows to be an evolution, and innate racial superiority an unjustifiable egotism common to the children of men. Ethnology and religion are alike in this, that those who hold to the partiality of nature deity believe theirs the favored class. I never know a foreordinationist in religion that did not count himself one of the elect; nor an advocate of racial superiority that did not think his the superior race. *Racial achievement means racial opportunity. Science knows no innately superior race.*[22]

Consequently, he saw his book, *American Civilization and the Negro: The Afro-American in Relation to National Progress,* as a scientific rebuttal to scientific racism. "The author aims to show that humanity is one in vices and virtues as well as blood; that the laws of evolution and progress apply equally to all; that there are no lethal diseases peculiar to the American Negro; that there are no debasing vices peculiar to the African; that there are no cardinal virtues peculiar to the European; that we are all sinners and have come short of the glories of civilization," he wrote in his introduction.[23] And like White writing more than a decade later, he also saw the project of refuting scientific claims about racial difference as serving larger goals: increasing "racial self-respect" and diminishing "racial antagonism."[24] While White lacked Roman's medical background, he had the benefit of a far broader audience for his work and access to insiders within the medico-scientific establishment whose whiteness lent them an added level of authority with white readers.

"Drawing upon Your Knowledge": Science as Friend and Foe

Toward that end, on August 31, 1927, White, then assistant secretary of the NAACP, wrote a two-page letter of introduction to Dr. Raymond Pearl (1879–1940), a professor of biology at Johns Hopkins University in Maryland. Pearl had recently critiqued eugenics, of which he had once been an outspoken proponent, in the popular periodical the *American Mercury* in November 1926. The publicity surrounding Pearl's apparent change of heart likely inspired White to approach him as a potential ally in his campaign against lynching, to which he attributed scientific theories of black inferiority as a contributing factor. Introducing himself through their mutual friend, H. L. Mencken, White turned quickly to the purpose of his correspondence. He explained that he was writing a book on lynching in the United States in which he was "attempting to treat lynching not only as an isolated phenomenon but also to fit it into its proper setting, which is made up of the social, economic, psychological and historical factors which cause this practice." Among those factors, one stood out as particularly relevant for a scientist like Pearl, long interested in biology and race. "One of the most important causes [of lynching], it seems to me," White noted, "is the theory or doctrine of inherent racial inferiority which serves some lynchers to feel they have not committed as grave an offense in killing a Negro as would be the case in putting a white man to death." If scientific racism was a contributing factor in lynching, then debunking its central claims was critical to combating racial violence, and White enlisted Pearl's help in doing so. He begged Pearl's patience for "drawing upon [his] knowledge in this fashion" before barraging Pearl with erudite questions relating to scientific theories of race as well as scientific methodologies.[25]

White's senior by fourteen years, Pearl was born in Farmington, New Hampshire, in 1879 and went on to receive an education typical for a driven Northern white man of comfortable means. Pearl obtained his bachelor's degree from Dartmouth College in 1899 and his PhD from the University of Michigan in 1902; he also spent time studying abroad at Leipzig and as a fellow in the Galton Laboratory at University College in London. He held a variety of lectureships in the United States and Europe but spent the bulk of his professional life at Johns Hopkins in Baltimore, where he built a career as a well-known and respected biologist, geneticist, and biometrician.[26] In Baltimore, he was also part of a lively circle of intellectuals who debated social issues. Pearl was an incredibly prolific scientist, penning or coauthoring 712 publications, 17 of which were books.[27] His research often funded by the Carnegie and Rockefeller Foundations, Pearl served on the editorial boards of numerous scientific journals and as editor of the *Quarterly Review of Biology* and *Human Biology*.[28] He published on topics ranging from animal husbandry to disease and longevity in human beings, but his work on eugenics, race, and population control was of particular interest among his contemporaries.

Historian Elazar Barkan observes that Pearl's "views on eugenics and race are especially interesting because he aspired to correlate his biological research to social questions, to combine his role as an expert, the absolute savant, to that of the intellectual and leader of public opinion."[29] While this is certainly true of Pearl, the same could be said about most racial scientists from Josiah Nott in the antebellum era to Pearl's contemporaries; the sciences of race, from ethnology to eugenics, were fundamentally applied sciences. Pearl was unique, however, in the enormous gap between his public writings and his private sentiments about race, demonstrated in his correspondence with friends and colleagues. Unlike scientists like Nott—or, later, Shufeldt, Lydston, and Daniel—Pearl came to critique racial science and race prejudice in his later professional publications, a rebuke that while relatively mild brought him the ire of his friends within eugenic circles. Meanwhile, he continued to express unabashed racism in private, indicating his apparent turn against eugenics was driven more by his irritation with its scientists and methodology than concern for racial equality.[30] Balkan's assessment of Pearl's ambivalence, a racist who nonetheless made insightful and influential criticisms of science's application to race, is borne out by his 1927 article, "The Biology of Superiority," and even to some extent his correspondence with Walter White.

Published in the popular periodical the *American Mercury,* "The Biology of Superiority" was Pearl's first and most well-known critique of eugenics. Pearl had been publishing articles on eugenics for decades, including "Breeding Better Men" in 1908, and advocated the creation of eugenic research centers at universities.[31] But in 1927, he attacked the field on several fronts, and while his criticisms were ostensibly about eugenics specifically, they had considerable resonance for racial sciences more broadly construed—something surely not lost on Walter White.[32] For one, Pearl asserted, eugenics was overly propagandistic. Propaganda had "always gone hand in hand with the purely scientific, from the very beginning of the development of eugenics," Pearl averred. But he feared the line between propaganda and science had "become almost inextricably confused, so that the literature of eugenics has largely become a mangled mess of ill-grounded and uncritical sociology, economics, anthropology, and politics, full of emotional appeals to class and race prejudices, solemnly put forth as science, and unfortunately accepted as such by the general public."[33] And it was indeed the general public Pearl reached with the article, which he published in a popular rather than purely scientific periodical—though it certainly did not escape the notice of his professional peers. He noted that no scientific man likes to think of himself as engaged in propaganda, and thus scientists soothe themselves by calling their endeavors "education, promoting the public welfare," which often fooled the public as well. "Propaganda is, however, a subtle and insidious reptile," he remarked, pointing to its inherent self-interest and "indifference to the truth" when the truth did not serve as an effective means to an end.[34]

Drawing on his vast knowledge of texts on eugenics, he then offered a remarkably perceptive summary of the chief components of eugenic thought: (1) All important human characteristics, "physical, mental, and moral," are determined primarily by heredity; offspring will always be similar to their parents; and other possible determining factors, such as environment, "are relatively unimportant from a racial point of view." (2) If such hereditary determinism is true and superior people will always have superior children while "inferior or defective people" will always have similarly inferior or defective children, then the "welfare of the race" depended on urging or even forcing superior people to have large families and inferior people to reproduce little if at all. And finally, (3) "some races of people are superior to other races, and . . . intermixture or even contact of the superior with the inferior should be prevented by exclusive immigration laws"—which cut to the core of scientific attitudes toward miscegenation that predated eugenics itself. Furthermore, he added, "superior people," in regards to "individuals, classes, or races," always seemed to mean either "my kind of people" or "people whom I happen to like." It is easy to see why such pointed and salient criticism would be so appealing to someone like Walter White and so threatening to white eugenicists, all the more so for having come from within their ranks.

Undermining racial prejudice, however, was not Pearl's major focus. Rather, the bulk of his article was devoted to attacking eugenics as based on faulty, outdated science. "Leaving aside all discussion of what might perhaps be called the broad humanitarian aspects of these eugenic theses," he insisted that such ideas—and their many social and legal applications—were built around the outdated science of "pre-Mendelian genetics." Throughout the article, he repeated that eugenics was largely premised on the belief that "like produces like," that a superior parent will produce a superior child, an idea he alternately called a "fallacy" and "folklore."

Despite his critique of eugenics' logic and methodologies, Pearl was still concerned with "the interest of the race," and his conversion was not as dramatic as it seemed. In private, he maintained connections to prominent figures in the eugenics movement and demonstrated deep ambivalence about race in his personal correspondence. Ultimately, Pearl's article was a call to reform eugenics rather than to dismiss it entirely. Specifically, he wanted to bring the field into accordance with established genetic principles toward the goal of "breeding great men." "It would seem to be high time that eugenics cleaned house, and threw away the old-fashioned rubbish which has accumulated in the attic," he concluded in the last line of the article.[35]

Still, it is not difficult to see why White chose to reach out to Pearl as a potential ally. While White does not refer to "Biology of Superiority" by name in his August 31, 1927, letter of introduction, it is likely that the article (published nine months earlier) prompted him to write to the scientist. For one, the publicity surrounding the article was swift and widespread, and White was quite well read. Furthermore, the *American Mercury* was a popular periodical, and White himself submitted an article, "I Investigate Lynching," to the journal in 1928, which ran in the January 1929 issue.[36] Most of all, "The Biology of Superiority" painted a picture of a well-established scientific discipline applied to issues of race and society that was both inherently biased and methodologically flawed. Such critiques were certainly not new to black intellectuals; C. V. Roman, for example, attacked ethnology on very similar grounds in 1916.[37] But the fact that Pearl was white and an insider in racial science gave his critique an added authority and salience among a white audience, precisely the people White wanted to sway against the practice of lynching. That Pearl had been a notable eugenicist, a field premised on scientific theories of racial difference and hierarchy, made him all the more appealing to White, for he wished white America would follow the prominent scientist in a broader cultural change of heart.

White's first letter to Pearl cited a variety of scientific and medical texts on racial difference, and White asked for Pearl's assessment of each study. Notably, he did not seek Pearl's advice on the scientific soundness of widespread claims about black sexuality that permeated both popular and scientific discourse on lynching. Instead, his questions about racial science focused on more general, seemingly gender-neutral claims about black inferiority, primarily in regards to intellectual capacity. Black women appeared only briefly as forced by economic circumstances to work, possibly risking their children's development in the process; the black man as rapist trope that White refuted elsewhere did not appear in his letter to Pearl at all, even though it had been enthusiastically perpetuated by scientists for decades.[38]

As ethnological science itself had done, White began with and honed in on the human head. He asked Pearl if several recent comparative studies of black and white brains were "sufficiently accurate or representative . . . to justify the drawing of definite conclusions as to Negro inferiority." White asked the same questions of similar studies conducted during the Civil War, which had continued to be employed by many scientists into the 1920s. "In brief, have sufficient measurements and studies of Negro brain capacity and structure been made to justify the sweeping statements of this sort?" White wondered. Exploring the "nature versus nurture" question that had been central to racial science since its inception in the nineteenth century, he suggested that "disuse rather than race" likely determined differences in brains. Similarly, he wondered if the "economic status of Negroes which forces both mother and father to work" might be responsible for the "old assumption that the brains of Negro children close around age fourteen" or if Pearl could cite any legitimate science that supported a biological explanation.[39] Finally, he queried Pearl about the scientific soundness of intelligence tests that had been employed to further malign the intellectual capacities or potential of the black race.[40] "It may be that the Negro is inferior but the evidence of progress he has made in the arts, sciences and other fields as his economic condition improves, and as the weight of oppressive social conditions is lifted seems to cast grave doubts upon any hasty assumption of inferiority," he demurred. "I very much want your candid opinion and trust you will be generous enough to give it to me."[41]

For the most part, Pearl seemed to do exactly that. White had raised an "important . . . but difficult problem," Pearl remarked before attempting to tackle White's numerous queries. Though he insisted it was "impossible to give categorical answers to such questions," he nonetheless "doubt[ed] very much if any of the comparisons of negro brain weights with white brain weights are worth much statistically." While Pearl expressed some concerns with the methodology of these studies, particularly the sample size, he maintained that even if a measurable difference in brain weight between the two races could be demonstrated, "very little importance" should be attached "to brain weight as evidence of intellectual capacity." Pearl also expressed skepticism about drawing "broad conclusions" from intelligence tests, "particularly relative to racial matters." As he had in "The Biology of Superiority," he critiqued scientists' tendency to hierarchically rank the races "because it all depends upon who decides upon the yardstick by which superiority shall be measured." Still, he admitted he believed the races to be biologically distinct. He further noted that such attempts at ranking were "futile" since racial groups have rarely if ever had "equality of opportunity." Determining racial superiority was a question "incapable of final resolution" because it could not be separated from "emotion, taste, and prejudices." Pearl apologized to White for not being more helpful and extended him an invitation to visit when he returned from France.[42]

White, however, found Pearl's response more than helpful. Indeed, he reprinted large portions of it in *Rope and Faggot*—with Pearl's approval. And White did in fact visit Pearl in Baltimore when he returned to America.[43] Moreover, White sought Pearl's input on a draft of his chapter on racial science, which Pearl read and returned with his comments. "On the whole I think it is excellent," Pearl declared. But he did advise, "You weaken your case greatly by calling [prominent racial scientists] names." Pearl cautioned him not to refer to one man employed as a "full professor in good standing in a good, respectable university" as a "near-scientist," even if his conclusions were wrong, and warned against implying another scientist was "only slightly more erudite and respectable than the Ku Klux Klan." White graciously thanked Pearl for his advice, replying, "You are right about the epithets and out they come."[44] The finished product reveals that White ignored the second portion of Pearl's advice by continuing to identify scientists by name whom he deemed only "slightly more erudite and respectable" than the Klan.[45]

On behalf of his publisher, White asked Pearl if he would be willing to read the final draft of *Rope and Faggot* and offer a brief endorsement, presumably for the back cover or publicity materials. Pearl declined, citing a busy schedule that would prevent him from attending to the matter in time. Nonetheless, White sent Pearl an inscribed copy of the book on March 27, 1929; Pearl responded within days with thanks and congratulations. "I read it last night in one sitting. What a dreadful indictment of the American people the book is. It will surely have a large sale," Pearl declared, before warning White that he should not "venture South of Baltimore in the future without a large and determined bodyguard." Ever persistent and mindful of the added authority his book would gain with an endorsement from a prominent white scientist like Pearl, White changed tacks and asked Pearl if the publisher could quote from his letter of congratulations. Again, Pearl declined.[46] Despite Pearl's refusal to endorse the book publicly, the two men continued to correspond congenially for years.

In the chapter of *Rope and Faggot* devoted to racial science, White began by quoting Herbert Adolphus Miller, a white professor of sociology at Ohio State University, who had given an address, "Science, Pseudo-Science and the Race Question," before the NAACP in 1925. Miller warned that science went all too often unquestioned; it had taken on the absolute authority previously reserved for religion. For this reason, science could be as dangerous as it could be instructive. White pointed out approvingly that he had also warned about "the appropriation of the scientific jargon by the totally unscientific who rationalize their prejudices and think that God intended it so because they can say it in scientific terms." Such was the case with lynching, White argued. As it had since the nineteenth century, science gave credence to and fueled existing racial prejudices, but more important, it influenced not just ideas about black people but how they were treated—or mistreated—in life.

White's use of Miller set the stage for the tenor and structure of the chapter as a whole. Throughout, he cited and summarized a variety of scientific studies by white men that endeavored to prove black inferiority, but he also strategically employed the critiques of more sympathetic white scientists like Pearl, whose letters he quoted extensively in the chapter. He discussed at length a number of well-known studies of brain weight, skulls, or intelligence tests that "proved" black inferiority and then identified holes in the study's logic or methodological problems, bolstering his critiques by citing other white scientists. In so doing, he represented scientific claims about black inferiority as under attack not just by himself and other African Americans but also by "scientists and scholars worthy of the name," knowing all too well the added authority these scientists' names would carry.[47]

These white scientists might not have gone as far as White would have liked in their challenges to claims about black inferiority, but he used them nonetheless. For example, he cited a Pearl article, "Variation and Correlation in Brain Weight," that conceded that brain weights did indeed differ by race, but "brain-weight and intelligence in the sense of mental capacity are probably not sensibly correlated."[48] Pearl and the other more sympathetic white scientists White cited stopped short of rejecting the existence of biological differences between the races, but their caution against extrapolating broadly about those differences was nevertheless appealing to the pragmatic White. However, he noted, "It is almost a tragic circumstance that such reasoned and temperate conclusions as these gain circulation at but a fraction of the speed of those which sow the seed of racial hatreds and antagonisms. It is also deplorable that especially in the United States prejudices . . . all but overwhelm those who counsel sanity and scientific accuracy."[49]

The Intellectual Quagmire of Gender and Race

Throughout his lengthy analysis of racial science in *Rope and Faggot,* White focused largely on claims about black intellectual capacity rather than the highly charged issues of sex and manhood that dominated both popular and scientific discourse on lynching. To be sure, many early twentieth-century scientists added fuel to white paranoia about miscegenation and widespread beliefs in black hypersexuality. But White did not address these claims in his discussion of racial science's influence on lynching. Instead, he challenged broad scientific claims about black physical and intellectual inferiority, which he insisted made lynching possible by devaluing black lives. Even his brief discussion of the "question which so sorely agitates American minds—intermarriage" focused on scientific claims about the brain size and capacities of the mulatto rather than the associated but more volatile issues of sex. He did, however, note the irony of scientific considerations of the mulatto, in which intermarriage with whites could be either the black race's destruction or its salvation, writing, "Here we have a bewildering example of much of the reasoning on the alleged inferiority of the Negro—first, the mulatto is 'deceptive,' 'dishonest,' 'inferior physically and mentally,' and 'dangerous' and he 'almost invariably' dies young the nearer he approaches the Caucasian"—claims that the biracial White, thirty-six years old and healthy at the time of the book's publication, must have found especially absurd. "And, second," he added, "the only hope of making anything at all of the poor Negro lies in intermarriage and crossing with other races!"[50] Mirroring antebellum ethnology in his focus on brains and skulls rather than sexed bodies, he argued that scientific claims that the black race was limited in its intellectual and social possibilities by inferior biology contributed to a general climate of racial animosity that required little spark to boil over into violence.

However, while he focused on relatively gender-neutral claims about black people in general, he identified the key players in the perpetuation of racist ideology—and its negation—as an entirely male group.[51] He was of course not entirely incorrect in this assessment; men had continued to dominate the racial sciences on both sides of the "Negro question." A number of women, including Alice Fletcher and Matilda Stevenson, had made names for themselves in cultural anthropology, a field that was presenting tremendous challenges to scientific racism. His omission of these female scientists may have had as much to do with his focus on biological studies of race—in which there were indeed virtually no women—as his general antipathy toward powerful women and his pragmatic emphasis on the scientists whose names would carry the most authority: white men.

White maintained that three overlapping and mutually informing groups of men perpetuated the notion of a natural racial hierarchy: the Klan, scientists who "attempt[ed] to prove Negro inferiority by brain weight or structure," and Nordicists, which he referred to as "the ballyhoo experts of a blue-eyed, blond-haired, dolichocelphalic [large head/brain] superman."[52] White likely wrote these words with a keen sense of irony: his varied accomplishments earned this blond-haired, blue-eyed African American the title of Renaissance man if not superman. For White, the role of the Klan in lynching was obvious and needed little elaboration. But racial scientists and Nordicists also played a "not unimportant part in the creation of a national psychology toward the Negro" and promoted violence against blacks. Furthermore, he believed, white race pride was as dangerous as theories of black inferiority. "Race pride and conceit" resulted in "the lawlessness and bigotry which find outlet in, among other ways, lynch-law."[53] He bemoaned the huge number of hearts and minds reached by all three groups of men, but he insisted it was essential to address the claims of the often overlooked—and thus all the more insidious—group, the scientists. After all, White noted, scientists reached and influenced large portions of society that would never align themselves with the Klan or formal white supremacy organizations.

White's tactic of poking holes in various studies asserting black inferiority proved to be fraught with ideological and rhetorical challenges. In the context in which he lived and worked and at a time in which ideas about the biological inferiority of black people persisted in popular thought even as they lost ground among mainstream scientists, some empirical investigations into these earlier yet still influential studies were necessary in order to debunk them once and for all—a project that others continued well into the late twentieth century. But like the nineteenth-century black ethnological writers previously discussed, Walter White faced the challenge of dismantling the master's house using the master's tools. Much like Pearl, he ended up implying that racial science was not inherently flawed; the science itself should simply be better. "It is certain that no conclusions regarding the Negro brain, whether those conclusions be absolute or relative, are worth very much until examination by unbiased and competent scientists is made of a sufficiently large number of Negro brains," White declared.[54] The whole scientific endeavor of comparing races need not be abandoned entirely, he seemed to say, remarking, "For whatever the result may be worth, it is not to be hoped that the methodology of the past will be abandoned and not only that qualified and unprejudiced experts may study a sufficient number of Negro brains to make their findings valuable and trustworthy."[55] In so doing, White lent a certain degree of legitimacy to the very science he set out to undermine.

White's omission of any discussion of scientific claims about black sexuality from his chapter on racial science, in contrast, was likely strategic. White was certainly aware that scientists had perpetuated the image of black men as a sexual threat. For one, in *Rope and Faggot,* he discussed scientific work on race dating from the 1870s to the 1920s, a period that coincided with racial scientists' fixation on "furor sexualis" among black men. Scientific claims about the supposedly enlarged genitalia and unrestrained sexual impulses of black men routinely appeared in the same journals from which White cited studies of brain weight and intelligence. Furthermore, claims about black hypersexuality were sometimes espoused by the same scientists White cited on issues of race and intelligence. For example, he discussed a comparison of black and white brains in a 1906 article, "Some Racial Peculiarities of the Negro Brain," written by University of Virginia physician Robert Bennett Bean and published in the *American*

Journal of Anatomy. However, White ignored Bean's assertion in the same article that black men were "passionate [. . .] under stimulation" and suffered from a "lack of self-control, especially in connection with the sexual relation."[56] Nor was this aspect of Bean's article mentioned in White's chapter on "Sex and Lynching." Even though he analyzes there the kind of rhetoric used by Bean, White did not associate it with Bean or any other scientist.

White's omission of scientists from his analysis of the construction and perpetuation of the most volatile stereotypes about black men was not simply symptomatic of White's general evasiveness or timidity about issues of sex. On the contrary, throughout *Rope and Faggot,* White's discussions of sex and its role in the lynch mob mentality were pointed and unflinching. Indeed, one suspects that Pearl's caution to White that, upon the book's publication, he never return to the South without bodyguards was shrewd advice. White described the South as "sex-obsessed" because the region's backwardness left its citizens wanting for entertainment, and the Southern brand of strict but emotional religiosity worked its followers into a frenzy with no permissible outlet. The problem was compounded by widespread ignorance and illiteracy, White maintained.[57] White women specifically were prone to "hysteria where Negroes are concerned . . . an aspect of the question of lynching which needs investigation by a competent psychologist."[58] Women accusing black men of rape in the South were not to be taken seriously; in his experience, these women were largely young girls "passing through the difficult period of adolescence," menopausal or postmenopausal women, women in unhappy marriages to "unattractive husbands," and "spinsters." With the large scope of the female lifespan and experience covered by White, his antilynching argument also seemed to reveal his thoughts about women in general: that few, if any, were to be taken seriously.

As for white men, White echoed Wells in arguing that lynching was an indictment of white rather than black manhood. To counter the connection of lynching with rape, White frequently pointed to the number of women and children lynched. With considerable irony and turning the rhetoric of civilized manhood on its head, White broke the number down by state, noting, "Mississippi leads in this exhibition of masculine chivalry, with sixteen women victims." He continued, "Tennessee and South Carolina mobs have bravely murdered seven women each. . . . Three of the twelve Texas victims were a mother and her two young daughters killed by a mob in 1918, when they 'threatened a white man.' Thus was white civilization maintained!"[59] White further argued that the "bogey of black rape" was a manifestation of white men's fears that black men would return in kind to white women the treatment white men had inflicted on black women throughout American history. White pointed out that the inheritance of slave status through the mother had given white men not only tacit permission but economic incentive for raping enslaved women and that since Emancipation, miscegenation continued to be only tolerated in one direction. "The man who attempts to maintain a fixed respect towards one group of women and indulges meanwhile in all manner of immoralities with another group may seek ever so hard to maintain such a balanced dual standard," he added. Moreover, "for more than two hundred years this moral deterioration has affected the Southern states, and from that decay arises the most terrifying of all aspects of the race problem to the white man."[60]

This terrifying aspect of the race problem was the prospect of black men breaking through the "dual standard" of the South and asserting the same male prerogatives to female bodies that existed under a traditional white patriarchy.[61] In fact, White noted, many lynchings occurred in the South because black men had tried to protect women in their own race, not because they threatened white women. Furthermore, white men's supposed repulsion for black physiognomy was more than disproved by high rates of miscegenation before, during, and after slavery, but "suspicion that the absence of repulsion applies to both sexes of both races" motivated the emasculation that often accompanied lynch violence. Lynching then represented white men's fears of sexual as well as economic and political competition from black men. Castration was also a frequent component of lynching because of "stories of Negro superiority in sex relations," White remarked, "and it did not matter whether or not that rumoured superiority existed in fact or fancy—the very violence of opposition by the mobbist seemed to lend credence to the truth of the assertion."[62]

Notably, White himself did not lend credence to such "stories" of black hypermasculinity or hypersexuality by attributing them to scientists. Instead, such claims appeared as rumors of indeterminate origin. "Of all the emotional determinants of lynching none is more potent in blocking approach to a solution than sex, and of all the factors, emotional or otherwise, none is less openly and honestly discussed," White wrote. "With the most intransigent Negrophobe it is possible to conduct a conversation of certain phases of the race question and do so with a measured calmness of manner," he maintained, and the brief correspondence he exchanged with scientist Robert Bennett Bean seemed to bear this out. "But when one approaches, however delicately or remotely, the question of sex or 'social equality,'" he added, "reason and judicial calm promptly take flight," replaced instead by "berserk rage."[63] To be sure, the trope of the black sexual predator pervaded popular as well as scientific discourse around lynching. But White likely feared that attaching the names of scientists to these explosive claims would lend them greater legitimacy even as he was trying to undermine them.

Walter White's engagement with racial science—and racial scientists—was complex and ambivalent. For White, scientists were both sinners and saviors in his campaign against racial violence. He enlisted the help of insiders like Pearl to fight science with science. But he could not escape the limitations of working within a field that, even at its most sympathetic to the black race, was premised on innate biological difference between the races. His relationship with Pearl provides a perfect example of the difficulties he faced. In the years that followed the publication of *Rope and Faggot* in 1929, the two men maintained their correspondence and White continued to seek additional information on the issue of differences in brain weight or size between the races. In 1934, Pearl published an article, "The Weight of the Negro Brain," in the journal *Science*. Therein, he reexamined the many racial studies of brains and crania dating back to Samuel Morton and the Civil War anthropometric studies.[64] Pearl often found fault with the studies' methodologies but nonetheless supported their conclusion that black brains were 92.1 percent the size of whites'. Subjecting previous measurements to his own biometric processes, he announced that his results corroborated those of Morton and others, and thus the only "reasonable" conclusion he could draw was that "the Negro brain is, on the average, from 8 to 10 percent lighter than the fairly comparable white brain."[65] Though he had noted in his first letter to White seven years earlier that he cautioned against drawing conclusions from racial measurements, he offered no such warning here. White must have been disappointed, but he made no mention of it in his letters to Pearl, which continued after the article's publication.

Ultimately, White's relationship with Pearl—and with racial science more generally—was pragmatic. In combating lynching, White sought to disprove widespread beliefs in black inferiority that had long been legitimized by science, but in so doing, he divorced from racial science its most dangerous claims about sex and gender. As White noted, "Sex with all its connotations so muddies the waters of reason that it is impossible to bring the conversation back to its more impassioned state."[66]

Walter White had an intimate knowledge of racial violence in America, which he devoted much of his storied career with the NAACP to combating. Long represented by historians as an organization man and contrasted with more grassroots activists like his predecessor in the antilynching battle, Ida Wells, White was indeed a pragmatic leader who often allied himself with powerful whites to further his cause. But he was also very shrewd, with a keen understanding of the complex machinations of American racism. An important component of his antilynching work and thought was his critical engagement with the nation's history of scientific racism.

In key public writings and personal correspondence, White argued that scientific theories about black inferiority—and threat—enabled and even provoked racial violence. White enlisted the assistance of white scientists, most notably Raymond Pearl of Johns Hopkins University, to chip away at biological justifications of racial inequality, but these alliances were not without problems. Scientists like Pearl usually stopped short of dismissing any biological difference between the races, even as they warned against reading too much into them. Still, White clearly recognized the authority to which scientists laid claim. It was likely this authority that led White to distance science from claims about black hypersexuality and focus instead on less volatile arguments about differences in brain size and intelligence. It is clear that White was well aware that scientists had long naturalized tropes about blackness, gender deviance, and sexual aggression, but by associating such claims with angry mobs and the ignorant masses rather than men of science, he stripped them of some of their power and legitimacy.

Though it is difficult to argue with his claim that scientific theories of black inferiority exacerbated lynch violence, White himself was not immune to some of the ideological traps of American racial thought.

Moreover, as he often found in his dealings with Pearl, attempting to fight science with science could engender its own problems and compromises, but ones he was ultimately willing to accept in the face of more immediate racial concerns that were, quite literally, a matter of life and death.

Epilogue

BY THE TIME WALTER WHITE published his autobiography in 1948, both scientific racism and racial violence were waning. While his scientific correspondent Raymond Pearl never gave up his belief that the races were biologically different, even if he warned publicly against drawing conclusions about superiority and inferiority, White argued until the end that race was fundamentally a fiction. White revisited the pernicious effect of racial science in the final pages of his memoir, published seven years before his death. "At the root of my anger and my frequent deep discouragement, is the knowledge that all race prejudice . . . is founded on one of the most absurd fallacies in all thought—the belief that there is a basic difference between a Negro and a white man," he lamented, adding, "There is no such basic difference." However, he believed there was cause for hope. Just as scientists had lent support and authority to human prejudice for more than a century, scientists also held the redemptive power to discredit and dismantle racism—even if Pearl himself had fallen short of White's objective in that regard. With mounting optimism, White observed, "More and more scientists, realizing the dire importance of the race problem to human welfare, are going out of their way to state unequivocally the falseness of the belief that such a difference exists. . . . Even the belief that racial markings are apparent in a Negro, no matter how light he is, has over and over been proved false."[1] Undermining one of the most central tenets of racial science—that the physical body could be read with scientific precision to reveal the character and capacities of its owner—White posited, "Suppose the skin of every Negro in America were suddenly to turn white. What would happen to all the notions about Negroes, the bases on which are built race prejudice and race hatred? What would become of their presumed shiftlessness, their cowardice, their dishonestly, their stupidity, their body odor? Would they not merge with the shiftlessness, the cowardice, the dishonestly, the stupidity, and the body odor of the whites? Would they not then be subject to individual judgment in matters of abilities, energies, honesty, cleanliness, as are whites? How else could they be judged?" As White understood, rather than an objective measure of racial difference, the body revealed only what the scientist expected to see in it.

As he drew his life story to a close, he described a memorable encounter he had with a white scientist, Sir Arthur Keith, the president of the British Association for the Advancement of Science, who he met in London several years earlier. "Before he spoke a word of greeting, he extended a long, bony hand in welcome and, abstractedly, continued to hold my hand as he examined my features intently," White recounted. Upon this awkward first meeting, Keith told the blond-haired, blue-eyed White, "The only way I can tell that you have Negro blood is by the shape of your eyes." Recalling the incident, White wrote, "Startled, I asked him, 'What is there in them which reveals my ancestry?' He did not reply immediately but continued his scrutiny of my face. It probably lasted no more than a minute, but the time seemed much longer. Sir Arthur then shook his head as though trying to rid it of cobwebs."

Time stood still for White in that loaded moment, a quintessential snapshot of scientific engagement with race. The white male scientist stood face to face with his black subject, struggling to read the man's physiognomy for some larger meaning. But unlike Samuel Morton's skulls or the thousands of Civil War soldiers measured and probed or the anonymous black bodies interspersed throughout R. W. Shufeldt's racial invectives, Keith's subject had the opportunity to speak back. And struck momentarily silent by White's challenge, Keith actually reconsidered his preconceptions about race. It is not insignificant that White chose an anecdote about a scientist to demonstrate the fiction of race in the final pages of his autobiography. Scientists had lent authority to human prejudice and fueled the fires of racial violence, White believed, but their ideological conversion, however flawed, embodied White's optimism that racism could also be overcome through scientific reason. It is also worth noting that White did not use Pearl, who appeared nowhere in White's autobiography, as an example here. Instead, White's brief encounter with Keith was all the more powerful rhetorically because Keith's change of heart about the logic of racial science could stand alone, untainted by subsequent disappointing words and actions that revealed his conversion to be incomplete. For White, the encounter represented a critical moment of dialogue and, ultimately, hope. In response to White's startled query, Arthur Keith answered, "No—I'm wrong. If you had not told me in one of your letters that you have Negro ancestry, I would have seen nothing. But because you did tell me, I *thought* I saw some indication. That's unscientific. Sit down, won't you—and let's talk."

Indeed, by the 1940s, the character of racial science and its function in society had changed considerably. For one, science no longer had the same degree of cultural authority over questions of race that it had in the nineteenth and early twentieth centuries. Eugenics, one of the more public and popular wings of racial science, for example, lost credibility in the United States after World War II, largely as a result of its association with the enemy—the Nazis, who took scientific racism to horrifying new extremes—but did not die entirely. Instead, negative eugenics, which emphasized preventing the "unfit" to reproduce through sterilization or, in the Nazi case, wholesale genocide, fell out of favor, at least publically, while positive eugenics, which encouraged careful but prolific reproduction among the "fit" culminated in the baby boom of postwar America.[2] In addition, growing numbers of scientists were questioning or openly challenging the long-held scientific belief in innate and permanent racial differences, differences that also implied natural inferiority and superiority.[3] In part, this change reflected a new generation of scientists, but as illustrated by White's interactions with Pearl and Keith, some individual scientists were also reconsidering their own previously held convictions. While scientists certainly did not abandon their search for quantifiable differences between the races after the 1930s and 1940s—quite the contrary, they conducted comparative studies on everything from intelligence to sexual behavior—they were no longer at the center of U.S. racial discourse and race relations.[4] In midcentury America, the nation's most contentious battles over race occurred in the courts.

Ultimately, though, one of the biggest challenges to scientific racism came from science itself, with the development of cultural anthropology as a discipline. Widely considered the founder of the field, anthropologist and Columbia University professor Franz Boas was no stranger to racism himself, having immigrated to the United States in part to escape anti-Semitism in his native Germany. Beginning in the 1890s, he argued that culture and environment, not biology, best explained human variation and amassed considerable evidence showing that physical characteristics varied greatly within each race, particularly across generations. Moreover, he suggested that racial discrimination, not inherent inferiority, had limited black advancement in the United States and insisted that a direct link between intelligence and race had never been proven by science, critiquing scientists' long-standing focus on head size.

However, Boas's views largely fell on deaf ears at the turn of the century, drowned out by still dominant biological paradigms of racial difference. For instance, Daniel Brinton, a professor of ethnology and archaeology at the Academy of Natural Sciences in Philadelphia, rebutted Boas's claims in a number of popular and scientific publications in which Boas was not given an audience at that point. Seemingly undaunted, over the next two decades, Boas attacked the logic of eugenics and, as part of a lengthy government study of immigration, proposed a fluid model of race in which the physical and mental character of groups of people changed as their environment changed. Through much of his life, Boas's work was met with controversy among scientists, but ultimately, his cultural model of race supplanted earlier biological and evolutionary models in American science—even if the general public remained less than committed to the ideal of racial equality.[5]

Still, carrying on his model of cultural relativism and explicitly antiracist agenda, a number of Boas's students at Columbia went on to be influential anthropologists in their own right, and, notably, the two most well known among them were women: Ruth Benedict and Margaret Mead. In its insistence that no race was innately or irrevocably inferior as well as its growing inclusion of female scientists, Boasian anthropology stood in sharp contrast to nineteenth-century racial science. Moreover, Benedict and Mead's anthropological work on race also enabled them to challenge social constraints placed on women in both their own professional lives and their critiques-by-comparison of America's gender system.[6] In this regard, they are part of not just a Boasian intellectual tradition but also a protofeminist genealogy of women who escaped the home by entering the (ethnographic) field, dating back to the late nineteenth century with Matilda Coxe Stevenson's work with the Zuni in the American Southwest, Alice Fletcher's ethnographic studies of the Plains Indians through the Peabody Museum of Archeology and Ethnology, and Mary French-Sheldon's racially charged expedition across East Africa.[7]

While female involvement in anthropology in and of itself does not appear to be a primary reason for the "retreat of scientific racism," women like Mead and Benedict are nonetheless noteworthy participants in the project of remaking race in the early to mid-twentieth century. After all, who scientists were—their personal backgrounds and cultural politics—shaped the science they produced. The scientists I have discussed herein have nearly all been male, reflecting the composition of racial science during the scope of the nineteenth and early twentieth centuries and my focus on proponents of biological frameworks of racial difference, who were very much an overwhelmingly male group.

Certainly, gender, as well as sexuality, featured prominently in Mead's and Benedict's professional writings on race. Mead in particular used her studies of "primitive" societies to critique the rigidity of "Western" gender roles and sexual mores. Mead's groundbreaking 1928 book, *Coming of Age in Samoa,* and her 1935 book, *Sex and Temperament in Three Primitive Societies,* both circulated widely within and outside of scientific circles, and Mead soon developed a lifelong public presence, speaking on a variety of contemporary issues and writing for numerous popular periodicals. While her work was intended to denaturalize gender and racial hierarchies and understand sexuality in its specific cultural contexts, this nuanced argument was likely lost on some in her wider audience, who could instead read her work as titillating and voyeuristic accounts that reaffirmed America's long-standing association between "primitiveness," hypersexuality, and deviance from white, middle-class gender norms. Indeed, gender and sexuality often served as markers of cultural difference in modern anthropology, just as they had previously served as markers of biological difference for ethnologists in the nineteenth and early twentieth centuries.[8]

To be sure, the scientific trend toward cultural rather than biological understandings of race did not mean an end to racism—within or outside the world of science. Enormously popular in mid-twentieth-century America, *National Geographic* brought anthropology—and near fetishistic imagery of scantily clad "primitives"—to an even wider audience, for example. Ostensibly celebrating human diversity, the periodical simultaneously reaffirmed traditional assumptions about "primitiveness" and revived old ethnological tropes, with frequent images of dark-skinned people engaged in physical labor and light-skinned people in more leisurely pursuits.[9] And in his 1965 *The Negro Family: The Case for National Action*—soon better known as the Moynihan Report—Daniel Patrick Moynihan, then assistant secretary for policy planning and research in the U.S. Labor Department, described the African American community as suffering from a kind of cultural pathology. Though he framed the problem in terms of "culture," his language and logic nonetheless echoed earlier biological paradigms of race and pathology. Moreover, gender and sexuality were central to Moynihan's pathology model. He blamed the "matriarchal culture" of black America for hindering the race's progress. According to Moynihan, black women were generally better educated and more often employed than their male counterparts, emasculating black men and driving them out of the home, which in turn left the community without strong male figures and black families in disarray. At the same time, he also called black women's sexual respectability into question by arguing they were more prone to illegitimate births, further undermining African American family and community structures.

Similarly, as American scientists in the nineteenth century focused more and more on gender and sexuality in their analyses of race, their writing also became increasingly political in its tone and far-reaching in its aim. Scientists in the long nineteenth century made policy recommendations, served as expert witnesses in criminal trials, endorsed political candidates, and became involved in the social hygiene and eugenics movements. In other words, racial scientists carefully cultivated a niche for their expertise and had an increasingly public function in U.S. society. They sought to place the authority over the nation's "race problem" in their own hands and saw their role as "social doctors" fixing America's ills.

This book has depicted an overlapping group of scientists who shaped ideas about race and human difference in America for more than a century and the world in which they lived. Between the antebellum period and the early twentieth century, these scientists constructed and maintained their own authority by bringing science to bear on the country's most pressing political issues and social problems. They represented a range of scientific disciplines, regional backgrounds, and political affinities, united by their belief that race was a biological entity. They sought to explain how the races had originated and made predications about where they were headed in the future. Most of all, they naturalized racial hierarchy in the present as transhistorical, rooted in the body, and permanent. The scientists saw their work as socially relevant, and, indeed, the imprint of it could be seen throughout American culture in a myriad of ways. Consequently, this book has sought to unpack the mutually informing relationship between racial science, politics, and culture around questions of race and gender. My fundamental concern has been the social function of science and the cultural context in which American scientists theorized about race.

In many ways, this has been as much a story about continuity over time as it is about change. Throughout the nineteenth century, science gained mainstream popularity and legitimacy in the United States, with many looking to the emerging sciences of race to explain and naturalize the social order. But the authority scientists held during their period had to be actively built and defended. Toward that end, scientists brought their expertise on race to bear on pressing issues in American society, but society was constantly changing. They constantly adapted their ideas about race—and its intersection with other categories, like gender and sexuality—to the new social, cultural, and political circumstances. But at the same time, several core beliefs about race remained relatively constant in science throughout this period: that human difference and destiny is located in the body and that differences of physicality explain and even demand social hierarchy.

In the end, part of me shares Walter White's optimism that science has the potential to unmake race, just as it had lead the project of its construction in the nineteenth century. But at the same time, even scientific studies that seemingly strive to challenge racial difference often stop short of arguing that race is entirely a fiction, particularly those studies that reach a relatively wide audience. Even the pragmatic White must have been somewhat disappointed by the incomplete conversion of his scientist friend, Raymond Pearl, on matters of race. Likewise, as a tool for dismantling the house it helped build, science remains an unfulfilled promise in the twenty-first century. Scientists are always a product of their own cultural context—even when they intend their work to challenge, subvert, or revolutionize the status quo—and in the United States, race continues to have considerable ideological power as a framework for categorizing and dividing human beings. In the late eighteenth century, Thomas Jefferson suggested that race was permanent and rooted in the body, though he expressed uncertainty as to precisely where; in the nineteenth century, some scientists embodied race in skulls, others in hair, and, by the end of the century, in the sexual organs.

Now scientists are looking further still into the body, to microscopic DNA, for the essence—or fallacy—of race. The featured article in the December 2003 issue of *Scientific American* questioned whether there was any biological basis for the concept of race, for example. It pointed out that scientists had uncovered an enormous amount of genetic variation within racial groups as currently defined, not just across them; that only superficial physical traits such as eye or hair color could be genetically linked to those groups; and that racial categories shifted from one culture to the next. Ultimately, however, the article seemed to suggest that racial categorization in and of itself was not the problem but that we needed additional or more precise categories and that racial groupings remained helpful for the treatment of diseases. Moreover, the issue's cover page underscores that the flurry of work on race, biology, and genetics at the turn of the *twenty-first* century may be driven at least in part by scientists' desire to reestablish their authority over a domain in which they had lost ground in the previous century. "Does Race Exist?" the title queried. "Science Has the Answer."[10]

Acknowledgments

THIS BOOK, DERIVED FROM MY dissertation, received generous support from several sources at Rutgers University. Graduate fellowships from the Department of History and an Excellence Fellowship from the Institute for Health, Health Care Policy, and Aging Research provided much-needed time for research early on in the project. The Institute for Research on Women also provided monetary support, but far more important, participation in its yearlong seminar on "Gender, Health, and Bodies," led by Beth Hutchison and Nancy Hewitt, offered the opportunity for interdisciplinary exchange of the best kind. My work and my sanity greatly benefited from the feedback I received from such a smart, friendly group of scholars. In addition, since its inception, the Center for Race and Ethnicity at Rutgers was a second home to me, and our "in-house" research group proved a most supportive environment in which to write and revise. Over the last two years of my graduate work and when I later returned as a postdoctoral fellow, the members of the group offered insightful commentary and practical advice on several chapters; I extend my thanks to Nadia Brown, Jill Campaiola, Simone Delerme, Jeffrey Dowd, Joe Gabriel, Richard Mizelle, Liz Reich, Rebecca Scales, Anantha Sudhakar, Sonja Thomas, Dora Vargha, Directors Keith Wailoo and Mia Bay, and Associate Director Ann Fabian. Now at Rochester Institute of Technology, Rebecca Scales also trudged out to a cold cemetery in upstate New York to procure several of the images, for which she has my gratitude.

I also had the great fortune of receiving a postdoctoral fellowship in the Gender Studies Department of Indiana University, which provided additional funding and precious time to devote to this project. No less important, my time at IU also brought me several sanity-sustaining friendships, including my fellow Rutgers history alums, Amrita Chakrabarti Myers, Khalil Gibran Muhammad, and Matthew Guterl, as well as Kim Williams and Jennifer Shaw. Sincere thanks are due to Jen Shaw in particular for turning her keen eye to several chapters, which are much improved for her efforts.

I have had the pleasure of conducting research in a number of excellent libraries and archives along the way, including the American Philosophical Society; Rutgers University Special Collections; the National Library of Medicine; and the New York Public Library, Manuscripts and Archives Division. The College of Physicians of Philadelphia Historical Library in particular has been a treasure trove of resources, with much help from Edward Morman, Lisa Gensel, and Richard Fraser. And last but certainly not least, the library staff at Rutgers, Indiana University, and the University of Kentucky have all been enormously helpful, particularly the interlibrary loan departments.

Feedback from a number of sources has shaped—and improved—this project. Deborah Gray White, Temma Kaplan, and Bonnie Smith commented on my work at critical early stages and have also offered encouragement and sage advice on many occasions. Krystal Frazier, Pamela Epstein, Emily Westkaemper, Jacqueline Castledine, Emily Sloat Shaw, Liviya Mendelsohn, Danielle McCole, Megan Clark, Janine Everett, and Candace Grand Pre have been enthusiastic readers and much-valued friends. Jessica Restaino, my constant cheerleader, writing partner, and rock-steady friend, has read drafts of every chapter—some more than once, bless her heart. She has not only improved my prose; she has made every stage of academia easier in innumerable ways, including an always-available couch to crash on. Thank you for being a friend, Jess. Undergraduate mentors and now dear friends, Maria Mitchell, Padmini Mongia, Misty Bastian, and Renée Sentilles have continued to offer advice and good cheer at every turn. I never would have chosen to be an academic without them. And I would not have had the opportunity to do so without the unflagging encouragement of my former coworkers (and honorary family for life) in the grants office of the School District of Philadelphia, particularly Ethel Goldberg, Leonard Rossio, and Marcia Schulman.

My dissertation committee was and continues to be an unmatched source of intellectual and professional support. Nancy Hewitt provided encouragement and wisdom at every stage; it is not difficult to see why no committee at Rutgers seems to be complete without her on it. Always encouraging me to find the “story” in my project, Ann Fabian has the rare talent of giving constructive criticism that left me more excited about my project than before. Keith Wailoo has made me think of science and medicine in entirely new ways while also providing some of the best practical advice around on navigating the academy. With spot-on feedback, Marc

Stein has been an incredibly generous, attentive, and cheerful outside reader. My advisor, Mia Bay, has read multiple versions of every chapter with patience, keen insight, and a critical eye. She has continuously pushed me to think big and write simply; my work is infinitely better for her efforts. That I so thoroughly enjoyed graduate school is largely to her credit.

Most recently, I have been blessed with some of the best colleagues one could imagine when I joined the Gender and Women's Studies Department at the University of Kentucky in 2011. My chair, Karen Tice, has not only been an invaluable mentor and friend; she provided crucial feedback on several drafts of the entire manuscript and much-needed encouragement and resources to complete it. I am forever in her debt. Carol Mason has also given me helpful feedback on the project, especially the introduction. Even more important, her friendship, and that of her partner, Rachel Elliot, has guaranteed that even amid the stress of a new job and a book deadline, I have laughed often. I also had the good fortune of receiving astute feedback on the introduction from the History Department's Works-in-Progress group. Thank you as well to my research assistant, Jenrose Fitzgerald, who provided much-needed help with the manuscript's many images, and my editor at the University of Minnesota, Jason Weidemann, for his patience and guidance through the process.

My family has supported me and my work in too many ways to list, not the least of which were frequent queries of "Aren't you done yet?" My late grandparents never missed an opportunity to brag as I completed my PhD, and I blame them for any lingering ego inflation on my part. My parents, Marilyn and Donald, always reminded me to take time out for rest, fun, and, of course, visiting family—and intervened heroically whenever I didn't follow their advice. Their boundless love and constant encouragement have sustained me throughout. I would be remiss if I didn't thank my furry family members as well—the irreplaceable Zora, my infamous feline "other half" for fifteen years, and our dog Delilah and new cat, Anya, who were always eager to remind me that there are more important things than writing. Finally, my wife, Kelly, has lived patiently with this project since we met. She has given me comments on every chapter, fed me a never-ending supply of delicious home-cooked meals to keep me going, and provided much-needed perspective at every turn. Her humor, warmth, and generosity are unparalleled, and I never would have finished this book without her.

· APPENDIX ·

Charting Racial Science

Data and Methodology

THE CHARTS AND GRAPHS IN chapters 1 and 2 are based on keyword searches of the *Index-Catalogue* of the National Library of Medicine (NLM). To chart changes in racial science between the antebellum period and the early nineteenth century, I searched this database for scientific texts on race. Now digitized, the library's first catalog was initially published in 1880 and comprises materials dating from the fifteenth century to the early twentieth, including journal articles, dissertations and theses, monographs, pamphlets, and reports.[1] I chose to use the National Library of Medicine rather than a broader repository like the Library of Congress because its stated purpose was to be a repository for medical and scientific publications, and in the late nineteenth century, the NLM was the largest medical library in the world. The *Index-Catalogue,* then, reflects contemporary understandings of what was considered science or medicine—or at least what was of interest to scientists and physicians. Texts by prominent figures like Josiah Nott, Samuel Cartwright, and John Van Evrie, who are often labeled pseudoscientists today, were featured in the *Index-Catalogue,* another indicator that their contemporaries did not view them as such.

To demonstrate change over time, I conducted separate searches for each decade from 1830 to 1930, utilizing various racial terms in keyword searches, and then went through each reference to determine, by its title and, when necessary, its subject heading, the text's primary focus, both in terms of theme and race or races. To be sure, the data that make up these charts is intended not be comprehensive but rather visually representative of the changes I discuss in more detail. It should also be noted that such a method only reflects the primary focus of each text as determined by the title and subject heading; only a thorough reading of every text referenced in the searches would be able to measure how often each theme appears within the body of each text, a method that would be near impossible when the number of citations measure in thousands.

The trends demonstrated by this measure are if anything more exaggerated when the content of representative texts from each time period are examined. For example, while 20 percent of the references from the 1880–1930 period focus on topics related to gender, sex, or sexuality, an examination of numerous texts from this period demonstrate that these themes were omnipresent within individual texts in the racial sciences during that period, even when those texts were ostensibly focused on another theme. Many of the texts that focused on racial taxonomy at the turn of the century used the degree of sexual differentiation within a race to distinguish it from other races, for instance. In other words, gender could be both a theme or a paradigm or framework with which to approach another theme within the racial sciences.

Gender more often than not overlapped with another category of measure. However, as demonstrating changes in gender and sex within racial science is the central task of the first two chapters of the book in particular, any reference that either focused solely on issues of sex/gender or took a gendered approach to another theme measured in the following approaches was counted toward the gender total, and each reference was counted toward only thematic column. I included within the parameters of the gender/sex category studies of diseases that were male or female specific, venereal diseases (since such studies tended to dwell much on sex and gender within afflicted populations), comparative studies (often using the brain or skull) between a specific race or races and women, discussions of race suicide and women's education and/or birth control, fertility, genital differences or abnormalities, assessments of masculinity across racial lines, childbirth, rape, and so on. Similarly, in the category of hybridity, I included any text dealing with race mixture or mixed race people themselves as well as texts about other people who complicated racial binaries—studies of African American albinos, for example.

Another change one sees in the racial sciences is that they become, quite literally, less black and white over time. Early ethnology tended to focus on blacks, whites, and, to some extent, Native Americans, but over time, with increasing immigration, other races became more present in racial science. The charts demonstrate this change, but this too is an imperfect measure, for it simply measures all references and does not show that the texts on African Americans, or on race in general, were more likely to be book length than those on other races and also seemed to be more widely read within and outside the scientific establishment based on the number of reprints, editions, and citations by political writers and social commentators. In addition, texts that were ostensibly focused on race in general, such as racial history and taxonomy texts, would themselves often focus disproportionately on Africans and African Americans after briefly defining and describing the other races. That the preceding chapters then focus more in depth on scientific discourse about (and to a lesser extent by) African Americans or the dynamics between blacks and whites is a reflection of the nature of nineteenth- and early twentieth-century racial science itself. Furthermore, the low percentage of texts focused on whites is both misleading and telling, for it measures only those texts where whites were the focus, not those in which they were present in texts comparing whites to another race or races. In those texts, the reference was counted toward the nonwhite category, not to reinforce white as normative, but rather to demonstrate the extent to which whites were utilized as such in comparative studies by scientists during this period.

Notes

Introduction

1. The incident, the ensuing controversy, and Lawrence's repeated apologies received considerable media coverage for months and were the subject of heated debate on college campuses and in newspaper opinion pages across the country. See, for example, "A Second Chance at Rutgers," *New York Times*, February 8, 1995, http://www.nytimes.com/1995/02/08/opinion/a-second-chance-at-rutgers.html.

2. "Online Focus: Women and Science," *PBS Online Newshour*, February 22, 2005, http://www.pbs.org/newshour/bb/science/jan-june05/harvard_02–22a.html. This site also includes a transcript of reactions from female faculty and scientists.

3. Barbara Miller Solomon, *In the Company of Educated Women: A History of Women and Higher Education in America* (New Haven: Yale University Press, 1986), 56–57.

4. Rebecca Jordan-Young, *Brain Storm: The Flaws in the Science of Sex Difference* (Boston: Harvard University Press, 2010). See also Anne Fausto-Sterling's *Sexing the Body: Gender Politics and the Constructions of Sexuality* (New York: Basic Books, 2000), particularly chapter 5 (115–45), which offers a concise overview of the various biological and psychological studies of gender and the brain, with an astute assessment of the cultural politics behind such studies.

5. See, for example, Roy Porter, "Born That Way? More from the Doctor Who Says Homosexuality May Be a Matter of Genetics, and from Some of His Colleagues," *New York Times*, August 11, 1996, BR8; Nicholas Kristof, "Gay at Birth," *New York Times*, October 25, 2003, A19; "The 'Gay Gene' Debate," *PBS Online Frontline*, February 2000, http://www.pbs.org/wgbh/pages/frontline/shows/assault/genetics/; and "A Finger on Sexuality," *BBC News*, March 29, 2000, http://news.bbc.co.uk/1/hi/sci/tech/695142.stm.

6. Sandi Doughton, "Born Gay? How Biology May Drive Orientation," *Seattle Times*, June 19, 2005, http://seattletimes.nwsource.com/html/localnews/2002340883_gayscience19m.html.

7. Bernard S. Talmey, "Notes on Homosexuality," *Medico-Legal Journal* 34, no. 8 (November/December 1917): 4.

8. On the genealogy and implications of the contemporary "nature/nurture" debate among scientists, see William Byne, Udo Schuklenk, Mitchell Lasco, and Jack Drescher, "The Origins of Homosexuality: No Genetic Link to Social

Change," and C. Phoebe Lostroh and Amanda Udis-Kessler, "Diversity and Complexity in Gay/Lesbian/Bisexual/Transsexual Responses to the 'Gay Gene' Debates," in *The Double-Edged Helix: Social Implications of Genetics in a Diverse Society,* ed. Joseph S. Alper, Catherine Ard, Adrienne Asch, John Beckwith, Peter Conrad, and Lisa Geller (Baltimore: Johns Hopkins University Press, 2002); Vernon Rosario, *Homosexuality and Science: A Guide to the Debates* (Santa Barbara: ABC-Clio, 2002); and Richard Pillard, "The Search for a Genetic Influence on Sexual Orientation," and Garland Allen, "The Double-Edged Sword of Genetic Determinism: Social and Political Agendas in Genetic Studies of Homosexuality, 1940–1994," in *Science and Homosexualities,* ed. Vernon Rosario (New York: Routledge, 1996).

9. See, for example, "Controversial Heart Drug," *PBS Online Newshour,* June 17, 2005, http://www.pbs.org/newshour/bb/health/jan-june05/heart_6–17.html.

10. On the cultural politics of race and disease, see Joseph Graves Jr., *The Emperor's New Clothes: Biological Theories of Race at the Millennium* (New Brunswick, N.J.: Rutgers University Press, 2001), 173–92; and Keith Wailoo, *Dying in the City of the Blues: Sickle Cell Anemia and the Politics of Race and Health* (Chapel Hill: University of North Carolina Press, 2001). On race, genetics, and DNA, see Alan Goodman, Deborah Heath, and M. Susan Lindee, eds., *Genetic Nature/Culture: Anthropology and Science Beyond the Two-Culture Divide* (Berkeley: University of California Press, 2003); and Alper et al., *The Double-Edged Helix.*

11. Evelyn Brooks Higginbotham, "African-American Women's History and the Metalanguage of Race," *Signs* 17, no. 2 (Winter 1992): 251–74.

12. Laura Briggs, "The Race of Hysteria: 'Overcivilization' and the 'Savage' Woman in Late Nineteenth-Century Obstetrics and Gynecology," *American Quarterly* 52, no. 2 (June 2000): 253.

13. See, for example, Ivan Hannaford, *Race: The History of an Idea in the West* (Washington, D.C.: Woodrow Wilson Center Press, 1996); Audrey Smedley, *Race in North America: Origin and Evolution of a Worldview* (Boulder: Westview Press, 1993); and Kim Hall, *Things of Darkness: Economies of Race and Gender in Early Modern England* (Ithaca, N.Y.: Cornell University Press, 1995).

14. On gender and science prior to the nineteenth century, see Thomas Laqueur, *Making Sex: Body and Gender from the Greeks to Freud* (Cambridge, Mass.: Harvard University Press, 1990); and Londa Schiebinger, *Nature's Body: Gender and the Making of Modern Science* (Boston: Beacon, 1993).

15. George M. Fredrickson, *The Black Image in the White Mind: The Debate on Afro-American Character and Destiny, 1817–1914* (Middletown, Conn.: Wesleyan University Press, 1971), xvii.

16. For more on the development of color prejudice, racism, and whiteness as a social category tied to citizenship, see Jill Lepore, *The Name of War: King Philip's War and the Origins of American Identity* (New York: Vintage, 1999);

Edmund Morgan, *American Slavery, American Freedom* (New York: W. W. Norton, 2003); Anthony Parent, *Foul Means: The Formation of a Slave Society in Virginia, 1660–1740* (Chapel Hill: University of North Carolina Press, 2006); Kirsten Fischer, *Suspect Relations: Sex, Race, and Resistance in Colonial North Carolina* (Ithaca, N.Y.: Cornell University Press, 2001); and Melissa Stein, "Race as a Social Construction," in *Black Women in America*, 2nd ed., vol. 3, ed. Darlene Clark Hine (New York: Oxford University Press, 2005), 1–12. For more on the cultural and legal construction of whiteness in U.S. history, see especially Matthew Frye Jacobson, *Whiteness of a Different Color: European Immigrants and the Alchemy of Race* (Cambridge, Mass.: Harvard University Press, 1998); Ian F. Haney López, *White by Law: The Legal Construction of Race* (New York: New York University Press, 1996); and David R. Roediger, *The Wages of Whiteness: Race and the Making of the American Working Class*, rev. ed. (London: Verso, 1999) among the foundational texts in what has become the burgeoning field of critical whiteness studies.

17. Thomas Jefferson, *Notes on the State of Virginia*, in *Thomas Jefferson: Writings*, ed. Merrill D. Peterson (New York: Library of America, 1984), 264–65.

18. Ibid., 270.

19. Alice Dreger, *Hermaphrodites and the Medical Invention of Sex* (Cambridge, Mass.: Harvard University Press, 2000).

20. Lisa Jean Moore and Adele Clarke, "Clitoral Conventions and Transgressions: Graphic Representations in Anatomy Texts, c.1900–1991," *Feminist Studies* 21 (Summer 1995): 255–301.

21. On the history of British racial science, for example, see George W. Stocking Jr., *Victorian Anthropology* (New York: Free Press, 1987); and Nancy Stepan, *The Idea of Race in Science: Great Britain, 1800–1960* (Hamden, Conn.: Archon Books, 1982). On early racial science more generally, see Andrew S. Curran, *The Anatomy of Blackness: Science and Slavery in an Age of Enlightenment* (Baltimore: Johns Hopkins University Press, 2011); Fred W. Voget, *A History of Ethnology* (New York: Holt, Rinehart and Winston, 1975); and David Bindman, *Ape to Apollo: Aesthetics and the Idea of Race in the 18th Century* (Ithaca, N.Y.: Cornell University Press, 2002).

22. See, for example, Winthrop Jordan, *White over Black: American Attitudes toward the Negro, 1550–1812* (Chapel Hill: University of North Carolina Press, 1968), 30–40; and Jennifer Morgan, "'Some Could Suckle over Their Shoulder': Male Travelers, Female Bodies, and the Gendering of Racial Ideology, 1500–1770," *William and Mary Quarterly* 3, no. 54 (January 1997): 167–92.

23. See Anne Fausto-Sterling, "Gender, Race, and Nation: The Comparative Anatomy of 'Hottentot' Women in Europe, 1815–1817," in *Deviant Bodies: Critical Perspectives on Difference in Science and Popular Culture*, ed. Jennifer Terry and Jacqueline Urla (Bloomington: Indiana University Press, 1995), 19–48; Z. S. Strother, "Display of the Body Hottentot," in *Africans on Stage: Studies in Ethnological Show*

Business, ed. Bernth Lindfors (Bloomington: Indiana University Press, 1999), 1–55; and Clifton Crais and Pamela Sully, *Sara Baartman and the Hottentot Venus: A Ghost Story and a Biography* (Princeton: Princeton University Press, 2008).

24. Michael Omi and Howard Winant, *Racial Formation in the United States from the 1960s to the 1990s* (New York: Routledge, 1994).

25. Allison Sneider, *Suffragists in an Imperial Age: U.S. Expansion and the Woman Question, 1870–1929* (New York: Oxford University Press, 2008); and Evelyn Nakano Glenn, *Unequal Freedom: How Race and Gender Shaped American Citizenship and Labor* (Cambridge, Mass.: Harvard University Press, 2004). To be sure, there has been an enormous—and enormously rich—body of historical literature that has interrogated the intersections of race and gender more generally. Prolific and influential work by historians Kathleen Brown, Jennifer Morgan, Kirsten Fischer, Deborah Gray White, and Gail Bederman, as well as legal scholar Dorothy Roberts, remain models of intersectional analysis, demonstrating race and gender as mutually constructive at critical moments in U.S. history. Likewise, a number of scholars—Shawn Michelle Smith's *American Archives: Gender, Race, and Class in Visual Culture* (Princeton: Princeton University Press, 1999) and Robyn Wiegman's *American Anatomies: Theorizing Race and Gender* (Durham, N.C.: Duke University Press, 1995) stand out here—have been taking a decidedly interdisciplinary approach to similar topics in U.S. history, and, in doing so, have raised exciting new theoretical questions about cultural representation, identity, and discursive practices with which I attempt to grapple in this book.

26. Michelle Alexander, *The New Jim Crow: Mass Incarceration in the Ages of Color-Blindness* (New York: New Press, 2012).

27. Anne Fausto-Sterling, "The Myth of Neutrality: Race, Sex, and Class in Science," *Radical Teacher* 19 (1981): 213.

28. G. Frank Lydston, "Sex Mutilations as a Remedy for Social Ills: With Some of the Difficulties in the Way of the Practical Application of Eugenics to the Human Race," printed as a booklet for the New York Academy of Medicine (October 27, 1917), first published in *New York Medical Journal*, April 6, 1912, 1.

29. By "sexed body," I mean the body marked as "female" or "male"; by "sexualized," I mean to connote the body under a sexually objectifying gaze.

30. Fredrickson, *The Black Image;* William Stanton, *The Leopard's Spots: Scientific Attitudes toward Race in America 1815–59* (Chicago: University of Chicago Press, 1960); Steven Jay Gould, *The Mismeasure of Man* (New York: W. W. Norton, 1996); John Haller Jr., *Outcasts from Evolution: Scientific Attitudes of Racial Inferiority, 1859–1900* (Urbana: University of Illinois Press, 1971); and Jordan, *White over Black.*

31. Joan W. Scott, "Gender: A Useful Category of Historical Analysis," *American Historical Review* 91, no. 5 (December 1986): 1053–75.

32. See in particular Nancy Leys Stepan, *"The Hour of Eugenics": Race, Gender, and Nation in Latin America* (Ithaca, N.Y.: Cornell University Press, 1996) and

The Idea of Race in Science: Great Britain, 1800–1960 (Hamden, Conn.: Archon, 1982).

33. Mia Bay, *The White Image in the Black Mind: African-American Ideas about White People, 1830–1925* (New York: Oxford University Press, 2000).

34. Fredrickson, *The Black Image*, 76–96.

35. As such, it speaks to a rich scholarship on race, masculinity, and citizenship—much of which has focused on the turn of the century or the civil rights era of the twentieth century—by analyzing a moment in which the legal category of citizenship was in flux with the threat and then reality of slavery's demise and scientists attempted to quantify the exact biological makeup of the paradigmatic man and citizen body. See, for example, Gail Bederman, *Manliness and Civilization: A Cultural History of Gender and Race in the United States, 1880–1917* (Chicago: University of Chicago Press, 1996); John Kasson, *Houdini, Tarzan, and the Perfect Man: The White Male Body and the Challenge of Modernity in America* (New York: Hill and Wang, 2002); E. Anthony Rotundo, *American Manhood: Transformations in Masculinity from the Revolution to the Modern Era* (New York: Basic Books, 1994); Michael Kimmel, *Manhood in America: A Cultural History*, 3rd ed. (New York: Oxford University Press, 2011); Dana Nelson, *National Manhood: Capitalist Citizenship and the Imagined Fraternity of White Men* (Durham, N.C.: Duke University Press, 1998); Steve Estes, *I Am a Man! Race, Manhood, and the Civil Rights Movement* (Chapel Hill: University of North Carolina Press, 2005); Kristin Hoganson, *Fighting for American Manhood: How Gender Politics Provoked the Spanish-American and Philippine-American Wars* (New Haven: Yale University Press, 2000); Eleanor Hannah, *Manhood, Citizenship, and the National Guard: Illinois, 1870–1917* (Columbus: Ohio State University Press, 2007); and Eric Sundquist, *Strangers in the Land: Blacks, Jews, Post-Holocaust America* (Cambridge, Mass.: Belknap Press of Harvard University Press, 2009).

36. Alys Eve Weinbaum's *Wayward Reproductions: Genealogies of Race and Nation in Transatlantic Modern Thought* (Durham, N.C.: Duke University Press, 2004) and Laura Briggs's *Reproducing Empire: Race, Sex, Science, and U.S. Imperialism in Puerto Rico* (Berkeley: University of California Press, 2002) are particularly instructive here, while Michele Mitchell's *Righteous Propagation: African Americans and the Politics of Racial Destiny after Reconstruction* (Chapel Hill: University of North Carolina Press, 2004) offers a fascinating take on the ways in which African Americans adapted—and subverted—pronatalist ideology toward their own community's survival and advancement, as well as the politics of black nationalism.

37. Lisa Duggan, *Sapphic Slashers: Sex, Violence, and American Modernity* (Durham, N.C.: Duke University Press, 2001); Siobhan Somerville, *Queering the Color Line: Race and the Invention of Homosexuality in American Culture* (Durham, N.C.: Duke University Press, 2000); Kevin Mumford, *Interzones: Black/White*

Sex Districts in Chicago and New York in the Early Twentieth Century (New York: Columbia University Press, 1997); Julian Carter, *The Heart of Whiteness: Normal Sexuality and Race in America, 1880–1940* (Durham, N.C.: Duke University Press, 2007); Sharon Patricia Holland, *The Erotic Life of Racism* (Durham, N.C.: Duke University Press, 2012); and Nancy Ordover, *American Eugenics: Race, Queer Anatomy, and the Science of Nationalism* (Minneapolis: University of Minnesota Press, 2003).

1. "Races of Men"

1. John Campbell, *Negro-Mania: Being an Examination of the Falsely Assumed Equality of the Various Races of Men* (Philadelphia: Campbell and Power, 1851), 9–10.

2. Ibid., 4.

3. Ibid., 458–61.

4. "Negro-mania" [book review], *Southern Quarterly Review* (January 1852), 154.

5. Prolific ethnologist John Van Evrie in particular made this claim frequently, discussed in more detail in chapter 2. As demonstrated by works like Edmund Morgan's *American Slavery, American Freedom* (New York: W. W. Norton, 1975) and David Brion Davis's *The Problem of Slavery in the Age of Revolution, 1770–1823* (Ithaca, N.Y.: Cornell University Press, 1975), slavery was essential to many white Americans' definition of freedom.

6. See, for example, Londa Schiebinger, *The Mind Has No Sex: Women in the Origins of Modern Science* (Cambridge, Mass.: Harvard University Press, 1989); Mary Creese, *Ladies in the Laboratory? American and British Women in Science, 1800–1900: A Survey of Their Contributions to Research* (Lanham, Md.: Scarecrow Press, 1998); and Margaret Alice, *Hypatia's Heritage: A History of Women in Science from Antiquity through the Nineteenth Century* (Boston: Beacon Press, 1986).

7. Data are based on citations in the National Library of Medicine (NLM) *Index-Catalogue*. See Appendix.

8. On biblical histories and race, particularly the "Curse of Ham" so often employed as an antiblack argument, see Jordan, *White over Black*, especially 3–44; Bay, *The White Image*, 26–55; and George M. Fredrickson, *The Black Image in the White Mind: The Debate on Afro-American Character and Destiny, 1817–1914* (Middletown, Conn.: Wesleyan University Press, 1971), 71–96.

9. For much of the basic chronology of the racial origins debate in American ethnology, I am indebted to Fredrickson's *The Black Image*, chapter 3 (71–96).

10. Charles L. Brace, *The Races of the Old World: A Manual of Ethnology* (New York: Charles Scribner, 1863), 1. Brace and Darwin also appeared to be friends, with Brace and his wife visiting Darwin's home in Down, England in the summer

of 1872, as described in *The Life and Letters of Charles Darwin, Vol. 2,* edited by his son, Francis Darwin (New York: Basic Books, 1959), 343.

11. For more on Smith's theories and influence, see Fredrickson, *The Black Image,* 72. For more on ideas about climate and racial difference, see William Stanton, *The Leopard's Spots: Scientific Attitudes toward Race in America 1815–59* (Chicago: University of Chicago Press, 1960), 7–10.

12. Charles Caldwell, *Autobiography of Charles Caldwell, M.D.* (New York: Da Capo Press, 1968), XXIII, 447. See also Reginald Horsman, *Race and Manifest Destiny: The Origins of American Racial Anglo-Saxonism* (Cambridge, Mass.: Harvard University Press, 1981), 117–18.

13. Fredrickson, *The Black Image,* 73.

14. Paul Erickson, "The Anthropology of Charles Caldwell, M.D.," *Isis* 72, no. 2 (June 1981): 252.

15. On Victorian gender ideology and women's presumed moral and spiritual natures, see Nancy F. Cott, "Passionlessness: An Interpretation of Victorian Sexual Ideology, 1790–1850," *Signs* 4, no. 2 (Winter 1978): 219–36; and Carroll Smith-Rosenberg, "The Female Animal: Medical and Biological Views of Woman and Her Role in Nineteenth Century America," *Journal of American History* 60, no. 2 (1973): 332–56.

16. Fredrickson, *The Black Image,* 74.

17. Ann Fabian, *The Skull Collectors: Race, Science, and America's Unburied Dead* (Chicago: University of Chicago Press, 2010), 13–14.

18. Ibid., 17–21.

19. Ibid., 23.

20. Ibid., 24–25.

21. Ibid., 26.

22. Ibid., 28–29.

23. Yet another variation of scientific studies of the human head was craniometry, a subset of anthropometrics, which measured the dimensions of the heads of living human beings, with largely the same purpose as craniology, to reflect intellectual capacity.

24. Joseph Barnard Davis, "On the Method of Measurements, as a Diagnostic Means of Distinguishing Human Races," *American Journal of Science and Arts* 29, no. 2 (May 1960): 329–30.

25. Fabian, *The Skull Collectors,* 36–39; and Steven Jay Gould, *The Mismeasure of Man* (New York: W. W. Norton, 1996), 82–83.

26. Steven Jay Gould, *Mismeasure of Man,* 85.

27. Ibid., 93–101.

28. Stanton, *The Leopard's Spots,* 39–41.

29. Fabian, *The Skull Collectors,* 70–91.

30. "The Candidate of Many Parties. A Phrenological Examination to Ascertain What His Political Principles Are," [1848], HarpWeek, *American Political Prints, 1766–1876,* http://loc.harpweek.com/LCPoliticalCartoons/DisplayCartoonMedium.asp?MaxID=89&UniqueID=65&Year=1848&YearMark=1846.

31. Samuel Morton, *Crania Americana: Or a Comparative View of the Skulls of the Various Aboriginal Nations of North and South America* (Philadelphia: J. Dobson, 1939), 1.

32. Ibid., 82.

33. Samuel Morton, *An Inquiry into the Distinctive Characteristics of the Aboriginal Race of America* (Philadelphia: John Penington, 1844). First read at the Annual Meeting of the Boston Society of Natural History, on April 27, 1842.

34. Fredrickson, *The Black Image,* 78.

35. Josiah C. Nott and George R. Gliddon, eds., *Types of Mankind, or Ethnological Researches, Based upon the Ancient Monuments, Paintings, Sculptures, and Crania of Races, and Upon Their Natural, Geographical, Philological, and Biblical History,* 8th ed. (Philadelphia: J. B. Lippincott, 1857), 87.

36. Michael O'Brian, *Conjectures of Order: Intellectual Life and the American South, 1810–1860, Volume 1* (Chapel Hill: University of North Carolina Press, 2004), 240.

37. Quoted in Fredrickson, *The Black Image,* 78.

38. The "American School" was largely associated with polygenesism, or the separate origin of the races, and *Types of Mankind* endeavored to harness "empirical evidence" from a variety of scientific disciplines in defense of the controversial theory.

39. Nott and Gliddon, *Types of Mankind,* 7. Though this study focuses on the United States, it is worth noting that there was a considerable amount of transatlantic exchange in racial thought between the U.S. and Europe during the nineteenth century. American ethnology was read in Europe, and a number of its practitioners read papers before the Royal Society in London, as well as other European venues. Several American ethnologists, including Josiah Nott, published articles in British ethnology or anthropology journals or had portions of their work reprinted therein. When an English edition of the popular French treatise, Arthur De Gobineau's *The Moral and Intellectual Diversity of Races,* was published in the United States in 1856, it included as appendices tables of Samuel Morton's crania measurements and an essay by Josiah Nott advocating the theory of polygenesis. Within the text itself, De Gobineau wrote not just about France and Europe but also at length about race relations in the United States as an excellent case study in racial difference and the racial order, a sentiment numerous other European ethnologists shared. See also Bernth Lindfors, ed., *Africans on Stage.*

40. Brace, *Races of the Old World*, 1.

41. I mean "himself" literally, for, as previously mentioned, I have not identified any female ethnologists writing in the first half of the nineteenth century.

42. Reginald Horsman, *Josiah Nott of Mobile: Southerner, Physician, and Racial Theorist* (Baton Rouge: Louisiana State University Press, 1987), 52.

43. Ibid., 57–58.

44. Ibid., 132–35.

45. Ibid., 58.

46. For more on Brace and this work with the Children's Aid Society, see Stephen O'Connor, *Orphan Trains: The Story of Charles Loring Brace and the Children He Saved and Failed* (Chicago: University of Chicago Press, 2004).

47. Quoted in Fredrickson, *The Black Image*, 79.

48. William Charles Wells, *Two Essays: One, upon Single Vision with Two Eyes; the Other, on Dew; a Letter to the Right Hon. Lloyd, Lord Kenyon; and an Account of a Female of the White Race of Mankind, Part of Whose Skin Resembles That of a Negro; with Some Observations on the Causes of the Differences in Color and Form between the White and Negro Races of Men* (London: Longman, Hurst, Rees, Orme and Brown, 1818).

49. Samuel Cartwright, "The Natural History of the Prognathous Race of Mankind," [1857] in *The Dred Scott Decision: Opinion of Chief Justice Taney, with an Introduction by Dr. J.H. Van Evrie. Also, an Appendix, containing an essay on the Natural History of the Prognathous Race of Mankind, originally written for the New York Day-Book, By Dr. S.A. Cartwright, of New Orleans* (New York: Van Evrie, Horton, 1863); Peter A. Browne, Esq., *The Classification of Mankind, by the Hair and Wool of Their Heads* (Philadelphia: A. Hart, 1850); and Robert Knox, *The Races of Men* (London: Henry Renshaw, 1850).

50. Moreover, the masculine generic in the English language is not a natural or transhistorical convention, but rather a relatively recent grammatical rule, tracing its origins to the eighteenth century, the result of efforts by British grammarians. One such grammarian, John Kirby, wrote in his 1746 text "Eighty-Eight Grammatical Rules" that because the male gender is "more comprehensive," "man" and "mankind" may be used to refer to all people. This, rule twenty-one of his eighty-eight grammatical rules, was then made a legal standard by an act of the British Parliament in 1850. On the masculine generic, see Ann Bodine, "Androcentrism in Prescriptive Grammar: Singular 'They,' Sex Indefinite 'He,' and 'He or She,'" *Language in Society* 4 (1975): 129–46; and Ann Weatheral, *Gender, Language and Discourse* (East Sussex: Routledge, 2002), 14–17 in particular.

51. While studies of skulls like Morton's often specified the sex of each skull, general ethnology texts often referred to such studies and talked about differences in cranial capacities along racial lines without regard for sex.

52. Fredrickson, *The Black Image*, 78.

53. Many ethnology texts pointed to three races in America: black, white, and red, in nineteenth-century ethnological language. Some compared all three at length, while others contained brief discussions of Native Americans before focusing on African Americans alone or in comparison to whites. My data is based on citations for scientific texts on race in the National Library of Medicine's *Index-Catalogue* (see Appendix).

54. Josiah C. Nott, *Two Lectures, on the Natural History of the Caucasian and Negro Races* (Mobile, Ala.: Dade and Thompson, 1844), 23.

55. It should be noted that studies that were explicitly focused on comparing the black and white races are included in the "black" category in the chart, rather than separated out as a distinct category. This is not intended to reinforce white as normative, but to demonstrate the extent to which whites were utilized as such in comparative studies in the 1830–50 period, for no texts appeared in the NLM search for this specific period that focused on the white race alone rather than whites as a point of comparison.

56. John Van Evrie, "Abolition Is National Death" (1866), in *Anti-Black Thought, Vol. 1: Anti-Abolition Tracts and Anti-Black Stereotypes,* ed. John David Smith (New York: Garland, 1993), 5.

57. Cartwright, "Natural History," 45–47.

58. Van Evrie, "Abolition Is National Death," 4; and Daniel G. Brinton, *Races and Peoples: Lectures on the Science of Ethnography* (Philadelphia: David McKay, 1901), 30.

59. "Samuel A. Cartwright and Family Papers Finding Aid," Louisiana and Lower Mississippi Valley Collections Special Collections, Hill Memorial Library, Louisiana State University Libraries (Baton Rouge: Louisiana State University Press, 2008), 4.

60. Cartwright, "Natural History," 45.

61. "The Negro—The Ovary Cells—Dr. Cartwright," *Georgia Blister and Critic* 1, no. 2 (April 1854): 38–39.

62. Italics mine. Cartwright's language here seems to be both conforming to and a step beyond the more common masculine generic conventions of using "him" as a gender neutral pronoun or "man" to mean all humanity. First he speaks of the "negro race" then questions whether "he" belongs to the same race as the "white man." He then questions whether the black man is the "son of Adam," not whether blacks and whites belong to the same race and are the "progeny" or "descendants" of "Adam and Eve" rather than Adam alone. One suspects, however, that his word choice was less a conscious decision to render women invisible than a reflection of the entirely male-centric world view that Cartwright shared with his colleagues and intended audience.

63. Samuel Cartwright, "Diseases and Physical Peculiarities of the Negro Race," *New Orleans Medical and Surgical Journal* (May 1852). Reprinted as "Cartwright

on the Diseases and Physical Peculiarities of the Negro Race," *Georgia Blister and Critic* 1, no. 5 (July 1854): 110.

64. Brace, *Races of the Old World,* 441. Of course, referring to God as male, as "Father," was hardly unique to ethnology, or even the nineteenth century in general, but rather represents a longstanding Judeo-Christian convention dating back to ancient scripture.

65. John Van Evrie, *Negroes and Negro "Slavery": The First an Inferior Race, The Latter Its Normal Condition* (New York: Van Evrie, Horton, 1861), 59.

66. O'Brian, *Conjectures of Order,* 334–35, 526–87.

67. "The Negro—The Ovary Cells," *Georgia Blister and Critic,* 38–39.

68. Samuel Morton, "Hybridity in Animals, Considered in Reference to the Question of the Unity of the Human Species," *American Journal of Science and Arts* 3, no. 7 (January 1847): 39.

69. Ibid., 40.

70. Nott, *Two Lectures,* 1.

71. Nott and Gliddon, *Types of Mankind,* 97.

72. Ibid., 399.

73. Slave women frequently bore the children of their white masters who, by laws in place in the United States since the late seventeenth century that mandated that all children born to black women would inherit their mother's status as slave or free, would only further increase the master's human property.

74. "Reviews—Types of Mankind, by Nott and Gliddon," *Georgia Blister and Critic* 1, no. 5 (July 1854): 112. The *Blister* excerpted this list of critical "facts" on hybridity from Nott and Gliddon, *Types of Mankind,* 373.

75. Nott and Gliddon, *Types of Mankind,* 373.

76. Ibid., 4.

77. Ibid., 398. By personal and professional circumstances, Nott is likely referring to his status as a slaveholder himself and his work in a medical clinic for slaves, which he co-founded.

78. Ibid.

79. See, for example, Van Evrie, "Abolition Is National Death" (1866), 4.

80. Ibid.

81. As historian Londa Schiebinger demonstrates, social issues around gender were implicit in Linnaeus's development of taxonomy, starting with his controversial placement of the genus "man" within the class "mammalia," one of the few groupings to be defined by female physiognomy (milk production) alone. Schiebinger situates his decision within the context of the French Revolution-era movement to encourage women to eschew wet nurses and breastfeed their own babies, which challenged the class dichotomies that had allowed wet nursing to thrive while simultaneously pushing women out of the public sphere where they had been gaining a more prominent role by reinforcing their "natural" maternal

function within the home ("Chapter 2: Why Mammals Are Called Mammals" in *Nature's Body,* 40–74). By the mid-nineteenth century, Linnaeus's taxonomy had been widely accepted in mainstream science. However, while Linnaeus used the female of the species to differentiate humans from other animals in the eighteenth century, antebellum ethnologists, situated within a new cultural and political backdrop, almost always used the male to differentiate *among* humans.

82. Nott and Gliddon, *Types of Mankind,* 53.

83. Linnaeus himself did not, however, view the races as different species of the genus man, but rather as variations within the same species.

84. Nott, *Two Lectures,* 7.

85. According to George M. Fredrickson, Northern acceptance of theories of polygenesis and black inferiority was part of a larger democratic ideology that, even in its celebration of the "common man," was quick to emphasize democracy as for whites only, defending "racial distinctions in an otherwise egalitarian society because, unlike the odious class divisions of Europe, they were based on the natural inequalities" (*The Black Image,* 91). For more on the controversies between religious and scientific accounts of racial origins, see Mason Stokes, "Someone's in the Garden with Eve: Race, Religion, and the American Fall," *American Quarterly* 50, no. 4 (December 1998).

86. Horsman, *Josiah Nott of Mobile,* 130–32.

87. Ibid., 82.

88. For more on the reception of *Types of Mankind,* and of Nott's work more generally, in the South, see O'Brian, *Conjectures of Order,* 215–52.

89. Quoted in Fredrickson, *The Black Image,* 81–82.

90. David Walker, *Appeal to the Colored Citizens of the World* (1829), repr. *"One Continual Cry"—David Walker's Appeal to the Colored Citizens of the World: Its Settings and Meanings,* ed. Herbert Aptheker (New York: Humanities Press, 1965), 68. Three editions of Walker's *Appeal* were published between 1829 and 1830, each with slightly different titles; the text cited herein is from the third and final edition. The capitalized emphasis on "men" is Walker's.

91. William S. Powell, ed., "David Walker, 1785–1830," *Dictionary of North Carolina Biography* (Chapel Hill: University of North Carolina Press, 1979–1996); accessed through *Documenting the American South,* http://docsouth.unc.edu/index.html.

92. Ibid., 78.

93. Ibid., 79.

94. On masculinity in black ethnology, see Bay, *The White Image,* especially 38–42 and 221–22. On black ethnological writers' assessment of whites (white men in particular) as brutal predators, a trope Bay coins the "Angry Saxon," see 45–55.

95. James W. C. Pennington, *A Textbook of the Origin and History &c. &c. of the Colored People* (Hartford: L. Skinner, 1841; reprint, Detroit: Negro History Press, 1969), 89 and 46. Quoted in Bay, *The White Image,* 52.

96. Hosea Easton, *A Treatise on the Intellectual Character, and Civil and Political Condition of the Colored People of the United States, and the Prejudice Exercised Toward Them* (Philadelphia: Historic Publications, 1969), 5.

97. Walker, *Appeal*, 62, 63.

98. Frederick Douglass, *The Claims of the Negro, Ethnologically Considered: An Address before the Literary Societies of Western Reserve College, at Commencement, July 12, 1854* (Rochester, N.Y.: Lee, Mann, 1854), 6–9.

99. Ibid., 20. Italics mine.

100. See, for example, Bay, *The White Image*, particularly 6, and 38–42.

101. Walker, *Appeal*, 63.

102. Powell, ed., "David Walker, 1785–1830," *Dictionary of North Carolina Biography*. See also Herbert Aptheker, ed., *"One Continual Cry"—David Walker's Appeal to the Colored Citizens of the World: Its Settings and Meanings* (New York: Humanities Press, 1965).

103. See Bay, *The White Image*, 38–74.

104. Easton, *Treatise*, 20.

105. Bay, *The White Image*, 38–74.

106. D. R. Goodwin, "The Unity of Language and of Mankind," *North American Review* 73, no. 152 (July 1851): 164, 168.

107. Ibid., 170.

108. Ibid., 164.

109. "Are All Men Descended from Adam," *Putnam's Monthly Magazine of American Literature, Science and Art* 5, no. 25 (January 1855): 79–88.

110. "Ethnology, or the Races of Men," *Scientific American* 12, no. 1 (November 23, 1856): 85.

111. Ibid.

112. Darwin–Lyell Correspondence (B D25.L)—Charles R. Darwin to Charles Lyell (June 2, 1847). American Philosophical Society (APS), Philadelphia, Pa.

113. Darwin–Lyell Correspondence (B D25.L)—Charles R. Darwin to Charles Lyell (September 23, 1860), 7–8. APS.

114. Ibid., 9.

115. Josiah Nott, "The Negro Race: Its Ethnology and History," a letter to Major-General O. O. Howard, Superintendent Freedmen's Bureau, reprinted in the *Mobile Daily Times*, February 22, 1866 (originally written November 25, 1865). In John David Smith, ed., *Anti-Black Thought, Volume Seven: Racial Determinism and the Fear of Miscegenation, Pre-1900* (New York: Garland, 1993), 10.

116. See Bay, *The White Image*, 13–74.

117. It is important to note that just as not all ethnology was explicitly proslavery, not all proslavery writing was ethnological. A number of prominent proslavery apologists such as George Fitzhugh looked to sociology and political theory to justify slavery, while others employed primarily economic arguments in support

of America's "peculiar institution." But what differentiated ethnology from other proslavery writing was the centrality of science, particularly the natural or biological sciences.

118. Morton, *Crania Americana*, 152.

119. Ibid., 82.

120. "Letter from Josiah Nott to E.G. Squier, dated May 24, 1851," General Correspondence folder, *E. G. Squier Papers, 1841–1888*, Library of Congress.

121. Samuel Stanhope Smith, *Essay on the Causes of the Variety of Complexion and Figure in the Human Species* (New Brunswick, N.J.: J. Simpson and Co, 1810), 249–52.

122. Charles Caldwell, *Thoughts on the Original Unity of the Human Race* (New York: E. Bliss, 1830), vii.

123. Ibid., vi–vii.

124. Caldwell himself does not mention owning slaves during his thirty-four years living in Kentucky in his autobiography, though he does reference growing up in a household in North Carolina with small enough numbers that his father and brothers often worked in the fields alongside them (*Autobiography*, 62). However, both of the following are among the scholarly sources that refer to Caldwell as a slave owner, and indeed a "Charles Caldwell" appears in the 1850 census records for Jefferson County, where he would have been living at the time: Paul Erickson, "The Anthropology of Charles Caldwell," *Isis* 72, no. 2 (June 1981): 254; and Adrian Desmond and James Moore, *Darwin's Sacred Cause: How a Hatred of Slavery Shaped Darwin's Views on Human Evolution* (New York: Houghton Mifflin Harcourt, 2009), 152.

125. Emma Brace, ed., *The Life of Charles Loring Brace, Chiefly Told in His Own Letters* (London: Sampson Low, Marston, 1894), 72. His daughter, Emma, collected, introduced, and provided context for an enormous body of letters, and the posthumous autobiography of sorts was published four years after his death.

126. Ibid., 116–17.

127. Brace, *The Races of the Old World*, 72.

128. Ibid., 308.

129. Ibid., 343.

130. Douglass, *Claims of the Negro*, 6.

131. Ibid., 14. Italics mine.

132. For more on paternalism and slavery, see Walter Johnson, *Soul By Soul: Life inside the Antebellum Slave Market* (Cambridge, Mass.: Harvard University Press, 1999), 19–44 and 107–12 in particular; Eugene Genovese, *Roll Jordan Roll: The World the Slaves Made* (New York: Vintage, 1972), 1–149; Fredrickson, *The Black Image*, 43–70; and Peter Kolchin, *American Slavery 1619–1877*, rev. ed. (New York: Hill and Wang, 2003), 93–168.

133. For more on Egypt's place in American racial thought, see Scott Trafton, *Egypt Land: Race and Nineteenth Century American Egyptomania* (Durham, N.C.: Duke University Press, 2004), especially chapter 1 (41–84), which includes an in-depth analysis of *Types of Mankind*. On *Types of Mankind*'s lasting impact on American ideas about Egypt and race, see Melani McAlister, *Epic Encounters: Culture, Media, and U.S. Interests in the Middle East, 1945–2000* (Berkeley: University of California Press, 2001), especially 144–45.

134. Horsman, *Josiah Nott of Mobile*, 130–31.

135. Nott and Gliddon, *Types of Mankind*, 246–71.

136. Ibid., 255, 260.

137. Egypt was not central only to white ethnologists' considerations of race, however. Black intellectuals throughout the nineteenth century frequently discussed Egypt as well, but toward opposite ends, to defend the black race and it origins (see Bay, *The White Image*, particularly chapters 38–116, and 187–218). Egypt continued to have a central place in black racial thought well into the twentieth century and continues today among some black leaders. While classicists have often ignored or dismissed claims by black thinkers that the ancient Egyptians were black, the publication of *Black Athena: The Afroasiatic Roots of Classical Civilization* (New Brunswick, N.J.: Rutgers University Press, 1987), by Martin Bernal, a white academic who made an extensive argument to that effect, was anything but ignored. On the ensuing firestorm the book raised in academia, see Jacques Berlinerblau's *Heresy in the University: The Black Athena Controversy and the Responsibilities of American Intellectuals* (New Brunswick, N.J.: Rutgers University Press, 1999).

138. Nott and Gliddon, *Types of Mankind*, 255.

139. As historian Jennifer Morgan's provocative article, "'Some Could Suckle over Their Shoulder': Male Travelers, Female Bodies, and the Gendering of Racial Ideology, 1500–1770" (*William and Mary Quarterly* 3, no. 54, January 1997, 167–92) illustrates, such descriptions and images had a long history in Western racial thought, even as the context changed. Black women's bodies, particularly their breasts, were characterized as both animalistic and erotic and were linked to physical and reproductive labor as a means of naturalizing and justifying chattel slavery. Though these racial tropes are implicit in Nott's examples—and familiar enough to his audience that Nott did not need to elaborate himself—Nott's more explicit point was about the permanence of racial difference and the naturally servile role of black people.

140. Not all ethnological writers who compared blacks to infants or children supported slavery. Indeed, some argued that slavery was an abuse of power and that because black people were innately childlike, they needed protection from exploitation. In other words, racist paternalism could be used to attack slavery as

well as defend it. More often in ethnology, however, such comparisons were made toward an overtly proslavery argument.

141. Samuel Cartwright, "On the Diseases and Physical Peculiarities of the Negro Race," 108–9.

142. Ibid., 109.

143. Of course, the new, romanticized image of childhood as a sheltered time of innocence and play described the reality of only the most well-to-do families. See, for example, Steven Mintz, *Huck's Raft: A History of American Childhood* (Cambridge, Mass.: Belknap Press, 2004). Comparisons between black adults and white children also beg the question of where black children figure in. Southern planters and proslavery scientists often described black children as clever and energetic, equal, sometimes even superior, to young whites in intelligence. But as white children grew physically and mentally, the skulls of black children fused early, stunting further development, some scientists (including Cartwright) argued. On the status of black children under slavery, see Lester Alsten, "Children as Chattel," in *Small Worlds: Children and Adolescents in America, 1850–1950,* ed. Elliott West and Paula Petrik (Lawrence: University Press of Kansas, 1992); and Wilma King, *Stolen Childhood: Slave Youth in Nineteenth Century America* (Bloomington: Indiana University Press, 1998).

144. *Georgia Blister and Critic* 1, no. 1 (March 1854): 12.

145. See, for example, Deborah Gray White, *Ar'n't I a Woman: Female Slaves in the Plantation South,* rev. ed. (New York: W. W. Norton, 1985), 70–77.

146. Samuel Cartwright, "On the Diseases and Physical Peculiarities of the Negro Race (cont.)," *Georgia Blister and Critic* 1, no. 7 (September 1854): 157.

147. Ibid., 156–57.

148. Ibid., 159.

149. *Georgia Blister and Critic* 1, no. 2 (April 1854): 38–39.

150. On the contentious political climate of the 1850s, see David Morris Potter, *The Impending Crisis, 1848–1861* (New York: Harper and Row, 1976); and George M. Fredrickson, *The Inner Civil War* (New York: Harper and Row, 1965), 7–52.

151. See, for example, Samuel Cartwright, "On the Diseases and Physical Peculiarities of the Negro Race," 110; and Samuel Cartwright, "Natural History," 45.

152. Cartwright, "Natural History," 46.

153. *Georgia Blister and Critic* 1, no. 3 (May 1854): 64.

154. Van Evrie, ed., *The Dred Scott Decision,* iv.

155. Ibid.

156. John Van Evrie, *Free Negroism,* 2nd ed. (New York: Van Evrie, Horton, 1863), 7.

157. Van Evrie, *Abolition and Secession* (New York: Van Evrie, Horton, 1862), 17.

158. The Free Soil Democrats and Whigs consolidated into the Free Soil Party for the 1848 election, nominating Martin Van Buren for President. "Marriage of the Free Soil and Liberty Parties," HarpWeek, *American Political Prints, 1766–1876,* http://loc.harpweek.com.

159. George M. Fredrickson, "Hinton Rowan Helper," in *American National Biography* (New York: Oxford University Press, 2000).

160. Hinton Rowan Helper, *The Impending Crisis of the South: How to Meet It,* ed. Earl Schenck Miers (New York: Collier Books, 1963), 19.

161. Ibid., 25.

162. Ibid., 256–57.

163. Ibid., 254–55.

2. An "Equal Beard" for "Equal Voting"

1. John Van Evrie, *Negroes and Negro "Slavery": The First an Inferior Race, The Latter Its Normal Condition* (New York: Van Evrie, Horton, 1861), 103–4.

2. Ibid., 121.

3. Benjamin Apthorp Gould, ed., *United States Sanitary Commission Memoirs—Volume 2, Statistical: Investigations in the Military and Anthropological Statistics of American Soldiers* (New York: Published for the U.S. Sanitary Commission, by Hurd and Houghton, 1869), 225–27.

4. Ibid., 218–19.

5. Ibid., 223–27. Italics mine.

6. Reid Mitchell, *The Vacant Chair: The Northern Soldier Leaves Home* (New York: Oxford University Press, 1993), 4.

7. Joseph Barnard Davis, "On the Method of Measurements, as a Diagnostic Means of Distinguishing Human Races," *American Journal of Science and Arts* 29, no. 2 (May 1960): 329.

8. For a detailed discussion of the scientific methodologies employed by the Sanitary Commission and the Provost-Marshal-General's Bureau, as well as the tensions between the two organizations, see John Haller Jr., *Outcasts from Evolution: Scientific Attitudes of Racial Inferiority, 1859–1900* (Urbana: University of Illinois Press, 1971), 19–34.

9. Gould, *Sanitary Commission Memoirs,* 227.

10. The PMGB, for example, listed the nationalities represented among Union soldiers in the introduction to its report for "curious" readers. Also, the Bureau sent questionnaires to all the physicians who examined soldiers for military service; among the questions was "What nationality presents the greatest physical aptitude for military service?" See J. H. Baxter, *Statistics, Medical and Anthropological, of the Provost-Marshal-General's Bureau, Derived from Records of the*

Examination for Military Service in the Armies of the United States during the Late War of the Rebellion, Volume 1 (Washington, D.C.: Government Printing Office, 1875), 13, 169. In their questionnaire responses, the physicians often elaborated at length on how and why each nationality fought in the war.

11. Bederman, *Manliness and Civilization,* 20.

12. On the ideological and legal linkages between citizenship rights, manhood, and "whiteness"—an ever-changing category—in the nineteenth century, see Matthew Frye Jacobson, *Whiteness of a Different Color: European Immigrants and the Alchemy of Race* (Cambridge, Mass.: Harvard University Press, 1998), 13–90; and Bederman, *Manliness and Civilization,* 21–31.

13. Bederman, *Manliness and Civilization,* 20. Italics mine.

14. See, for example, Nancy Isenberg, *Sex and Citizenship in Antebellum America* (Chapel Hill: University of North Carolina Press, 1998).

15. On the varied and often deeply personal reasons why Union soldiers fought in the Civil War, see James McPherson, *For Cause and Comrades: Why Men Fought in the Civil War* (New York: Oxford University Press, 1998).

16. Quoted in Haller, *Outcasts from Evolution,* 21.

17. Historians have shown increasing attention to the experiences of and attitudes toward black soldiers during the Civil War. See, for example, James McPherson, *The Negro's Civil War: How American Negroes Felt and Acted During the War for the Union* (New York: Pantheon Books, 1965); Joseph Glatthaar, *Forged in Battle: The Civil War Alliance of Black Soldiers and White Officers* (New York: Free Press, 1990); John David Smith, ed., *Black Soldiers in Blue: African American Troops in the Civil War Era* (Chapel Hill: University of North Carolina Press, 2002); and Ira Berlin, Joseph Reidy, and Leslie Rowland, eds., *Freedom's Soldiers: The Black Military Experience in the Civil War* (New York: Cambridge University Press, 1998).

18. Gould, *United States Sanitary Commission Memoirs—Volume 2, Statistical,* 231.

19. Gould, *United States Sanitary Commission Memoirs—Volume 2, Statistical,* 221.

20. George C. Comstock, "Biographical Memoir: Benjamin Apthorp Gould, 1824–1896," in *National Academy of Science, Volume XVII, 7th Memoir* (1922); and "Gould, Benjamin," in *Appletons' Cyclopædia of American Biography, Volume II,* ed. James Grant Wilson and John Fiske (New York: D. Appleton, 1887).

21. Comstock, "Biographical Memoir," 161–62.

22. Lauren Julius Harris and Jason B. Almerigi, "Probing the Human Brain with Stimulating Electrodes: The Story of Roberts Bartholow's (1874) Experiment on Mary Rafferty," *Brain and Cognition* 70, no. 1 (2009), 93.

23. Ibid., 92–93.

24. "Chiefs of the Medical Department, U.S. Army 1775–1940, Biographical Sketches," *Army Medical Bulletin* 52 (1940), 62–65, compiled by James M. Phalen, Colonel, Medical Corps, U.S. Army and available through the U.S. Army Medical

Department: Office of Medical History website, http://history.amedd.army.mil/surgeongenerals/J_Baxter.html; *Southern Practitioner: An Independent Journal Devoted to Medicine and Surgery* 12, no. 9 (September 1890), 401.

25. Baxter, *Statistics,* VI.

26. Ibid.

27. Ibid., 3. The quotation was from Carl Vogt, and Lambert Adolphe Quetelet, to whom Vogt referred, was a Belgian philosopher, mathematician, and scientist who conducted extensive anthropometric research in Europe earlier in the nineteenth century. His work was cited extensively in the reports of both the PMGB and the SC.

28. Ibid., 20–21.

29. Ibid., 15.

30. Ibid., III–IV.

31. Ibid., 11.

32. "Surgeons' Reports—Dixi Crosby," in Baxter, *Statistics,* 189.

33. Baxter, *Statistics,* LIV.

34. Ibid., VI.

35. Ibid., 189. On scientific understandings of intersex conditions, as well as the relative social flexibility of the "hermaphrodite," in the nineteenth century, see Alice Dreger, *Hermaphrodites and the Medical Invention of Sex* (Cambridge, Mass.: Harvard University Press, 2000).

36. As Benjamin Gould notes repeatedly in the second volume of the Sanitary Commission's reports, the Secretary of War, Edwin Stanton, under whose authority the PMGB studies were conducted, denied the SC access to the Bureau's data and resources.

37. Gould, *United States Sanitary Commission Memoirs—Volume 2, Statistical,* 246.

38. Ibid., 247. On the influence of Quetelet's data and methodology on the Sanitary Commission's work, see Haller, *Outcasts from Evolution,* 21–22.

39. As historian Sarah Igo observes in reference to Quetelet's foundational search for the "average man" in the 1830s, "the drive to determine the average was part empirical quest, part cultural preoccupation." Sarah E. Igo, *The Averaged American: Surveys, Citizens, and the Making of a Mass Public* (Cambridge, Mass.: Harvard University Press, 2007), 19. Making no mention of the SC and PMGB's quest for the "average man" during the Civil War, however, she concludes, "scientific characterizations of the 'average' or 'typical' American were a striking phenomenon of the new [twentieth] century" (11).

40. Gould, *United States Sanitary Commission Memoirs—Volume 2, Statistical,* 568.

41. Ibid., 568. It is worth noting that the 25th Army Corps served in Texas after the Civil War officially ended; many of the black soldiers expected to be mustered out of the army but were instead sent to patrol the Mexican border. See, for example, Ira Berlin, Barbara J. Fields, Steven F. Miller, Joseph P. Reidy, and Leslie S.

Rowland, eds., *Free at Last: A Documentary History of Slavery, Freedom, and the Civil War* (New York: New Press, 1992), 511–13. This is noteworthy because this particular physical investigation fell outside the general chronology of the Sanitary Commission's anthropometric studies, a final, somewhat unofficial attempt to interrogate the manhood of black bodies, even more of an issue once the Civil War had ended and the question of citizenship rights was of considerable consequence.

42. Gould, *United States Sanitary Commission Memoirs—Volume 2, Statistical,* 569.

43. "Surgeons' Reports—Joseph H. Streeter," in Baxter, *Statistics,* 201.

44. "Surgeons' Reports—Alex Burbank," in Baxter, *Statistics,* 173.

45. "Surgeons' Reports—Dr. Robert B. Carswell," in Baxter, *Statistics,* 183.

46. "Surgeons' Reports—Dr. John L. Sullivan," in Baxter, *Statistics,* 210.

47. "Surgeons' Reports—Dr. R. McC. Lord and Dr. George Douglas; Dr. George Douglas," in Baxter, *Statistics,* 237 and 250, 270.

48. "Surgeons' Reports—Dr. R. McC. Lord," in Baxter, *Statistics,* 237.

49. "Surgeons' Reports—Dr. H.S. Chubbuck," in Baxter, *Statistics,* 280–81.

50. "Surgeons' Reports—Dr. George Douglas," in Baxter, *Statistics,* 270.

51. "Surgeons' Reports—Dr. William C. Roberts," in Baxter, *Statistics,* 251.

52. "Surgeons' Reports—Dr. C.L. Hubbell," in Baxter, *Statistics,* 261. Nott summarized his own conclusions, and the prevailing sentiment among ethnologists in general, on the inferior health and vitality of the mulatto in *Types of Mankind* (373).

53. Roberts Bartholow, "Sanitary Memoirs of the War—Chapter First: The Various Influences Affecting the Physical Endurance, the Power of Resisting Disease, etc., of the Men Composing the Volunteer Armies of the United States," in *United States Sanitary Commission Memoirs—Volume 1, Medical: Contributions Relating to the Causation and Prevention of Disease, and to Camp Diseases,* ed. Austin Flint (New York: Published for the U.S. Sanitary Commission, by Hurd and Houghton, 1867), 4. It is worth clarifying that for Bartholow "the term American applies, of course, to the composite race now inhabiting the continent, and not to the aborigines."

54. Ibid., 3.

55. Ibid., 4.

56. Ibid., 4–5.

57. Gould, *United States Sanitary Commission Memoirs—Volume 2, Statistical,* 221.

58. Bartholow, *United States Sanitary Commission Memoirs—Volume 1, Medical,* 5.

59. Ibid., 5. "Scrofulous" referred to a condition that included inflamed lymph nodes, likely a form of tuberculosis, and was frequently cited by ethnologists and

medical professionals as symptomatic of degraded physiology—often associated with the "lower races"—rather than a product of poor living conditions and inadequate health care.

60. Harris and Almerigi, "Probing the Human Brain," 92–93.

61. Ibid., 97.

62. My description of Bartholow's experiments and the ensuing fallout is drawn from the accounts in both Susan Lederer, *Subjected to Science: Human Experimentation in America before the Second World War* (Baltimore: Johns Hopkins University Press, 1997), 7–9; and "The History of Neurosurgery in Cincinnati," an exhibit presented by the University of Cincinnati Department of Neurosurgery and Mayfield Clinic, April 2009, http://www.mayfieldclinic.com/PDF/HistoryNeurosurg_web.pdf.

63. Harris and Almerigi, "Probing the Human Brain," 112.

64. "Chiefs of the Medical Department, U.S. Army 1775–1940, Biographical Sketches," *Army Medical Bulletin* 52 (1940): 62–65, compiled by James M. Phalen, Colonel, Medical Corps, U.S. Army and available through the U.S. Army Medical Department: Office of Medical History website, http://history.amedd.army.mil/surgeongenerals/J_Baxter.html; *Southern Practitioner: An Independent Journal Devoted to Medicine and Surgery* 12, no. 9 (September 1890): 401.

65. Comstock, "Biographical Memoir," 166–67.

66. Ibid., 169.

67. Baxter, *Statistics,* 23.

68. See, for example, Josiah Strong, *Our Country: Its Possible Future and Its Present Crisis* (New York: Baker and Taylor, 1891), 218–19.

69. On the immigrant experience in the Civil War and its impact on group and individual assimilation, see, for example, Martin Öfele, *True Sons of the Republic: European Immigrants in the Union Army* (Westport, Conn.: Praeger, 2008); Dean Mahin, *The Blessed Place of Freedom: Europeans in Civil War America* (Washington, D.C.: Brassey's, 2002); and William Burton, *Melting Pot Soldiers: The Union's Ethnic Regiments* (Ames: Iowa State University Press, 1988).

70. Margaret Humphreys, *Intensely Human: The Health of the Black Soldier in the American Civil War* (Baltimore: Johns Hopkins University Press, 2008), 146.

71. Gould, *United States Sanitary Commission Memoirs—Volume 2, Statistical,* vii.

72. Josiah Nott to Joseph Leidy, October, 18, 1866, Box 5, Ser. 1.1. Subfolder—Nott, Josiah Clark, Joseph Leidy Papers (Library of the College of Physicians of Philadelphia, Philadelphia, Pa.).

73. The Reconstruction Amendments have been the subject of a tremendous amount of scholarly literature. See, for example, Alfred Avins, ed., *The Reconstruction Amendments' Debates: The Legislative History and Contemporary Debates in Congress on the 13th, 14th, and 15th Amendments* (Richmond: Virginia Commission on Constitutional Government, 1967); Earl Maltz, *Civil Rights, the Constitution,*

and Congress, 1863–1869 (Lawrence: University Press of Kansas, 1990); and David A. J. Richards, *Conscience and the Constitution: History, Theory, and Law of the Reconstruction Amendments* (Princeton: Princeton University Press, 1993). Senate .gov offers a useful brief overview of the amendments' history in their web article, "Landmark Legislation: The Civil War and Reconstruction Amendments to the Constitution," which I utilize here; see http://www.senate.gov/artandhistory/history/common/generic/CivilWarAmendments.htm.

74. On the relationship between the women's and black suffrage movements, see especially Faye E. Dudden, *Fighting Chance: The Struggle over Woman Suffrage and Black Suffrage in Reconstruction America* (New York: Oxford University Press, 2011).

75. Eric Foner's foundational work on Reconstruction is an essential starting point for scholarship on the subject; see in particular *Nothing but Freedom: Emancipation and Its Legacy* (Baton Rouge: Louisiana State University Press, 1983); *Reconstruction: America's Unfinished Revolution, 1863–1877* (New York: Harper & Row, 1998); and *A Short History of Reconstruction* (New York: Harper Perennial, 2000). W. E. B. DuBois's *Black Reconstruction in America: An Essay toward a History of the Part Which Black Folk Played in the Attempt to Reconstruct Democracy in America, 1860–1880* (New York: Harcourt Brace, 1935) is also a classic; more recent scholarship has further challenged accounts of African Americans as passive recipients of Reconstruction-era reforms, including Steven Hahn, *A Nation under Our Feet: Black Political Struggles in the Rural South from Slavery to the Great Migration* (Cambridge, Mass.: Belknap Press of Harvard University Press, 2005); and James Lowell Underwood and W. Lewis Burke Jr., eds., *At Freedom's Door: African American Founding Fathers and Lawyers in Reconstruction South Carolina* (Columbia: University of South Carolina Press, 2000). Historians have also begun to consider the gendered aspects of Reconstruction, including Laura F. Edwards, *Gendered Strife and Confusion: The Political Culture of Reconstruction* (Urbana: University of Illinois Press, 1997); Jane Dailey, *Before Jim Crow: The Politics of Race in Postemancipation Virginia* (Chapel Hill: University of North Carolina Press 2000); Mary Farmer-Kaiser, *Freedwomen and the Freedmen's Bureau: Race, Gender, and Public Policy in the Age of Emancipation* (New York: Fordham University Press, 2010); and LeeAnn Whites, *Gender Matters: Civil War, Reconstruction, and the Making of the New South* (New York: Palgrave Macmillan, 2005).

76. Josiah Nott, "The Negro Race," 26.

77. Isenberg, *Sex and Citizenship*, xii.

78. Amy Dru Stanley, "Conjugal Bonds and Wage Labor: Rights of Contract in the Age of Emancipation," *Journal of American History* 75 (September 1988): 471–500.

79. Nott, "The Negro Race," 10.

80. Josiah Clark Nott, "Instinct of Races" (1866), in *Anti-Black Thought, Vol. 7: Racial Determinism and the Fear of Miscegenation, Pre-1900,* ed. John David Smith (New York: Garland, 1993), 58.

81. Ibid., 47.

82. Ibid.

83. Horsman, *Josiah Nott of Mobile,* 291.

84. Ibid., 294.

85. Ibid., 296.

86. Ibid., 325.

87. Edwin G. Burroughs and Mike Wallace, *Gotham: A History of New York City to 1898* (New York: Oxford University Press, 1999), 883–99; and Robert C. Kennedy, "On This Day: August 1, 1863," *HarpWeek Online* (New York Times, 2001), http://www.nytimes.com/learning/general/onthisday/harp/0801.html.

88. Burroughs and Wallace, *Gotham,* 926.

89. D. H. Jacques, "Colored People: Considered Scientifically and Socially," *Phrenological Journal and Science of Health* (December 1877): 402.

90. See, for example, Jacques, "Colored People."

91. Hinton Rowan Helper, *Nojoque: A Question for a Continent* (New York: Carleton, 1867), 16.

92. Helper, *Nojoque,* frontispiece.

93. Ibid., vii, vi.

94. Ibid., 14–15.

95. Ibid., 16.

96. Helper apparently went beyond relying on secondhand testimony in his consideration of the "peculiar and distinguishing characteristics of the negro" (*Nojoque,* 16). While residing in New York in the 1860s, a relative who was attending the University Medical School allowed Helper on several occasions to "accompany him to the dissecting rooms" where he inspected black bodies, quite literally, for himself (*Nojoque,* 22).

97. Ibid., 51.

98. Ibid., 54–55.

99. Hinton Rowan Helper, *The Negroes in Negroland, The Negroes in America, and Negroes Generally* (New York: G. W. Carleton, 1868).

100. Ibid., VIII.

101. Ibid., VIII–X.

102. Isenberg, *Sex and Citizenship,* 7.

103. John Van Evrie, "The Six Species of Men" (1866), repr. John David Smith, ed., *Anti-Black Thought 1863–1925—Volume One: Anti-Abolition Tracts and Anti-Black Stereotypes* (New York: Garland, 1993), 138.

104. Nancy Tomes, *The Gospel of Germs: Men, Women, and the Microbe in American Life* (Cambridge, Mass.: Harvard University Press, 1998), 104, 126–27. It is also

important to note that Van Evrie was by no means the first scientist to attach racial significance to the beard. For example, in *Nature's Body*, Schiebinger discusses the beard as racial marker in eighteenth-century European science; see especially 120.

105. Quoted in John D'Emilio and Estelle B. Freedman, *Intimate Matters: A History of Sexuality in America* (New York: Harper and Row, 1988), 190. See also Bederman, *Manliness and Civilization.*

106. Van Evrie, "Six Species of Men," 137.

107. I use "white" in parentheses here to indicate that when white ethnologists did use women as a point of comparison, the context and language they used indicates they generally meant white women specifically, even though they tended to simply refer to "women." When they were writing about black women, they specified this by referring to "negro women" or "negresses." So just as ethnological writ often used language that conflated men with all humanity, so too did they use language that made whiteness the norm for womanhood.

108. Nancy Leys Stepan, "Race and Gender: The Role of Analogy in Science," *Isis* 77, no. 2 (June 1986): 263. Stepan's argument is most easily applied to postbellum thought, a point she does not emphasize but is relevant here. Of the primary sources she cites that compare women and the "lower races," all but one were written after 1860.

109. John Van Evrie, *Subgenation: The Theory of the Normal Relation of the Races, An Answer to Miscegenation* (New York: John Bradburn, 1864), iv.

110. Ibid., 30. Italics mine.

111. Ibid., 25.

112. Ibid., 27–29. Italics mine.

113. Ibid., 32–33.

114. Ibid., 30.

115. Ibid., 42–43.

116. Ibid., 44.

117. John Van Evrie, *White Supremacy and Negro Subordination; or, Negroes a Subordinate Race, and (So-Called) Slavery Its Normal Condition* (New York: Van Evrie, Horton, 1868), 185. *White Supremacy* is largely a reprint of his 1861 *Negroes and Negro "Slavery"* under a new title. He added new illustrations and sixty pages to the end of the book, however, in the form of appendices and a new conclusion that incorporated the reality of Emancipation into Van Evrie's continued defense of chattel slavery.

118. Frederick Douglass, "Sources of Danger to the Republic: An Address Delivered in St. Louis, Missouri, on 7 February 1867," in *The Frederick Douglass Papers. Series 1—Speeches, Debates, and Interviews, Volume 4 1864–1880,* ed. John Blassingame and John R. McKivigan (New Haven: Yale University Press, 1991), 153.

119. Frederick Douglass, "We Are Here and Want the Ballot-Box: An Address Delivered in Philadelphia, Pennsylvania, on 4 September 1866," in Frederick Douglass Papers. *Series 1, Volume 4, ed. Blassingame and McKivigan,* 123–24.

120. Ibid., 126.

121. Ibid., 131–32.

122. On tensions over race, particularly the issue of black male enfranchisement, within the women's suffrage movement, see Louise Michele Newman, *White Women's Rights: The Racial Origins of Feminism in the United States* (New York: Oxford University Press, 1999), 56–85.

123. Frederick Douglass, "Let No One Be Excluded from the Ballot Box: An Address Delivered in Albandy New York, on 20 November 1866," in Frederick Douglass Papers. *Series 1, Volume 4, ed. Blassingame and McKivigan,* 147–48.

124. Frederick Douglass, "Women's Rights Are Not Inconsistent with Negro Rights: An Address Delivered in Boston, Massachusetts, on 19 November 1868," in Frederick Douglass Papers. *Series 1, Volume 4, ed. Blassingame and McKivigan,* 183.

125. Ibid., 182.

126. For a comprehensive overview of black disenfranchisement in the United States, see especially Richard Valelly, *The Two Reconstructions: The Struggle for Black Enfranchisement* (Chicago: University of Chicago Press, 2004); and Michael Perman, *Struggle for Mastery: Disfranchisement in the South, 1888–1908* (Chapel Hill: University of North Carolina Press, 2000).

127. "Died" (death notices), *New York Times,* May 20, 1896, 5; and Mount Hope Cemetery Roll, January 1, 1893–December 31, 1906, 677. I thank Eric Nystrom of the History Department at Rochester Institute of Technology (RIT) for tracking down the information about Van Evrie's burial and the local history surrounding him, and Rebecca Scales, also of RIT's History Department and a graduate school colleague of mine, for photographing Van Evrie's and Susan B. Anthony's graves for me.

128. "Known Descendants of National Rochester," GenWeb, Monroe County, N.Y., http://mcnygenealogy.com/n-roch.htm.

129. Horsman, *Josiah Nott of Mobile,* 313–32.

130. Harriet Washington, *Medical Apartheid: The Dark History of Medical Experimentation on Black Americans from Colonial Times to the Present* (New York: Doubleday, 2006), 61–74.

131. Horsman, *Josiah Nott of Mobile,* 333–35.

132. Ibid., 304.

133. On the attitudes of Northerners more generally toward Reconstruction, as well as their role in its ultimate demise, see Heather Cox Richardson, *The Death of Reconstruction: Race, Labor, and Politics in the Post-Civil War North, 1865–1901* (Cambridge, Mass.: Harvard University Press, 2004).

134. On gender, evolutionary theory, and civilization discourse in American culture more generally at the turn of the century, see in particular Bederman, *Manliness and Civilization,* and Newman, *White Women's Rights.*

135. Brinton, *Races and Peoples,* 18.

136. Ibid., 19.

137. Ibid., 37–38.

138. On the integration of evolutionary theory into American race science specifically in the late nineteenth century, see Fredrickson, *The Black Image,* 228–55; John S. Haller, *Outcasts from Evolution*; and Joseph Graves Jr., *The Emperor's New Clothes: Biological Theories of Race at the Millennium* (New Brunswick, N.J.: Rutgers University Press, 2001), 53–104.

139. Richard Hofstadter, *Social Darwinism in American Thought,* rev. ed. (Boston: Beacon Press, 1955), 4–5.

140. Richard A. Brand, MD, "Frederick L. Hoffman, LLD (Hon), 1865–1946," *Clinical Orthopaedics and Related Research* 467, no. 10 (October 2009): 2489–90, http://www.ncbi.nlm.nih.gov/pmc/articles/PMC2745441/.

141. Frederick L. Hoffman, *Race Traits and Tendencies of the American Negro* (New York: Macmillan, 1896), v–vi.

142. Humphreys, *Intensely Human,* 152. Humphreys describes these health hardships throughout her medical history. References to the Civil War studies as evidence of African Americans' biological inferiority and their propensity for disease could be found in a range of popular and scientific journals during the Redemption period as well. See, for example, William Z. Ripley, "Acclimatization," *Appleton's Popular Science Monthly* (March 1896): 662–75; and Edward Eggleston, *The Ultimate Solution of the American Negro Problem* (Boston: Gorham Press, 1913), 224.

143. Frederick L. Hoffman, "Army Anthropometry and Medical Rejection Statistics," A Consolidation of Papers read before the National Academy of Science, Philadelphia, November 21, 1917, and the American Statistical Association, Philadelphia, December 28, 1917, Courtesy of Indiana University Libraries.

144. Ibid., viii.

145. On the impact of *Race Traits and Tendencies* both within and beyond science and medicine, see, for example, Fredrickson, *The Black Image,* 249–52; Thomas Gossett, *Race: The History of an Idea in America* (New York: Oxford University Press, 1997), 281–82; Khalil Gibran Muhammad, *The Condemnation of Blackness: Race, Crime, and the Making of Modern Urban America* (Cambridge, Mass.: Harvard University Press, 2010), 35–87; and Megan Wolff, "The Myth of the Actuary: Life Insurance and Frederick L. Hoffman's *Race Traits and Tendencies of the American Negro,*" *Public Health Reports* 121, no. 1 (January/February 2006): 84–91.

146. R. W. Shufeldt, *The Negro: A Menace to American Civilization* (New York: R. G. Badger, 1907), 116.

147. Kalman Lambrecht, "In Memoriam: Robert Wilson Shufeldt, 1850–1934," *Auk: A Quarterly Journal of Ornithology* 52, no. 4 (October 1935): 359.

148. R. W. Shufeldt, "Personal Adventures of a Human Skull Collector," *Medical Council* 15, no. 4 (1910): 123–27.

149. Mark V. Barrow, *A Passion for Birds: American Ornithology after Audubon* (Princeton: Princeton University Press, 2000), 63.

150. Ibid., 63–67.

151. Lambrecht, "In Memoriam," 359–61.

152. Shufeldt, *The Negro,* 112–13.

153. Ibid., 113.

154. Ibid., 115.

155. William Lee Howard, "The Negro as a Distinct Ethnic Factor in Civilization," *Medicine* (June 1903): 423–24.

156. Ibid., 423. Italics mine.

157. Howard A. Kelly, "Howard, William Lee (1860–1918)," *American Medical Biographies,* ed. Howard A. Kelly, MD and Walter L. Burrage, MD (Baltimore: Norman, Reddington, 1920): 567–68.

158. Ibid., 567. A similar version of this same passage, from a later edition of *American Medical Biographies* (then under the title of *The Dictionary of American Medical Biography*) is also quoted in *The American New Woman Revisited: A Reader, 1894–1930,* ed. Martha H. Patterson (New Brunswick, N.J.: Rutgers University Press, 2008), 279, as part of an editorial introduction to one of Howard's medical journal articles.

159. Howard, "The Negro," 423–24.

160. Ibid., 423.

161. Ibid., 423–25.

162. John William Leonard, ed., *Woman's Who's Who of America, 1914–1915* (New York: American Commonwealth, 1915), 491.

163. Ellen Barret Ligon, MD, "The White Woman and the Negro," *Good Housekeeping* (November 1903), 428.

164. Shufeldt, *The Negro,* 121–22. Emphasis on "of our own race" and "our own women" mine.

165. Ligon, "The White Woman," 428–29.

166. Ibid., 429.

167. On the sexual exploitation of black women, and its role in maintaining white supremacy, during the Jim Crow period, see Darlene Clark Hine, "Rape and the Inner Lives of Black Women in the Middle West: Preliminary Thoughts on the Culture of Dissemblance," *Signs* 14 (1989): 912–20; and Danielle McGuire, *At the Dark End of the Street: Black Women, Rape, and Resistance—A New History of the Civil Rights Movement from Rosa Parks to the Rise of Black Power* (New York: Alfred A. Knopf, 2010).

168. Ligon, "The White Woman," 429.

169. Dirk Schultheiss and Rainer M. Engel, "G. Frank Lydston (1858–1923) Revisited: Androgen Therapy by Testicular Implantation in the Early Twentieth Century," *World Journal of Urology* 21, no. 5 (November 2003): 358; and Frank Parsons Norbury, M.D., "The Reviewer's Table—*Over the Hookah: The Tales of a Talkative Doctor*" (book review), *Medical Fortnightly* 11, no. 2 (January 15, 1897): 48.

170. Not surprisingly, Lydston's experiment is recounted in many contemporary books, both popular and scholarly. For a brief sampling of the latter, see Schultheiss and Engel, "G. Frank Lydston (1858–1923) revisited"; and Angus McLaren, *Impotence: A Cultural History* (Chicago: University of Chicago Press, 2007), 182–207.

171. Daniel J. Kevles, *In the Name of Eugenics: Genetics and the Uses of Human Heredity* (Cambridge, Mass.: Harvard University Press, 1985), xiii.

172. Befitting its enormous popularity in the United States, there is a vast body of scholarship on eugenics in U.S. science and culture, to which I cannot give full justice here. A small sampling of useful contributions includes Edwin Black, *War against the Weak: Eugenics and America's Campaign to Create a Master Race* (New York: Four Walls Eight Windows Press, 2003); Paul A. Lombardo, *Three Generations, No Imbeciles: Eugenics, the Supreme Court, and Buck v. Bell* (Baltimore: Johns Hopkins University Press, 2008); Rebecca M. Kluchin, *Fit to Be Tied: Sterilization and Reproductive Rights in America, 1950–1980* (New Brunswick, N.J.: Rutgers University Press, 2009); and Wendy Kline, *Building a Better Race: Gender, Sexuality and Eugenics from the Turn of the Century to the Baby Boom* (Berkeley: University of California Press, 2005).

173. G. Frank Lydston, "The Negro—America's Special Problem in Eugenics," *Southern California Practitioner* 29, no. 1 (January 1914): 334.

174. Ibid., 335.

175. Julius F. Taylor, "Predicts the End of the Negro Race—G. Frank Lydston Points to 'White-Blacks in Chicago—Calls It a Disgrace, Says Jack Johnson Should Be in 'Morgue' Instead of Roaming the Streets," *Broad Ax* (February 8, 1913): 1.

176. Lydston, "The Negro," 336.

177. Jacobson, *Whiteness of a Different Color,* 41.

178. Ibid., 43.

179. Ibid., 45–46.

180. Ibid., 46–47.

181. Charles L. Brace, *The Dangerous Classes of New York and Twenty Years' Work among Them* (New York: Wynkoop and Hallenbeck, 1872), 30.

182. On the numbers of immigrants, and their countries of origin, in this period, see Mae M. Ngai, *Impossible Subjects: Illegal Aliens and the Making of Modern America* (Princeton: Princeton University Press, 2004), 272–73.

183. S. T. Joshi, ed., *Documents of American Prejudice: An Anthology of Writings on Race from Thomas Jefferson to David Duke* (New York: Basic Books, 1999), 509; and "Francis Amasa Walker, 1840–1897," *Statisticians in History*, American Statistical Society, https://www.amstat.org/about/statisticiansinhistory/index.cfm?fuseaction=biosinfo&BioID=35.

184. Francis Walker, "Restriction of Immigration," *Atlantic Monthly* 77, no. 6 (June 1896): http://www.theatlantic.com/magazine/archive/1896/06/restriction-of-immigration/306011/ (The Atlantic Monthly Group, 2013).

185. On public discourse on race at the turn of the century, particularly in New York City, in the context of both increasing immigration and migration of Southern blacks to the urban North, see Matthew Pratt Guterl's excellent *The Color of Race in America, 1900–1940* (Cambridge, Mass.: Harvard University Press, 2002).

186. Hofstadter, *Social Darwinism*, 170–71.

187. Lothrop Stoddard, *The Rising Tide of Color against White World Supremacy* (1920), reprint by Indo-European Publishing (Los Angeles, 2011), 8.

188. Ibid., 4. Italics mine.

189. Sheena Morrison and Elizabeth Fee, "Charles V. Roman: Physician, Writer, Educator, Historian (1864–1934)," *American Journal of Public Health* 100, no. S1 (April 2010): S69.

190. C. V. Roman, *American Civilization and the Negro: The Afro-American in Relation to National Progress* (Philadelphia: F. A. Davis, 1916), 40.

191. Ibid., 25.

192. Ibid., v.

193. Ibid., 4.

3. Inverts, Perverts, and Primitives

1. G. Frank Lydston, "Materialism versus Sentiment in the Study of the Causes and Correction of Crime," *Addresses and Essays*, 2nd ed. (Louisville, Ky.: Renz and Henry, 1892), 94. The essay derives from a speech Lydston gave before the Kentucky State Medical Society.

2. Ibid., 106.

3. While the terms "hermaphroditism" and "hermaphrodite" have fallen out of favor today, with "intersex" the preferred usage today, I use the former terminology here to denote the scientists' language and the context in which it was originally used. I use the terms "mulatto" and "sexual perversion" here in the same vein. "Inversion" was a popular concept in late nineteenth- and early twentieth-century science and medicine. It referred to a range of gender deviance and may or may not include physical indicators. And indeed, scientists looked hard for and had a rather amorphous definition of what such indicators could be—from

a woman's "square face" or "flattened labia" to a "coarse voice." Scientists often described inversion as a female soul trapped in a male body or vice versa. "Inversion" and "psychical hermaphrodism" generally referred to the same thing in scientific discourse during this period. However, scientists' concept of "inversion" generally stopped short of complete physical intersexuality, which they usually defined as a distinct category, "hermaphroditism" or "pseudo-hermaphroditism." In most scientists' usage at the time, hermaphroditism referred to individuals with functional male and female genitalia and pseudohermaphroditism to individuals with some mixture of primary and secondary sex characteristics of both sexes but with one sex predominating and/or diminished functionality of the other; however, to confuse matters even more, some scientists used the terms interchangeably or did not use the latter at all. Homosexual acts or relationships were seen as natural but not universal components of inversion; many scientists argued that not all homosexuals were inverts and the term "homosexual" was indeed in usage by the 1880s, as was the term "lesbian."

4. G. Frank Lydston, "A Lecture on Sexual Perversion, Satyriasis, and Nymphomania," *Addresses and Essays, 2nd ed.* (Louisville, Ky.: Renz and Henry, 1892), 257.

5. G. Frank Lydston, "Materialism versus Sentiment in the Study of the Causes and Correction of Crime," *Addresses and Essays, 2nd ed.* (Louisville, Ky.: Renz and Henry, 1892) 102.

6. Janice Irvine, *Disorders of Desire: Sex and Gender in Modern American Sexology* (Philadelphia: Temple University Press, 1991), 2.

7. Gail Bederman, *Manliness and Civilization: A Cultural History of Gender and Race in the United States, 1880–1917* (Chicago: University of Chicago Press, 1996), 1–44.

8. Lydston, "Lecture on Sexual Perversion," 257. Like other physicians of his era, Lydston distinguished "spurious" hermaphrodites, whose bodies appeared in some way opposite their "real" sex (which a clever doctor could ostensibly discern), from the more rare "true hermaphrodites" who possessed both male and female gonads. The cook, for example, suffered from severe hypospadias, an abnormality in which the urethral opening is located on the underside of the penis.

9. For example, Chicago physician and psychiatrist James Kiernan identified black men's supposed proclivity for raping white women as a highly dangerous form of perversion, while Louisiana physician Augustin Himel pointed to "young white men cohabiting with negro women" as a sexual transgression of critical interest to both psychologists and lawmen. James Kiernan, "Sexual Perversion and the Whitechapel Murders," *Medical Standard* 4, no. 6 (December 1888) 171; and Augustin Himel, "Some Minor Studies in Psychology, with Special Reference to Masturbation," *New Orleans Medical and Surgical Journal* 60, no. 6 (December 1907): 440–41.

10. See, for example, G. Frank Lydston, "Aberrant and Imperfect Differentiation of Sex," in *The Surgical Diseases of the Genito-Urinary Tract: Venereal and Sexual Diseases,* rev. ed. (Philadelphia: F. A. Davis, 1905), 517.

11. See especially George Chauncey, "From Sexual Inversion to Homosexuality: The Changing Medical Conceptualization of Female 'Deviance,'" in *Passion and Power: Sexuality in History,* ed. Kathy Peiss and Christina Simmons (Philadelphia: Temple University Press, 1989); and Eve Kosofsky Sedgwick, *The Epistemology of the Closet* (Berkeley: University of California Press, 1990).

12. Jennifer Terry, *An American Obsession: Science, Medicine, and Homosexuality in Modern Society* (Chicago: University of Chicago Press, 1999), 27–39 and 74–119.

13. Lydston articulated an influential model of the causes and symptoms of "furor sexualis" in his 1893 publication, *Sexual Crimes among the Southern Negroes,* which the next chapter will take up in detail. Miscegenation and "furor sexualis" were among the many topics Shufeldt addressed in his antiblack publications.

14. On turn-of-the-century discourse on miscegenation, see Avtar Brah and Annie Coombes, eds., *Hybridity and Its Discontents: Politics, Science, Culture* (New York: Routledge, 2000); Abby Ferber, *White Man Falling: Race, Gender, and White Supremacy* (Lanham, Md.: Rowman and Littlefield, 1998); Kevin Mumford, *Interzones: Black/White Sex Districts in Chicago and New York in the Early Twentieth Century* (New York: Columbia University Press, 1997); Joel Williamson, *New People: Miscegenation and Mulattoes in the United States* (New York: New York University Press, 1984); and Fredrickson, *The Black Image in the White Mind,* 288–82.

15. For a detailed analysis of the concept of "race suicide," see Bederman, *Manliness and Civilization,* 199–215; and Louise Michele Newman, *White Women's Rights: The Racial Origins of Feminism in the United States* (New York: Oxford University Press, 1999), 144–49. On "race suicide" in black thought, see Michele Mitchell, *Righteous Propagation: African Americans and the Politics of Racial Destiny after Reconstruction* (Chapel Hill: University of North Carolina Press, 2004).

16. There is an extensive body of scholarship that examines the construction of homosexuality and, specifically, the role of science in that complex project. Jonathan Ned Katz, for example, argues that not only was homosexuality invented in the nineteenth century (largely by sexologists), but so was heterosexuality, the latter defined against the former (*The Invention of Heterosexuality,* Chicago: University of Chicago Press, 2007). While the scholarly treatments of European sexology are too numerous to list here, several works have also addressed early American scientific discourse on homosexuality. Ronald Bayer's *Homosexuality and American Psychiatry: The Politics of Diagnosis* (Princeton: Princeton University Press, 1987) and Henry Minton's *Departing from Deviance: A History of Homosexual Rights and Emancipatory Science in America* (Chicago: University of Chicago Press, 2001) address homosexuals' engagement with the scientists who study them.

Foundational articles by George Chauncey, "From Sexual Inversion to Homosexuality," and Henry Abelove, "Freud, Male Homosexuality, and the Americans" (in *The Lesbian and Gay Studies Reader*, eds. Henry Abelove, Michele Barale, and David Halperin, New York: Routledge, 1993, 381–96) have shed considerable light on the gender politics at play in turn-of-the-century American sexual science. In recent years, groundbreaking scholars such as Lisa Duggan, Siobhan Somerville, and Jennifer Terry have also begun to examine the complex relationship between the sciences of race and the sciences of gender and sexuality during this period. See Lisa Duggan, *Sapphic Slashers: Sex, Violence, and American Modernity* (Durham, N.C.: Duke University Press, 2001), especially 123–92; Siobhan Somerville, *Queering the Color Line: Race and the Invention of Homosexuality in American Culture* (Durham, N.C.: Duke University Press, 2000), 15–38: and Terry, *An American Obsession*, 1–73.

17. Indeed, even in historical scholarship that is otherwise focused on the United States, early work on sexuality by American scientists like Lydston, Kiernan, and Shufeldt remains largely overshadowed by that of their European counterparts. For example, though focused on the development of sexology in the context of American race relations, Somerville writes primarily about European sexologists in her text—that is, she juxtaposes U.S. ethnology with European sexology. Similarly, Irvine traces the rise of sexology in Europe, without any mention of Americans writing in the same period, in her introduction to *Disorders of Desire*, and begins the body of her text on American sexology with the Kinsey studies in the 1940s and 1950s. Irvine is hardly alone, however. It has become a standard narrative to locate early sexology in Europe, with U.S. contributions to the field beginning with Kinsey. Chauncey's "From Sexual Inversion to Homosexuality," and Terry's *An American Obsession*—itself impressively contextualized within the backdrop of U.S. race relations, racial science, and social hygiene—constitute notable exceptions. Terry in particular makes a number of astute comparisons between this U.S. literature and European sexology, though she too positions American scientists as writing largely in response to their European counterparts (74–119).

18. See Richard von Krafft-Ebing, *Psychopathia Sexualis*, rev. ed., trans. F. J. Rebman (New York: Medical Art Agency, 1992), 48, 218, 344, 438, and 471; and Lydston, *Surgical Diseases*, 530–31.

19. Allan McLane Hamilton, "The Civil Responsibility of Sexual Perverts," *American Journal of Insanity* 4 (April 1896): 504. Charles Gilbert Chaddock was an American neurologist who translated an early edition of Krafft-Ebing's *Psychopathia Sexualis* into English.

20. See, for example, James Kiernan, "Sexual Perversion and the Whitechapel Murders," *Medical Standard* 4, no. 6 (December 1888): 170.

21. It is also worth noting that the vast majority of sexological writing in the United States appeared in article form in medical and psychiatric journals, unlike

European sexologists who often published book-length studies, as Terry also observes (75). In "The Medico-Legal Consideration of Perverts and Inverts" (*Pacific Medical Journal* 48, 1905, 385–93), R. W. Shufeldt hinted at the likely reason for this difference: American scientists could face obscenity charges for their explicit discussions of sexuality, and thus their work was largely targeted to other physicians.

22. Charles E. Rosenberg, *The Trial of the Assassin Guiteau: Psychiatry and the Law in the Gilded Age* (Chicago: University of Chicago Press, 1995), 143.

23. Ibid., 143–50.

24. "Noted Alienist Dies; James G. Kiernan was Expert for the Defense in Guiteau Trial," *New York Times,* July 3, 1923, 13.

25. Chad Heap, with an introduction by George Chauncey, *Homosexuality in the City: A Century of Research at the University of Chicago* (Chicago: University of Chicago Library, 2000), 13, 9.

26. Ibid., 6.

27. William I. Thomas, "The Psychology of Modesty and Clothing," *American Journal of Sociology* 5 (1899): 246–62.

28. "Review: *The Races of Europe. A Sociological Study.* By William Z. Ripley. New York: D. Appleton, 1899," *American Journal of Sociology* 5 (1899): 265.

29. James G. Kiernan, "Sexology—Increase in American Inversion. Birth Control Among Primitive Peoples," *Urologic and Cutaneous Review* 20 (1916): 44–49.

30. For more on neurasthenia, see Bederman, *Manliness and Civilization,* 14 and 84–88.

31. On "romantic friendships," see Carroll Smith-Rosenberg's classic essay, "The Female World of Love and Ritual: Relations between Women in Nineteenth-Century America," *Signs* 1 (1975): 1–29.

32. Scientists were far more undecided over what to make of the feminine women with whom these "inverts" often developed relationships. Many scientists viewed them as passive victims of aggressively masculine predators, though over time scientists began to see them as suffering from sexual pathology as well. See Leila Rupp, *A Desired Past: Same-Sex Love in America* (Chicago: University of Chicago Press, 1999), 79–84; Terry, *An American Obsession,* 97–199; and Chauncey, "From Sexual Inversion to Homosexuality," 87–117.

33. James Kiernan, "Psychological Aspects of the Sexual Appetite," *Alienist and Neurologist* 12 (1891): 188.

34. Kiernan makes the exact same argument—verbatim—in two different articles, published in 1888 and 1891. See Kiernan, "Sexual Perversion and the Whitechapel Murder," 130, and "Psychological Aspects of the Sexual Appetite," 195 respectively.

35. C. P. Oberndorf, "Diverse Forms of Homosexuality," *Urologic and Cutaneous Review* 33, no. 8 (August 1929): 519. Along similar lines and published in the

same issue of the *UCR*, see also John F. W. Meagher, "Homosexuality: Its Psychobiological and Psychopathological Significance," 507.

36. "Dr. C. P. Oberndorf, Psychiatrist, Dies; Columbia Ex-Professor, Also a Noted Psychoanalyst, Set Up Clinic at Mt. Sinai," *New York Times*, May 31, 1954, 13.

37. Bernard S. Talmey, "Notes on Homosexuality," *Medico-Legal Journal*, 34, no. 8 (November/December 1917): 4.

38. "Dr. Barnard S. Talmey" (death notice), *New York Times*, July 1, 1926, 23; and Barnard S. Talmey, *Woman: A Treatise on the Normal and Pathological Emotions of Feminine Love, for Physicians and Students of Medicine and Jurisprudence*, 2nd ed. (New York: Practitioners, 1908).

39. James Kiernan, "Sexual Perversion and the Whitechapel Murders," 171; and William Lee Howard, "Sexual Perversion in America," *American Journal of Dermatology and Genito-Urinary Diseases* 8 (1904): 10.

40. Howard, "Sexual Perversion in America," 9.

41. William Lee Howard, "Sexual Perversion," *Alienist and Neurologist* 17, no. 1 (January 1896): 6.

42. Shufeldt, "Consideration of Perverts and Inverts," 391.

43. "Dr. Isador Coriat, Pioneer in Psychoanalysis of Stammering, Dies at 67," *New York Times*, May 27, 1943, 25; and *Roster of the American Psychoanalytic Association* (New York: Association, 1968), 123.

44. Isador Coriat, "Homosexuality: Its Psychogenesis and Treatment," *New York Medical Journal* (March 22, 1913): 590. On the influence of Freud and psychoanalysis on American sexology, see Abelove, "Freud, Male Homosexuality, and the Americans," 381–96.

45. Shufeldt, 391.

46. Howard, "Sexual Perversion," 6.

47. Meagher, "Homosexuality," 506.

48. "Obituary: John F.W. Meagher," *Journal of Nervous & Mental Disease* 75, no. 1 (January 1932): 111.

49. Meagher, "Homosexuality," 512.

50. See, for example, Benjamin Karpman, "The Sexual Offender," *Psychoanalysis and Psychoanalytic Review* 10 (1923): 289–96; James G. Kiernan, "Sexology—Increase in American Inversion," 44–49; C. P. Oberndorf, "Homosexuality," *New York State Journal of Medicine* 22, no. 4 (April 1922): 176–80; Howard, "Sexual Perversion," 1; Talmey, "Notes on Homosexuality," 3; and William Lee Howard, "Psychical Hermaphroditism: A Few Notes on Sexual Perversion," *Alienist and Neurologist* 18 (1897): 115. Indeed, as Thomas Laqueur demonstrates in *Solitary Sex: A Cultural History of Masturbation* (New York: Zone Books, 2003), masturbation was an obsession in nineteenth-century American and European culture, with scientists linking the practice to a variety of physical and mental disorders as well as moral decay.

51. See Meagher, "Homosexuality," 506–7; and Augustin J. Himel, "Minor Studies in Psychology," 439–52.

52. Karpman, "The Sexual Offender," 289. Karpman also penned a book-length publication of the same name, in addition to nine other books, primarily on criminal psychoanalysis.

53. "Benjamin Karpman, M.D.," *American Journal of Psychiatry* 119 (May 1963): 1119; and "Ben Karpman, St. Elizabeths Psychotherapist Dies," *Psychiatric Quarterly* 36, no. 1 (December 1962): 796.

54. Kiernan, "Sexual Perversion and the Whitechapel Murders," 172.

55. Oberndorf, "Homosexuality," 180.

56. Himel, "Some Minor Studies in Psychology," 440.

57. Nayan Shah's *Contagious Divides: Epidemics and Race in San Francisco's Chinatown* (Berkeley: University of California Press, 2001) contains a fascinating discussion of medical experts who characterized syphilis as endemic to the Chinese, spread among the immigrant population by female prostitutes and men driven to sodomy in part because of the scarcity of women in America's Chinatowns, then transmitted to whites through the presence of Chinese "houseboys" as domestic laborers as well as through prostitution. On scientists' connections between homosexuality and venereal disease more generally, see Terry, *An American Obsession*, 94–105.

58. Howard, "Sexual Perversion in America," 1.

59. See, in particular, Kiernan, "Psychological Aspects of the Sexual Appetite," 188–219.

60. James Kiernan, "Psychical Treatment of Congenital Sexual Inversion," *Review of Insanity and Nervous Diseases* 5 (1894/1895): 293.

61. Howard, "Sexual Perversion," 1.

62. Howard, "Sexual Perversion in America," 9.

63. Lydston most definitely fits in this group. See also R. W. Shufeldt, "Judicial Ignorance of Sexual Crimes," *Pacific Medical Journal* 50 (1907): 79–82; and "The Medico-Legal Consideration of Perverts and Inverts," *Pacific Medical Journal* 48 (1905): 385–93.

64. Oberndorf, "Homosexuality," 179–80. In a similar vein, see also Meagher, "Homosexuality," 518.

65. Heap, *Homosexuality in the City*, 9–13; and Kiernan, "Psychical Treatment of Congenital Sexual Inversion," 293.

66. Hamilton, "Civil Responsibility of Sexual Perverts."

67. "Memoirs of a Famous Alienist: Dr. Allan McLane Hamilton's 'Recollections' Throw Light on Some of the Historic Murder Trials in Which He Took Part," *New York Times*, January 14, 1917.

68. "Medico-Legal Contributions," *American Journal of Neurology and Psychiatry* (1882): 323.

69. Shufeldt, "Consideration of Perverts and Inverts," 386.

70. Ibid., 393.

71. See, for example, James Kiernan, "Sexual Perversion and the Whitechapel Murders," *Medical Standard* 4, no. 5 (November 1888): 129–30 and 4, no. 6 (December 1888): 170–72; Kiernan, "Psychological Aspects of the Sexual Appetite," 188–219; Lydston, *Surgical Diseases,* 512–67; Howard, "Sexual Perversion," 1–6; and Coriat, "Homosexuality: Its Psychogenesis and Treatment," 589–94.

72. For example, under the heading of "sexual perversion," Lydston included "(a) those having a predilection (affinity) for their own sex; (b) those having a predilection for abnormal methods of gratification with the opposite sex; (c) those affected with bestiality." G. Frank Lydston, "A Lecture on Sexual Perversion, Satyriasis and Nymphomania," *Philadelphia Medical and Surgical Reporter* (September 7, 1889), reprinted in Lydston, *Addresses and Essays* (Louisville, Ky.: Renz and Henry, 1892), 246. Lydston's schematic here represents somewhat of a break with other scientific models, which often lumped same-sex affinity and bestiality together as the same "perversion." Over two decades later, Coriat invoked a similar model in the *New York Medical Journal,* stating, "The sexual aberrations in man, or the deviations from the normal sexual impulse, may be divided into two types, the inversions or deviations in relation to the sexual object choice, such as homosexuality, and the deviations in reference to the sexual aim, such as fetichism [*sic*], sadism, and masochism" ("Homosexuality: Its Psychogenesis and Treatment," 589).

73. The theory that homosexuality was invented as an exclusive and specific identity in the eighteenth and nineteenth centuries—though the behavior patterns predated the terminology that described it as such—has been argued in several influential texts, including Randolph Trumbach, *Sex and the Gender Revolution: Heterosexuality and the Third Gender in Enlightenment London* (Chicago: University of Chicago Press, 1998), Katz, *The Invention of Heterosexuality,* and George Chauncey's *Gay New York: Gender, Urban Culture, and the Making of the Gay Male World, 1890–1940* (New York: Basic Books, 1994), among others. The construction of homosexual identity(ies) has also been the subject of considerable theoretical work, including foundational texts like Michel Foucault, *The History of Sexuality Volume 1* (New York: Vintage Books, 1990); Judith Butler, *Gender Trouble: Feminism and the Subversion of Identity* (New York: Routledge, 1999); and Sedgwick, *Epistemology of the Closet.*

74. See, for example, Somerville, *Queering the Color Line,* especially 15–21; Rupp, 73–100; Chauncey, "From Sexual Inversion to Homosexuality," 57–69; Lucy Bland and Laura Doan, eds., *Sexology Uncensored: The Documents of Sexual Science* (Chicago: University of Chicago Press, 1999), 41–45; and Terry, *An American Obsession,* 43–45.

75. In discussing this schematic in *The Surgical Diseases of the Genito-Urinary Tract,* Lydston explained, "When the above classification first appeared, it was not

cordially received, its practicality being overlooked. Its recent adoption by Havelock Ellis, who asserts its superiority to Krafft-Ebing's classification, is extremely gratifying" (530–31). For his part, Kiernan felt the schematic was unnecessarily complex ("Psychological Aspects of the Sexual Appetite," 198–99).

76. Lydston, *Surgical Diseases,* 246, 244.

77. Ibid., 244. The translation of "Lusus Naturae" is a "sport of nature" or "freak of nature."

78. Kiernan, "Sexual Perversion and the Whitechapel Murders," 130.

79. Lydston, *Surgical Diseases,* 246.

80. Hamilton, "Civil Responsibility of Sexual Perverts," 504–5.

81. Howard, "Psychical Hermaphroditism," 114.

82. Meagher, "Homosexuality," 505.

83. Oberndorf, "Diverse Forms of Homosexuality," 519.

84. Hamilton, "Civil Responsibility of Sexual Perverts," 505.

85. See, for example, Kiernan, "Sexual Perversion and the Whitechapel Murders," 171; Kiernan, "Psychological Aspects of the Sexual Appetite," 202; and Karpman, "The Sexual Offender," 296.

86. "Perversions and Hypergenitalism," *Journal of the American Medical Association* 80, no. 18 (1923): 1346.

87. Kiernan, "Sexual Perversion and the Whitechapel Murders," 171.

88. Ibid. Scientists' link between black men and "sexual furor" or "furor sexualis" is discussed at length in the next chapter.

89. On the similarities between representations of black men and lesbians in popular and scientific discourse, see in particular, Duggan, *Sapphic Slashers.*

90. The 1922 text, *The Female-Impersonators,* written by self-professed invert Ralph Werther and sold to a medical and legal audience only, offers a particular revealing example. Drawing on common scientific ideas about homosexuality and inversion, he presented himself as a biological anomaly who should not be persecuted for his deviance. Included in the text's index were the terms "Beard growth of author," "Beardless men (natural)," and "Beard scanty;" there were ten citations for these three terms combined.

91. Terry, *An American Obsession,* 36.

92. Lydston, *Surgical Diseases,* 516; and Douglas McMurtrie, "Principles of Homosexuality and Sexual Inversion in the Female," *American Journal of Urology* 9 (1913): 150.

93. Kiernan, "Sexual Perversion and the Whitechapel Murders," 172.

94. Kiernan, "Psychological Aspects of the Sexual Appetite." One fetish or act in particular appears several times in Kiernan's paper, and in numerous medical and psychiatric texts during the same period: men who inserted various objects, often described as "wood splinters," into their urethra for sexual gratification. The potentially dangerous effects were likely one reason why so many scientists took

note of it, and indeed, many of the reporting physicians encountered the problem in their clinical practices when patients came in with accompanying injuries. However, I suspect that part of the reason these case studies were so often included amid discussions of inversion was because of their penetrative element, of male bodies (self) penetrated rather than penetrating, which disrupted Victorian gender expectations. It is perhaps not such a leap since several physicians characterized masturbation in general as a kind of homosexual act.

95. They were most likely white and Protestant, generally the norm in scientific work of this period, assumed unless otherwise noted. Antisemitism and anti-Catholic sentiment ran high in the U.S. during this period, as did nativism (these phenomena were related of course, since much of the immigration was at that point coming from Southern and Eastern Europe); Lydston even penned a book titled *That Boogey Man, the Jew* (Kansas City: Burton, 1921). Cases that described individuals from these groups, then, tended to be highlighted as such. For example, Kiernan included a case study of a necrophiliac priest, the man's occupation adding to the shock value and playing to Protestant Americans' fears about Catholicism. While he generally did not explicitly make connections between their "perversion" and their race/ethnicity in his case studies, it is likely the connection would have been implicit for the reader.

96. Kiernan, "Psychological Aspects of the Sexual Appetite," 218–19.

97. James Kiernan, "An Increase of American Inversion," *Urologic and Cutaneous Review* 20 (1916): 44–49; and Howard, "Sexual Perversion in America," 9. Italics mine.

98. Howard, "Psychical Hermaphroditism," 113–14.

99. On these urban same-sex subcultures and attempts by police and vice squads to regulate them, especially those that fostered cross-racial erotic encounters, see Mumford, *Interzones,* 53–92 and 133–56. On male urban subcultures—and their policing—during this period more generally, see Chauncey, *Gay New York,* 131–301.

100. Charles Hughes, "Postscript to Paper on 'Erotopathia'—An Organization of Colored Erotopaths," *Alienist and Neurologist* 14, no. 4 (October 1893): 731–32.

101. See, for example, Mumford, *Interzones,* 75; Terry, *An American Obsession,* 95–96; Duggan, *Sapphic Slashers,* 164–71; Vern Bullough, *Science in the Bedroom: A History of Sex Research* (New York: Basic Books, 1994), 93; Rupp, *A Desired Past,* 241; and Jonathan Katz, *Gay American History: Lesbians and Gay Men in the U.S.A.* (New York: Avon Books, 1976), 66–67, where Hughes's brief piece is reprinted in its entirety. Curiously, in Jennifer Terry's discussion of Hughes's text, in which she focuses on many of the same aspects as I have done here, she claims that "these spectacles were carried on by 'colored erotopaths,' who gathered annually, providing entertainment for each other and for similarly perverse whites," even though the piece itself makes no mention of white men as participants or observers.

102. Another possible reason for sexologists' overwhelming focus on whites is that American sexology texts often centered around patient case studies; whites of some financial means were most likely to seek and be able to afford to visit these physicians and psychiatrists in their private practice. Many of these sexologists also saw patients in an institutional context, at state hospitals for the mentally ill, but tended to describe these patients as more generally and pervasively insane, in which homosexual behavior could simply be one of many manifestations rather than their dominant diagnosis.

103. Psychologist Margaret Otis, who published an article on lesbian relationships in boarding schools in the *Journal of Abnormal Psychology* in 1913, is a notable exception to the male dominion over American sexology in the late nineteenth and early twentieth centuries. And social worker and penal reformer Katherine Bement Davis's groundbreaking study, *The Factors in the Sex Life of Twenty-Two Hundred Women* (1929), funded by the Rockefeller Foundation, in many regards paved the way for the work of Alfred Kinsey in the 1940s and 1950s and well as other female sexologists like Shere Hite later in the century. Both Otis and Davis were, however, white.

104. At least one scientist—Irving Rosse of Washington, D.C.—speculated that homosexuality and inversion could be found in somewhat greater numbers among black men or that they were at least more inclined to act out their proclivities (see Katz, *Gay American History,* 62). But this too seemed to be a minority opinion compared to the number of scientists who remarked on the high incidence of homosexuality among the intellectual and cultured population of white society.

105. Meagher, "Homosexuality," 508. Perhaps even stranger is that his negative assessment of homosexual men's whistling abilities was echoed in other sexology texts.

106. Ibid., 517.

107. Irving C. Rosse, "Homosexuality in Washington, D.C." (1892), excerpted in Katz, *Gay American History,* 65.

108. Charles Hughes, "Homo Sexual Complexion Perverts in St. Louis" (1907), excerpted in Katz, *Gay American History,* 75–76.

109. See Margaret Otis, "A Perversion Not Commonly Noted" (1913), reprinted in Katz, *Gay American History,* 101–5. See also Somerville, *Queering the Color Line,* 34–37, which contains an extended and insightful analysis of the Otis piece.

110. On vice squads' regulation of queer interracial sex, see especially Mumford, *Interzones,* 53–92.

111. Katz, *Gay American History,* 76.

112. Meagher, "Homosexuality," 508.

113. Ibid., 516.

114. Ibid., 513.

115. Ibid., 516.
116. E. S. Shepherd, "Contribution to the Study of Intermediacy," *American Journal of Urology and Sexology* 14, no. 6 (June 1918): 241–42.
117. Ibid., 243.
118. Ibid., 246.
119. Werther—also a pseudonym—used the name "Jennie June" as his female persona in the "sexual underworld" of New York. For the sake of simplicity, I refer to Werther/Lind/June as "Werther" throughout because he wrote two of his three books under that name and often refers to himself as such therein. His real name is unknown.
120. See, for example, Bernard Talmey, "Notes on Homosexuality" in the November/December 1917 issue, discussed previously.
121. Alfred Herzog, introduction to *The Female-Impersonators* by Ralph Werther (New York: Arno Press, 1975), vii.
122. "Dr. Alfred W. Herzog; Versatile Physician Was Editor of Medico-Legal Journal," *New York Times*, May 31, 1933, 17.
123. Alfred Herzog, introduction to *Autobiography of an Androgyne*, by Ralph Werther (New York: Arno Press, 1975), i.
124. Ibid., ii.
125. Notably, several of the photographs in his books were taken by R. W. Shufeldt, another scientist to whom Werther reached out and one who, like Lydston, inhabited the nexus of racial and sexual science.
126. Ralph Werther, *Autobiography of an Androgyne* (New York: Arno Press, 1975), 3.
127. Ralph Werther, *The Female-Impersonators* (New York: Arno Press, 1975), 49.
128. Werther, *Autobiography*, 27–28.
129. Ibid., 31.
130. Ibid., 41.
131. Minton, *Departing from Deviance*, 23–24.
132. Ibid., 24.
133. As the next chapter will take up, the late nineteenth and early twentieth centuries saw the eugenic sterilization of the mentally ill, physically handicapped, poor, and racial minorities, for example, while prominent scientists like Lydston recommended castration for black men as a solution to America's "lynching problem."
134. Werther, *Autobiography*, 74. It is worth underscoring here that Werther found—and absorbed—American as well as European texts on homosexuality during his visits to the medical library in the early 1890s.
135. Ralph Werther, *Riddle of the Underworld* (manuscript), 14, in the *Victor Robinson Papers, 1898–1947* (MS C 28), History of Medicine Division, National Library of Medicine, Washington, D.C. It is unclear whether *Riddle of the*

Underworld was ever published as a book, as was clearly Werther's intention. The contract between Werther and Victor Robinson (included with Robinson's edited copy of the *Riddle* manuscript in his papers at the NLM) for the text's serialization in *Medical Life* in 1921 referenced its planned publication the following year and *The Female-Impersonators* contained a detailed advertisement for it; however, I found no record of it as a published book.

136. Minton, *Departing from Deviance,* 24.

137. Werther, *Riddle,* 18.

138. Minton, *Departing from Deviance,* 26.

139. Ibid., 25.

140. Herzog, introduction, in *Autobiography,* xi.

141. Minton, *Departing from Deviance,* 27. See also Werther, *Autobiography,* 96–97.

142. See, for example, Werther, *Autobiography,* 96–97.

143. Werther, *Autobiography,* 1.

144. Ibid., 3.

145. Werther, *Riddle,* 1.

146. Werther, *The Female-Impersonators,* 266–68.

147. See, for example, Werther, *Riddle,* 6; and *The Female-Impersonators,* 49.

148. Werther, *Autobiography,* 10–13.

149. Ibid., 28.

150. Werther, *The Female-Impersonators,* 45–46.

151. Werther, *Riddle,* 3–6. Underlining is Werther's.

152. Ibid., 17.

153. Ibid., 3–4.

154. Ibid., 17.

155. Werther, *The Female-Impersonators,* 195–196.

156. Werther, *Autobiography,* 23.

157. Werther, *The Female-Impersonators,* 20–21. Werther is using homosexuality and "androgynism" relatively interchangeably here. In addition, he is using "bisexual" to mean "androgyne" or "intermediate," not in the current sense of the word, which indicates sexual object choice.

158. Werther, *Riddle,* 30. Note that the pagination here refers to the thirtieth page of the manuscript; the typed number at the top of the page reads "7" because Werther often restarted the pagination at the beginning of individual chapters (but unfortunately not consistently, creating even more difficulties for citation).

159. See Century Company Records. Series 1: General Correspondence. Box 91 [R. W. Shufeldt] and Century Company Records. Series 1: General Correspondence. Box 48 [William Lee Howard], Manuscripts and Archives Division, Humanities and Social Sciences Library, New York Public Library. William Lee Howard, *The Perverts* (New York: G. W. Dillingham, 1901).

160. F. E. Daniel, *Recollections of a Rebel Surgeon (and Other Sketches), or in the Doctor's Sappy Days* (Austin, Tex.: Von Boeckmann, Schutze, 1899) and *The Strange Case of Dr. Bruno* (Austin, Tex.: Von Boeckmann, Schutze, 1906), respectively; and G. Frank Lydston, *Panama and the Sierras: A Doctor's Wander Days* (Chicago: Riverton Press, 1900), *Over the Hookah: The Tales of a Talkative Doctor* (Chicago: F. Klein, 1896), and *The Blood of the Fathers: A Play in Four Acts* (Chicago: F. Klein, 1912), respectively.

161. Werther, *Riddle,* 16.

162. Werther, *The Female-Impersonators,* 49.

163. Lydston, "Sexual Perversion and Inversion," in *Impotence and Sterility, with Aberrations of the Sexual Function and Sex-Gland Implantation* (Chicago: Riverton Press, 1917), 31.

4. Unsexing the Race

1. On spectacle lynchings, see Grace Elizabeth Hale, *Making Whiteness: The Culture of Segregation* (New York: Vintage, 1999), 199–240.

2. See Gail Bederman, *Manliness and Civilization: A Cultural History of Gender and Race in the United States, 1880–1917* (Chicago: University of Chicago Press, 1996), 45–76.

3. Hunter McGuire and G. Frank Lydston, *Sexual Crimes among the Southern Negroes* (Louisville: Renz and Henry, 1893), 1.

4. James O. Breeden, "McGuire, Hunter Holmes," http://www.anb.org/articles/12/12-00600.html; *American National Biography Online,* February 2000. American Council of Learned Societies, published by Oxford University Press.

5. Hunter McGuire, *Address of the President before the Southern Surgical and Gynecological Association, at the Meeting Held in Nashville, Tennessee November 13, 1889* (Philadelphia: Wm. J. Dornan, 1890), 13–14. It is worth noting that McGuire received medical training in Philadelphia, the country's medical capital in the nineteenth century. But as tensions mounted between the North and South in the late 1850s, McGuire led a contingent of nearly 250 Southern medical students in protest out of Philadelphia.

6. McGuire and Lydston, *Sexual Crimes,* 2–4.

7. Ibid., 17.

8. Ibid., 7–16.

9. Ibid., 5–6.

10. Ibid., 21.

11. Ibid., 22.

12. Gary Taylor, *Castration: An Abbreviated History of Western Manhood* (New York: Routledge, 2002), 37–38. As Taylor explains, in harem societies, eunuchs

were the guardians of female chastity, a position of considerable importance in cultures that emphasized patrilineal bloodlines of inheritance and rule. In other contexts, eunuchs castrated after puberty were considered lusty, desirable lovers—they could still perform sexually without risk of pregnancy. And castration could also be associated with religious devotion, with men who had transcended temptations of the flesh and worldly concerns (41–43).

13. Taylor, *Castration,* 76–77.

14. See Winthrop Jordan, *White over Black: American Attitudes toward the Negro, 1550–1812* (Chapel Hill: University of North Carolina Press, 1968), 154–55; and Diane Sommerville *Rape and Race in the Nineteenth Century South* (Chapel Hill: University of North Carolina Press, 2004).

15. Dorothy Roberts, *Killing the Black Body: Race, Reproduction, and the Meaning of Liberty* (New York: Vintage Books, 1997), 28.

16. G. Frank Lydston, "Sex Mutilations as a Remedy for Social Ills: With Some of the Difficulties in the Way of the Practical Application of Eugenics to the Human Race," printed as a booklet for the New York Academy of Medicine (October 27, 1917), first published in *New York Medical Journal,* April 6, 1912, 2.

17. Quoted in Philip R. Reilly, *The Surgical Solution: A History of Involuntary Sterilization in the United States* (Baltimore: Johns Hopkins University Press, 1991), 28.

18. Reilly, *The Surgical Solution,* 28–29.

19. Ibid., xiii. Many of the patients were told they were being operated on for medical reasons, most often appendectomy, and few were informed after the fact that they had been sterilized while under the knife. On involuntary sterilization and eugenics in America, see also Harry Bruinius, *Better for All the World: The Secret History of Forced Sterilization and America's Quest for Racial Purity* (New York: Alfred A. Knopf, 2006).

20. Reilly, *The Surgical Solution,* 49. On eugenic discourse in the south, see Edward Larson, *Sex, Race, and Science: Eugenics in the Deep South* (Baltimore: Johns Hopkins University Press, 1995). Larson shows that while eugenic discourse was alive and well in the south, sterilization was not put into law the way it was in other parts of the country early in the twentieth century. Eugenics was largely a progressive movement, espoused by elite, often urban reformers, and many Southerners remained suspicious. This began to change in the South in the 1930s; most notably, as Johanna Schoen discusses in *Choice and Coercion: Birth Control, Sterilization, and Abortion in Public Health and Welfare* (Chapel Hill: University of North Carolina Press, 2005), North Carolina passed a state sterilization act in 1929 and performed growing numbers of sterilizations after World War II as other states saw sharp drops.

21. F. E. Daniel, "Castration of Sexual Perverts," *Texas Medical Journal* 27, no. 10 (April 1912): 378.

22. While we tend to use the word "castration" mostly in reference to men, the term can also refer to the surgical removal of women's ovaries—a fairly common linguistic usage in the nineteenth century. It is also possible that by saying he castrated the fourteen girls, he was referring to clitoridectomy; while not as common as the ovariectomies (also called oophorectomy) so often performed in the nineteenth century to cure "hysteria," among other disorders, clitoridectomy had its supporters in Western medicine and was most often prescribed for masturbation. See, for example, Elizabeth Sheehan, "Victorian Clitoridectomy: Isaac Baker Brown and His Harmless Operative Procedure," in *The Gender/Sexuality Reader,* ed. R. N. Lancaster and M. di Leonardo (New York: Routledge Press, 1997), 325–34; G. J. Barker-Benfield, *The Horrors of the Half-Known Life: Male Attitudes toward Women in Nineteenth-Century America* (New York: Routledge, 2000); and Deborah Kuhn McGregor, *From Midwives to Medicine: The Birth of American Gynecology* (New Brunswick, N.J.: Rutgers University Press, 1998).

23. Reilly, *The Surgical Solution,* 29, 38.

24. For example, while Puerto Rico was under U.S. control, up to a third of its women were sterilized, some forcibly, during the twentieth century. See Annette B. Ramirez de Arellano and Conrad Seipp, *Colonialism, Catholicism, and Contraception: A History of Birth Control in Puerto Rico* (Chapel Hill: University of North Carolina Press, 1983); Erica M. Gibson-Rosado, *The Sterilization of Women in Puerto Rico under the Cloak of Colonial Policy: A Case Study on the Role of Perception in U.S. Foreign Policy and Population Control* (Baltimore: Johns Hopkins University Press, 1993); Eileen J. Suárez Findlay, *Imposing Decency: The Politics of Sexuality and Race in Puerto Rico, 1870–1920* (Durham, N.C.: Duke University Press, 1999); and Laura Briggs, *Reproducing Empire: Race, Sex, Science, and U.S. Imperialism in Puerto Rico* (Berkeley: University of California Press, 2002).

25. F. E. Daniel, "Castration of Sexual Perverts" [Originally titled "Should Insane Criminals or Sexual Perverts Be Permitted to Procreate"], *Texas Medical Journal* 27, no. 10 (April 1912): 369. Daniel's 1893 article was reprinted in the *TMJ* in 1912; my page numbers herein correspond to the 1912 reprint.

26. Ibid., 378.

27. Ibid., 376.

28. Ibid., 378.

29. Ibid., 380.

30. Ibid., 369.

31. F. E. Daniel, "The Cause and Prevention of Rape," *Texas Medical Journal* 19 (May 1904): 452–62; and "Elements of Decay in American Civilization," *Texas Medical Journal* 25 (July 1909): 1–9. Daniel's views on black men's poor character and sexual aggression changed little between "Castration of Sexual Perverts" (1893) and these two later articles, though he did qualify the specifics of his

support for castration in these later articles. In 1904s, "The Cause and Prevention of Rape," he argued that castration should be reserved for lesser sexual offenses like attempted rape and masturbation, whereas rape (in which penetration had occurred) and/or murder should be punished by—presumably legal—execution.

32. Lydston, too, returned to the "castration or lynching" question in several other publications.

33. "San Diego Biographies—Peter Charles Remondino," *San Diego Historical Society*, http://www.sandiegohistory.org/bio/remondino/remondino.htm.

34. P. C. Remondino, "Questions of the Day: Negro Rapes and Their Social Problems," *National Popular Review: A Journal of Preventative Medicine and Applied Sociology* 4, no. 1 (January 1894): 4.

35. Ibid., 3.

36. Ibid., 5.

37. Bederman, *Manliness and Civilization*, 46.

38. Ibid., 49.

39. For more on Wells's speaking tours of England and Scotland, see Bederman, *Manliness and Civilization*, 45–75; Mia Bay, *To Tell the Truth Freely: The Life of Ida B. Wells* (New York: Hill and Wang, 2010), 109–90; Paula Giddings, *A Sword among Lions: Ida B. Wells and the Campaign against Lynching* (New York: Harper Paperbacks, 2009), 253–310; and Linda O. McMurry, *To Keep the Waters Troubled: The Life of Ida B. Wells* (New York: Oxford University Press, 2000), 188–205.

40. Remondino "Questions of the Day," 3–4.

41. Ibid., 4.

42. Ibid., 3.

43. Notably, Lydston himself was an outspoken advocate for circumcising all male infants, arguing that circumcision was hygienic, healthful, and would diminish the risk of masturbation. Lydston was in good company at the turn of the century in touting the benefits of circumcision, tying it to both sexual and social hygiene goals. The actual rate of circumcision increased steadily throughout the late nineteenth century, from approximately 15 percent in the 1890s to more than 40 percent in the 1920s, according to Edward Wallerstein, *Circumcision: An American Health Fallacy* (New York: Springer, 1980). Ironically, in contrast to Remondino's proposed targeting of black men in particular, circumcision in practice became a marker of class status and whiteness (see David Gollaher, *Circumcision: A History of the World's Most Controversial Surgery* [New York: Perseus Books, 2000], 106). Facing an enormous influx of new inhabitants, "so it came about that the foreskin, viewed as dangerous by the medical profession, commonly came to indicate ignorance, neglect, and poverty. As white middle-class Gentiles adopted circumcision, those left behind were recent immigrants, people of color, the poor, and other at the margins of respectable society," Gollaher writes (108).

44. "The Solution of the Negro Rape Problem," *Boston Medical and Surgical Journal* 130 (February 1, 1894): 126.

45. "Circumcision for the Correction of Sexual Crimes among the Negro Race," *Maryland Medical Journal* (February 10, 1894): 345.

46. John David Smith, *Black Judas: William Hannibal Thomas and The American Negro* (Chicago: Ivan R. Dee, 2002), 157.

47. Quoted in Smith, *Black Judas*, 157.

48. Smith, *Black Judas*, 158–59.

49. William Hannibal Thomas, *The American Negro: What He Was, What He Is, and What He May Become* (New York: Macmillan, 1901), 234.

50. Thomas, *The American Negro*, 236. As discussed earlier, Lydston stated in 1893, "A few emasculated negroes scattered around through the thickly-settled negro communities of the South would really prove the conservation of energy, as far as the repression of sexual crimes is concerned." Meanwhile, in his paper, "Should Insane Criminals Be Allowed to Procreate," published later the same year, F. E. Daniel voiced his support of Lydston's proposition as follows: "Dr. Lydston says, and very truly, that a hanging or even a burning is soon forgotten; but a negro buck at large amongst the ewes of his flock, minus the elements of his manhood, would be a standing terror to those of similar propensities."

51. Thomas, *The American Negro*, 230–33.

52. Ibid., 223.

53. Smith, *Black Judas*, 226–28.

54. The latter reprinted large portions of the earlier text under new chapter titles, with some additional text and illustrations, and as we will see, some noteworthy revisions.

55. Shufeldt, *The Negro*, 62. With a few minor changes in language, he repeated the same thing in *America's Greatest Problem: The Negro* (Philadelphia: F. A. Davis, 1915), 145–47.

56. Shufeldt, *The Negro*, 65; and *America's Greatest Problem*, 157.

57. Shufeldt, *America's Greatest Problem*, 157–58.

58. Ibid., 156.

59. Shufeldt, *The Negro*, 66.

60. Shufeldt, *America's Greatest Problem*, 150.

61. Ibid., 152 and 160.

62. Shufeldt, *The Negro*, 75.

63. Shufeldt, *America's Greatest Problem*, 151; and *The Negro*, 68.

64. Quoted in John Haller Jr., *Outcasts from Evolution: Scientific Attitudes of Racial Inferiority, 1859–1900* (Urbana: University of Illinois Press, 1971), 57.

65. See, for example, G. Frank Lydston, *Diseases of Society: The Vice and Crime Problem* (Philadelphia: J. B. Lippincott, 1904), 421–26.

66. Lydston, *Diseases of Society*, 424.

67. G. Frank Lydston, "Sex Mutilations," 7. Capitalization is Lydston's.

68. Ibid., 22–23.

69. "Dr. G. Frank Lydston" (death notice), *New York Times,* March 15, 1923, 19.

70. Lydston, "Sex Mutilations," 1.

71. Reprinted in Ralph Ginzburg, *100 Years of Lynching* (Baltimore: Black Classic Press, 1962), 113–14.

72. Hale, *Making Whiteness,* 234.

73. Letters to the Times, "A Riverside Woman's Plan of Salvation," *Los Angeles Times,* Nov. 20, 1895, 7. ProQuest Historical Newspapers.

74. Cited in Philip Dray, *At the Hands of Persons Unknown: The Lynching of Black America* (New York: Random House, 2002), 144.

75. Quoted in Dray, *Lynching of Black America,* 144.

76. William H. Fleming, "Race Problem Reviewed, Its Evils and Its Curses: A Diagnosis of the Rape Spirit and the Mob Spirit," *Macon Weekly Telegraph/Macon Daily Telegraph,* October 12, 1906, 4. The article was later reprinted as "The South's Race Troubles: Ex-Congressman Fleming of Georgia Issues and Open Letter to People of the South—A Calm Appeal—Criticizes Graves, Hoke Smith, and Dixon," *State* (Columbia, S.C.), October 21, 1906. http://www.newsbank.com. *America's Historical Newspapers Online,* NewsBank/American Antiquarian Society, 2004.

77. "To 'Fit the Crime,'" *Macon Weekly Telegraph/Macon Daily Telegraph,* October 20, 1907, 4. http://www.newsbank.com. *America's Historical Newspapers Online,* NewsBank/American Antiquarian Society, 2004.

78. "Cured One Negro: He Made Indecent Proposals to a White Girl in South Carolina," *Fort Worth Morning Register,* October 6, 1899, 1. *America's Historical Newspapers Online,* NewsBank/American Antiquarian Society, 2004.

79. Robert L. Zangrando, *The NAACP Crusade against Lynching, 1909–1950* (Philadelphia: Temple University Press, 1980), 6–7. According to Zangrando's data, in 1893, 34 whites were lynched and 118 blacks, for a total of 152.

80. The total (black and white) number of lynchings annually dipped below ten for the first time in 1932 (since 1882, the first year of Zangrando's data), and stayed in the single digits from 1936 on.

81. Hale, *Making Whiteness,* 201.

82. Henry Smith's lynching, anticipated for days as he was hunted down, was immediately reported across the country. By turns sympathetic to the victim or to the mob, newspaper accounts describe Smith's torture in agonizing detail, but none indicate that castration was among the horrors inflicted upon him. See the following from *America's Historical Newspapers Online* (NewsBank/the American Antiquarian Society, 2004, http://www.newsbank.com), "Met an Awful Fate: Little Myrtle Vance's Murderer Burned at the Stake," *Daily Inter Ocean* (Chicago, Ill.), February 2, 1893, 2; "Burned at the Stake," *Albuquerque Democrat,* February 2,

1893, 1; "Burned at the Stake," *Daily Charlotte Observer*, February 2, 1893, 1; "At the Stake—Smith Was Roasted," *Columbus Enquirer-Sun*, February 2, 1893, 1; "Horror of Horrors," *Dallas Morning News*, February 2, 1893, 1; and "Burned! Awful Revenge for a Most Horrible Crime," *Knoxville Journal*, February 2, 1893, 1. In addition, while spectacle lynchings—with their carnivalesque atmosphere, huge crowds, parading of the dead body and macabre souvenirs, and often advertisement of the event beforehand so many could attend—increased after 1893, Hale explains, "not all southern lynchings fit this new and evolving pattern. More often, small groups of white men hunted down and shot or hanged their African American victims after an argument over the year-end sharecroppers' settle or to send a message to other timber or turpentine camp laborers not to demand any better. These lynchings in the night claimed many more victims than the open-air spectacles of torture that drew such large crowds" (*Making Whiteness*, 201).

83. Dray, *Lynching of Black America*, 93–94.

84. For more on Hose and Washington, see Hale, *Making Whiteness*, 209–22.

85. Dray, *Lynching of Black America*, 268–69.

86. For a detailed account of the Neal lynching and its aftermath, see James McGovern, *Anatomy of a Lynching: The Killing of Claude Neal* (Baton Rouge: Louisiana State University Press, 1982). As Robert Ingalls demonstrates, Florida was the "lynching capital" of America during this period; see Ingalls, *Urban Vigilantes in the New South: Tampa, 1882–1936* (Knoxville: University of Tennessee Press, 1988).

87. Robert L. Zangrando, ed., *Papers of the NAACP [microfilm]—Part 7, The Anti-Lynching Campaign, 1912–1955* (Frederick, Md.: University Publications of America, 1987), Series A: Anti-Lynching Investigative Files, Reel 4: 0631. At the time, White and the NAACP kept the name of their white informant a secret, for his own safety and the integrity of the investigation; the source was later revealed to be Howard Kester, a college professor, preacher, and activist committed to racial justice.

88. "The Marianna, Florida Lynching: A Report of an Investigation Made for the National Association for the Advancement of Coloured People, 69 Fifth Avenue, New York, by a White Southern College Professor into the Killing of Claude Neal by a Mob on October 26, 1934," 3. *Papers of the NAACP—Part 7*, Series A, Reel 4: 0649. In the letter to Walter White accompanying his preliminary report, Kester recounted, "Last night I talked for one hour and forty minutes with a member of the mob which lynched Claude Neal . . . I was quite nauseated by the things which apparently gave this man the greatest delight to relate." *Papers of the NAACP, Part 7*, Series A, Reel 4:0635.

89. "The Marianna, Florida Lynching," 3–4. In their discussions of the Neal lynching, most historians have relied on and quoted heavily from Kester's report; the lynch mob member's description of Neal's castration, in particular, has been

quoted in numerous scholarly works, including Hale, *Making Whiteness;* Dray, *At the Hands of Persons Unknown*; McGovern, *Anatomy of a Lynching*; Dora Apel, *Imagery of Lynching: Black Men, White Women, and the Mob* (New Brunswick, N.J.: Rutgers University Press, 2004); Christopher Waldrep, *Lynching in America: A History in Documents* (New York: New York University Press, 2006); Mason Boyd Stokes, *The Color of Sex: Whiteness, Heterosexuality, and the Fictions of White Supremacy* (Durham, N.C.: Duke University Press, 2001); Felipe Smith, *American Body Politics: Race, Gender, and Black Literary Renaissance* (Athens: University of Georgia Press, 1998); and McKay Jenkins, *The South in Black and White: Race, Sex, and Literature in the 1940s* (Chapel Hill: University of North Carolina Press, 1999), among others.

90. NAACP memo, "Roosevelt and Cumming Get Gruesome Lynch Report," *NAACP Papers, Part 7,* Series A, Reel 4: 0666.

91. McGovern, 139–40. For an extensive analysis of lynching's transformation in the 1930s, see Jessie Daniel Ames, *The Changing Character of Lynching: Review of Lynching, 1931–1941* (Atlanta: Published by Commission on Interracial Cooperation Inc., 1942). The white founder and director of the Association of Southern Women for the Prevention of Lynching, Ames sought to challenge the stereotype of white women as passive objects in need of protection, toward the goal of ending the lynchings so often committed in their name. For more on Ames, see Jacquelyn Dowd Hall, *Revolt against Chivalry: Jessie Daniel Ames and the Women's Campaign against Lynching* (New York: Columbia University Press, 1993).

92. Historians have also pointed to a shift from spectacle lynchings to "legal lynchings" after the Neal case, whereby law enforcement officials used deadly force on black prisoners under the guise of escape attempts or other forms of resistance; see in particular, Steven Lawson, David Colburn, and Darryl Paulson, "Florida's Little Scottsboro," in *Civil Rights Crossroads: Nation, Community, and the Black Freedom Struggle* (Lexington: University Press of Kentucky, 2003).

93. Charles Payne, *I've Got the Light of Freedom: The Organizing Tradition and the Mississippi Freedom Struggle* (Berkeley: University of California Press, 1995), 13–14.

94. Quoted in Payne, *I've Got the Light,* 14.

95. W. Fitzhugh Brundage, *Lynching in the New South: Georgia and Virginia, 1880–1930* (Urbana: University of Illinois Press, 1993), 66. Brundage argues that historians "should be cautious in assuming that the ritual [of castration] was a central, even defining, element of lynchings for rape . . . Extant evidence, admittedly open to question, suggests that mass mobs mutilated one in three black victims lynched for alleged sexual offenses in Georgia and one in ten blacks lynched in Virginia" (66). However, he does not elaborate on or cite this evidence and his focus on Georgia and Virginia presents challenges to generalizing on lynching throughout the country. For one, he focused primarily on the more covert,

"under the cloak of darkness" lynchings than the more public spectacle lynchings in which torture, including castration, was an essential part of the spectacle. Also, he uses Virginia as a point of comparison with Georgia because of the former's relatively low incidents of lynching. And of the cases I discuss here in which castration *did* feature prominently in the lynch ritual—and its reporting—only one occurred in Georgia, with two from Mississippi, one from Florida, one from Tennessee, and three from Texas. In addition, he focuses here only on lynchings for alleged sex crimes; in several of the cases I discuss in which the lynch victim was castrated, he had been charged with nonsexual offenses, such as robbery (Walker), or incurred the mob's wrath for trying to prevent a rape rather than committing one himself (Burt Smith). Unlike Brundage, Philip Dray maintains, "For reasons of delicacy, direct allusions to castration were left out of most contemporary lynching accounts, the general term mutilation being substituted, but there can be little doubt that it was often the centerpiece of the entire lynching ritual" (*Lynching of Black America,* 82).

96. Literary critic Trudier Harris, for example, observes, "This element [castration] in particular would capture the imagination of many generations of black writers who, in their works, made the elements surrounding the lynching of black men a ritual which would be repeated in literary works for more than a hundred years" (5–6). Trudier Harris, *Exorcising Blackness: Historical and Literary Lynching and Burning Rituals* (Bloomington: Indiana University Press, 1984).

97. Quoted in Richard Wormser, *The Rise and Fall of Jim Crow* (New York: St. Martin's Press, 2003), 75.

98. Ida B. Wells-Barnett, "Southern Horrors: Lynch Law in All Its Phases," reprinted in *Ida B. Wells-Barnett on Lynchings* (New York: Arno Press, 1969), 19.

99. Richard Wright, *12 Million Black Voices* (New York: Thunder Mouth's Press, 1941), 46.

100. Wright, *12 Million Black Voices,* 88–89.

101. Wells-Barnett, "Southern Horrors," 4.

102. Ibid., 5.

103. August Meier, introduction to *Ida Wells-Barnett on Lynchings* by Ida B. Wells-Barnett (New York: Arno Press, 1969), iii.

104. Dray, *Lynching of Black America,* 81.

5. Walter White, Scientific Racism, and the NAACP Antilynching Campaign

1. Walter White, *A Man Called White* (New York: Arno Press, 1969), 9–11.

2. Walter White, *Rope and Faggot: A Biography of Judge Lynch* (New York: Alfred A. Knopf, 1929), viii–ix.

3. Ibid., 115–16.

4. On Pearl's participation in the Harlem hospital investigative committee, see Correspondence—Series 1, Box 21: NAACP Folder #2 of the *Raymond Pearl Papers,* American Philosophical Society (APS) Philadelphia, Pa. Professional and social meetings—missed and kept—between the two men are described in NAACP Folders 1–3.

5. On the relationship between Wells and White in Wells's later years, and the influence of Wells's earlier activism on White and the NAACP more generally, see Mia Bay, *To Tell the Truth Freely: The Life of Ida B. Wells* (New York: Hill and Wang, 2010), 311–19.

6. White, *Rope and Faggot,* 55.

7. Ibid., viii.

8. Walter White was a complex figure who offers a fascinating lens into a number of historical issues—racial identity, passing, black activism in the first half of the twentieth century, gender tensions within the NAACP, and interracial alliances and tensions, among others—many of which are outside the scope of my study here. For more on White's life, see White, *A Man Called White* and Kenneth Robert Janken, *Walter White, Mr. NAACP* (Chapel Hill: University of North Carolina Press, 2006). Janken's biography of White was originally published as *White: The Biography of Walter White, Mr. NAACP* (New York: New Press, 2003); all page numbers herein refer to the 2006 University of North Carolina Press edition.

9. Janken, *Walter White, Mr. NAACP,* 20.

10. Ibid., 20–24. On White's conflicts with and treatment of women with the NAACP, see also 48–49.

11. White, *A Man Called White,* 13–14.

12. Janken, *Walter White, Mr. NAACP,* 2.

13. Ibid., xiii–xiv.

14. White, *A Man Called White,* 11.

15. As discussed previously, black ethnologists often characterized the white race as hypermasculine, aggressive warmongers and celebrated the black race's "feminine" qualities—moral, spiritual, peaceful—tropes Mia Bay has termed the "Angry Saxon" and the "Redeemer Race," respectively.

16. White, *A Man Called White,* 12.

17. On Franz Boas and his influence on American racial and scientific thought, see Vernon J. Williams, *Rethinking Race: Franz Boas and His Contemporaries* (Lexington: University Press of Kentucky, 1996); Lee Baker, *From Savage to Negro: Anthropology and the Construction of Race, 1896–1954* (Berkeley: University of California Press, 1998), 99–126; and Bay, *The White Image in the Black Mind,* 187–229. In particular, Williams argues that Boas, a product of his times, was himself conflicted and sometimes contradictory on issues of race, but nonetheless precipitated important changes in scientific and social thought.

18. Baker, *From Savage to Negro,* 100.

19. Frederick Douglass, *The Claims of the Negro, Ethnologically Considered: An Address before the Literary Societies of Western Reserve College, at Commencement, July 12, 1854* (Rochester, N.Y.: Lee, Mann, 1854), 16.

20. Ibid., 15.

21. C. V. Roman, *American Civilization and the Negro: The Afro-American in Relation to National Progress* (Philadelphia: F. A. Davis, 1916), 5.

22. Ibid., 21.

23. Ibid., v.

24. Ibid., vii.

25. Walter White to Raymond Pearl (August 31 1927). *Raymond Pearl Papers,* APS Philadelphia, Pa. Correspondence—Series 1, Box 21: NAACP Folder #1.

26. According to the Merriam-Webster dictionary, the words *biometry* and *biometrician* could be traced to 1901; *biometry* is defined as "the statistical analysis of biological observations and phenomena."

27. "Guide to the Major Collections of the American Philosophical Papers—Raymond Pearl Papers," American Philosophical Society, Philadelphia, Pa. http://www.amphilsoc.org/library/guides/glass/pearl.html.

28. Elazar Barkan, *The Retreat of Scientific Racism: Changing Concepts of Race in Britain and the United States between the World Wars* (Cambridge: Cambridge University Press, 1992), 210–11.

29. Barkan, *Retreat of Scientific Racism,* 209.

30. For a detailed description of Pearl's public conversion against eugenics side-by-side with his private racism, see Barkan, *Retreat of Scientific Racism,* 210–20.

31. Barkan, *Retreat of Scientific Racism,* 212.

32. Raymond Pearl, "The Biology of Superiority," *American Mercury* 12 (November 1927): 257–66.

33. Ibid., 260.

34. Ibid., 260.

35. Ibid., 266.

36. Walter White, "I Investigate Lynchings," *American Mercury* 16, no. 61 (January 1929): 83. See also White's correspondence in *The Papers of the NAACP,* Part 7: The Anti-Lynching Campaign, 1912–1955; Series A: Anti-Lynching Investigative Files, 1912–1953; Reel 3:0765–0773.

37. In *American Civilization and the Negro,* Roman argued that the words of racial theorists became deeds in the streets (58). Clearly familiar with the "furor sexualis" model in racial science, he also noted, "The half-scientific rubbish and historical mendacity that seek to parade sexual excesses and moral delinquencies as peculiar racial vices, are a travesty upon learning, a perversion of justice, and a degradation of human reason" (4).

38. While on the decline by the 1920s, even the "castration or lynching" question specifically still appeared on occasion in medical journals. See, for example, J. H. James, "Asexualization: A Remedy for Crime and Criminality," *Minnesota Medicine* 3 (1926).

39. It was a common belief in ethnology dating back to the antebellum era that black and white children were comparable in intellect until adolescence, when whites continued to develop but black brains remained stunted in perpetual childhood.

40. The intelligence testing to which White refers here is discussed in depth in Steven Jay Gould, *The Mismeasure of Man* (New York: W. W. Norton, 1996).

41. Walter White to Raymond Pearl (August 31 1927). *Raymond Pearl Papers,* APS, Philadelphia, Pa. Correspondence—Series 1, Box 21: NAACP Folder #1.

42. Raymond Pearl to Walter White (September 20, 1927). *Raymond Pearl Papers,* APS. Correspondence—Series 1, Box 21: NAACP Folder #1.

43. Pearl mentions White's visit in a letter dated April 28 1928, *Raymond Pearl Papers,* APS. Correspondence—Series 1, Box 21: NAACP Folder #1.

44. Raymond Pearl to Walter White (September 20 1927), and Walter White to Raymond Pearl (May 19 1928). *Raymond Pearl Papers,* APS, NAACP Folder #1.

45. White, *Rope and Faggot,* 118–19.

46. White to Pearl (January 31 1929); Pearl to White (February 1 1929); Pearl to White (April 3 1929); telegram from White to Pearl (April 11 1929); and Pearl to White (April 11 1929), *Raymond Pearl Papers,* APS, NAACP Folder #1.

47. White, *Rope and Faggot,* 114–15.

48. Raymond Pearl, "Variation and Correlation in Brain Weight," *Biometrika* 4, quoted in White, *Rope and Faggot,* 127.

49. White, *Rope and Faggot,* 117–18.

50. Ibid., 132–33.

51. I do not intend to imply that scientific studies of race and intelligence were always implicitly or explicitly gender neutral. Indeed, during Reconstruction and Redemption, scientists often compared black men and white women to argue that neither had the intellectual capacity necessary for the franchise. However, the mostly early twentieth-century studies to which White referred tended to discuss the races in more general terms and some did not break down their data by sex.

52. White, *Rope and Faggot,* 121.

53. Ibid., 123.

54. Ibid., 134–35.

55. Ibid., 136–37.

56. Robert Bennett Bean, "Some Racial Peculiarities of the Negro Brain," *American Journal of Anatomy* 5, no. 4 (1906), 409.

57. White, *Rope and Faggot,* 56–61.

58. Ibid., 57.

59. Ibid., 227.
60. Ibid., 62–63.
61. Ibid., 62–69.
62. Ibid., 67–68.
63. Ibid., 54. On White's correspondence with Bean, see 124.
64. Raymond Pearl, "The Weight of the Negro Brain," *Science* 80 (November 9, 1934): 431–34.
65. Ibid., 433.
66. White, *Rope and Faggot*, 54.

Epilogue

1. Walter White, *A Man Called White* (New York: Arno Press, 1969), 363–64.
2. Wendy Kline, *Building a Better Race: Gender, Sexuality and Eugenics from the Turn of the Century to the Baby Boom* (Berkeley: University of California Press, 2005).
3. See especially Elazar Barkan, *The Retreat of Scientific Racism: Changing Concepts of Race in Britain and the United States between the World Wars* (Cambridge: Cambridge University Press, 1992).
4. On racial science in the second half of the twentieth century, see Joseph Graves Jr., *The Emperor's New Clothes: Biological Theories of Race at the Millennium* (New Brunswick, N.J.: Rutgers University Press, 2001), 140–72.
5. Lee Baker, *From Savage to Negro: Anthropology and the Construction of Race, 1896–1954* (Berkeley: University of California Press, 1998), 101–7, 123.
6. Barkan, *Retreat of Scientific Racism*, 127.
7. For more on Matilda Coxe Stevenson, see Will Roscoe, *The Zuni Man-Woman* (Albuquerque: University of New Mexico Press, 1991). On Mary French-Sheldon and Alice Fletcher, see *White Women's Rights: The Racial Origins of Feminism in the United States* (New York: Oxford University Press, 1999), 102–15 and 121–31, respectively.
8. According to historian Louise Michele Newman, Mead "broke with an earlier tradition of evolutionary or Victorian anthropology that supported Anglo-Americans' definitions of themselves as a superior race because of their supposedly unique, race-specific biological forms of sexual difference" (159). The cultural concept of race could engender its own claims to superiority, however. She argues that "while Mead challenged Anglo-Saxons' beliefs in their inherent biological superiority to primitive peoples, she did not challenge their belief in the cultural superiority of Western civilization . . . For Mead, primitive societies provided Americans with conceptual alternatives to reflect on, but she never advocated that the United States remake itself in the image of the primitive." Undeniably, Mead's work marked an important turning point in science,

but nonetheless promoted a racial paradigm that, while more fluid, still emphasized differences between groups of people. As Newman quite rightly points out, "opposition movements retain residues of that which they oppose" (159).

9. Catherine A Lutz and Jane L. Collins, "The Color of Sex: Postwar Photographic Histories of Race and Gender in *National Geographic Magazine,*" in *The Gender/ Sexuality Reader*, ed. Roger N. Lancaster and Micaela di Leonardo (New York: Routledge, 1997), 291–306. In their fascinating discussion of the racial politics at play in *National Geographic,* Lutz and Collins further argue that amid the social upheaval of civil rights, Black Power, and race riots in the 1950s and 1960s, the magazine's images also served to render race "safe" by placing various groups in their "natural" environments in isolation from each other.

10. Michael J. Bamshad and Steve E. Olson, "Does Race Exist?," *Scientific American* (December 2003).

Appendix

1. "About Index-Catalogue," National Library of Medicine, http://www.nlm.nih.gov/hmd/indexcat/abouticatalogue.html.

Index

MELISSA N. STEIN is assistant professor of gender and women's studies at the University of Kentucky. She completed postdoctoral fellowships at Indiana University and the Rutgers University Center for Race and Ethnicity.